COMMUNITIES OF BELIEF

For Julia

ἀλλά σ' ἔγω θέλω
ὄμμναισαι, σὺ δὲ λάθεαι
ὄσσα μόλθακα καὶ κάλ' ε'πάσχομεν.

SAPPHO

Communities of Belief

CULTURAL AND SOCIAL TENSION IN EARLY MODERN FRANCE

ROBIN BRIGGS

CLARENDON PRESS · OXFORD
1989

Oxford University Press, Walton Street, Oxford OX2 6DP

Oxford New York Toronto
Delhi Bombay Calcutta Madras Karachi
Petaling Jaya Singapore Hong Kong Tokyo
Nairobi Dar es Salaam Cape Town
Melbourne Auckland

and associated companies in
Berlin Ibadan

Oxford is a trade mark of Oxford University Press

Published in the United States
by Oxford University Press, New York

British Library Cataloguing in Publication Data

Briggs, Robin
Communities of belief: cultural and
social tension in early modern France.
1. France. Religious beliefs, ca. 1500–
1700. Sociological perspectives
I. Title
306'.6'0944
ISBN 0–19–821981–4

Library of Congress Cataloging in Publication Data

Briggs, Robin.
Communities of belief.
Includes index.
1. Church and social problems—France—
History—17th century. 2. Social conflict—
France—History—17th century. 3. Witchcraft—
France—History—17th century. I. Title.
HN39.F8B75 1988 303.6'0944 88–19700
ISBN 0–19–821981–4

Typeset by Litho Link Limited, Welshpool, Powys
Printed in Great Britain
at the University Printing House, Oxford
by David Stanford
Printer to the University

Preface

A collection of essays seems to demand some kind of explanation or apology, if only to reassure the reader that what is being offered is something more than the mere repackaging of shopsoiled goods. Of the ten pieces in this book, only three are reprinted essentially in the form in which they originally appeared. Two others have been substantially revised and expanded; the remaining five are entirely new, although some of them have their origin in papers given to a variety of very welcoming audiences over the past few years. They have been selected on the basis that they do possess a degree of thematic unity, whose nature is set out in the introduction.

The oldest of the essays, 'The catholic puritans', was a contribution to the *festschrift* for my undergraduate tutor Christopher Hill, *Puritans and Revolutionaries,* edited by Keith Thomas and Donald Pennington (Oxford, 1977). 'Witchcraft and popular mentality in Lorraine, 1580–1630', a paper given at a conference in Zürich, was published in *Occult and Scientific Mentalities in the Renaissance,* edited by Brian Vickers (Cambridge, 1984). '*Idées* and *mentalités*: the case of the catholic reform movement in France', another conference paper, this time given at the Maison Française in Oxford, appeared in *History of European Ideas,* 7 No. 1 (1986). I am very grateful to Brian Vickers and Bruno Neveu, not only for inviting me to participate in these conferences, but also for general discussions and stimulus over the years. Two other chapters originated from contributions to conferences. 'The church and the family in seventeenth-century France' was given to the Society for Seventeenth-century French Studies, and an abbreviated version was published in their *Newsletter,* 4 (1982); the present essay has been completely rewritten. 'Ill will and magical power in Lorraine witchcraft', delivered in French amidst the haunting grandeur of the Abbaye de Fontevraud, appeared in *Histoire des faits de sorcellerie,* edited by Jean de Viguerie (Angers, 1985). It has been translated and expanded for the present volume. I should like to thank Jean de Viguerie for his hospitality and help on that occasion, and René and Suzanne Pillorget, whose many kindnesses included the suggestion that I should attend the meeting with them.

My first general acknowledgement must be to All Souls College, which has for many years afforded me exceptional conditions for research, from which I have often received special research grants, and where my colleagues have given me so much moral and intellectual support. I have also been the recipient of generous financial assistance from the Oxford Faculty of Modern History and from the British Academy. A great many of the ideas in the book originate from the lectures and tutorials given in Oxford over the years; I am sure that few of my pupils realize how much I have learned, or come to question, in the course of teaching them. Special thanks are due to two of them; John Saxton, for his archival work in Nancy, and David Parrott, who helped obtain microfiches of rare books. It was an honour and a pleasure to be a Visiting Professor at the Sorbonne (Paris IV) in 1981, when early versions of the essays on popular revolt and the confessional were given. Among my hosts at that time I should particularly like to thank François Crouzet, André Corvisier, Pierre Chaunu, Jean Meyer, François Caron, and Viviane Barrie-Curien. It is impossible to list all the French friends who have helped me at one time or another, but I must record my debts to Emmanuel Le Roy Ladurie, with whom I started my serious research in French social history, to Yves and Nicole Castan, who fired me with enthusiasm for the *histoire des mentalités,* and to Guy Cabourdin for his generous help with my work in Lorraine. A special word of thanks is due to the staff of the *salle de travail* at the Archives départementales de la Meurthe-et-Moselle at Nancy, who were friendly and helpful far beyond the call of duty as I ordered up great piles of documents, then demanded hundreds of metres of microfilm. My work on witchcraft has gained immeasurably from the generosity of Al Soman, in lively discussions over many years.

At the risk of being invidious, I shall single out some of my friends and colleagues in England who have been particularly helpful. Keith Thomas has always been an inspiration, both for his own work and for the unfailing interest he has taken in mine. Robert Franklin, Richard Smith, Paul Slack, and Jenny

Wormald have all given me enormous stimulus and encouragement. Jack McManners, Olwen Hufton, and Peregrine Horden have read and criticized parts of the book, although I have not always taken their advice, and they bear no responsibility for its shortcomings. My fumbling attempts to incorporate certain ideas drawn from psychoanalysis would have been even more hazardous without the help of my analytical friends Frances Campbell, Julia Segal, and Daphne Nash (who has also nobly read the proofs). I am deeply grateful to my editor, Ivon Asquith, for being so patient with an errant author while retaining his enthusiasm for the project, and to Janet Godden for her deft and imaginative sub-editing of the text.

Many of the topics I have chosen to write about can only be tackled seriously with some reference to one's own personal experience. The dedication records my greatest debt, to my first wife Julia, who suggested the idea of the book in the first place, discussed so many of its ideas with me, and removed innumerable infelicities from my various drafts. Time and again what I have written reflects my life with her, and with Jonathan, Simon, and Jeremy. Although these words are written at a time when a partnership of some twenty years has ended, that misfortune cannot diminish my gratitude for all that I have been given for so long, and on which I have drawn so much.

Contents

Introduction

THE essays in this volume are all concerned with the beliefs and behaviour of the people of France (and sometimes the regions just outside the kingdom proper) in the early modern period. They attempt to combine the approaches of religious history, social history, and the history of ideas to explore some difficult and teasing questions, many of which recent French historians would classify as belonging to the 'histoire des mentalités'. This useful if necessarily vague term has attracted some splenetic British comments, sometimes being dismissed as a flashy manifestation of French intellectual fashion, which ought to be banned from the historical vocabulary. No doubt there is much to be said for good roast beef rather than French kickshaws and gewgaws, but this does not make the latter totally meretricious. While I have in practice been very sparing in using the concept of 'mentalités', I am not one of those modern rigorists who would like to see it put on the intellectual Index. How people conceptualized their society and its value systems, whether in relation to supernatural powers or to their neighbours, seems to me a very proper, indeed an essential, concern of the historian. It is notoriously difficult to handle such questions in abstract general terms, and I have found it most practical to approach them through a group of more narrowly focused studies. This avoids any dubious claims to universality, tending instead to emphasize the diversity familiar to students of both this period and social history in general.

The historian who ventures into this uncertain terrain faces serious methodological problems. The categories and concepts in general use tend towards binary simplifications: sacred and profane, élite and popular, religion and superstition, individual and communal. These are powerful and valuable intellectual tools, so long as we recognize that they represent ideal types, almost never found in their pure forms. In the real world we do not find sharp dividing lines, but numerous overlapping entities, rather in the style of a Venn diagram. These essays tend to employ the binary model, not least because they are so often concerned with conflict in various forms; while I have quite frequently attempted to subvert my own over-simplifications, the reader should try to inject a further dose of scepticism. The difficulties of achieving accurate descriptions and identifying problems, great as they are, are significantly less than those of reconstructing cause-and-effect relationships. In discussing some of these questions I have looked back with awe at the prodigious learning, encompassing the history of several civilizations, which Max Weber brought to bear on related topics. His work emphasizes how a search for explanations drives one ever back and outwards, into comparisons with remote times and places. I have attempted no such intellectual fireworks, leaving numerous untidy loose ends along the frontiers between the medieval, early modern, and modern periods. The general picture is certainly one which portrays the sixteenth and seventeenth centuries as a period of crucial change and upheaval, the watershed between two very different worlds of behaviour and feeling. This does not imply, however, a conception of the middle ages as long immobile centuries lacking in dynamism or conflict. Virtually all the structural shifts of the early modern period have vitally important medieval antecedents, and if they have often been no more than hinted at in these essays, this should not be taken to imply that I think them unimportant.

Inevitably I have written a good deal about Christianity, and here I do share a disability with Weber in being 'religiously unmusical'. There can be no disguising the fact that I have approached the subject from the viewpoint of an unbeliever, who regards religious beliefs as projections of human preoccupations and interests. This leads to a predilection for reductionist and functional interpretations, which are liable to be irritating, even offensive, to those who believe that religion is divinely inspired. I can only say that I have not set out to offend, nor to use historical evidence in support of a personal and private attitude to such matters, but to be as dispassionate and analytical as possible. In any case my discussions should make it clear that I regard belief and dogma as having independent strength and significance, behind which believers can very properly locate the hand of God. The historian is virtually bound to adopt the classic early modern device of treating only 'second causes', on the assumption that religion and society have always interacted, and that this interaction can be studied as a process in its own right. To explain many of the characteristics of seventeenth-century Christianity as responses to social pressures and changes is not necessarily to diminish the heroic figures of 'le siècle des saints'; only by placing them in their context can we hope to understand why they were so influential among many less exceptional men and women.

The ambiguity of the title is quite deliberate, intended to suggest at least some of the themes which run through the essays. Shared beliefs did help to bind French society together, even if they were interpreted differently by their holders; they were the ultimate legitimation for the institutional role of the catholic church, studied in the essay on church and state. Communities, whether local, social, or religious, played a vital part in conflicts over belief and behaviour, providing the encouragements or constraints which largely determined individual action. The first part, Rebels, Deviants, and Victims, concentrates primarily on history seen from the bottom up, giving pride of place to the humbler members of society. The three studies of witchcraft emphasize the crucial role of the village community in regulating the identification and persecution of this very particular class of deviants. Popular revolts involved deviance of a much more public and obvious kind, in the political rather than the religious sphere, but these episodes are particularly revealing of both communal attitudes and divisions. While the ordinary people are still very much present in the second part, Agencies of Control, the emphasis here is more on the intellectual and institutional superstructure, and its relationship to society as a whole. Evolving clerical attitudes towards families and the imposition of moral standards through confession brought confrontations with alternative values which had deep communal roots. Some further aspects of this clash of values are evoked in both the analysis of the puritanical elements in Jansenism and rigorism and in the general essay on *idées* and *mentalités* in the catholic reform movement. The thematic unity should not be exaggerated, for these are separate studies, but I hope that they provide a range of vantage-points from which the reader will get some general sense of the forces working both for and against change in French society at this time. Above all, they try to show how that society can only be understood through a close analysis both of its component parts and of the ideas and beliefs through which they expressed themselves.

Citations from contemporary texts, very numerous in some of the essays, pose difficult problems. I have chosen to cite most of the French texts in the original, in order to preserve the flavour which is inevitably lost in translation. They are not always easy to construe, so I have also provided my own English versions, which pretend to nothing more than basic utility. Some longer extracts from printed works have been given in translation only, where this seemed appropriate. Spelling and accentuation are an intractable problem; most of the oddities of the original have been retained, but I have undertaken some discreet modernization where this made the sense easier to follow. Some of the quotations which I have taken from secondary sources have of course already been fully modernized.

I

Rebels, deviants, and victims

I

Witchcraft and the community in France and French-speaking Europe

THE COURTS AND THE RISE OF PERSECUTION

MORE than three centuries have passed since Colbert and Louis XIV put an end to the legal persecution of witches in France in the 1670s; by the same period trials had largely died out in the French-speaking regions beyond the eastern border of the kingdom proper. Even at this distance, however, the historian who reads the trial records cannot hope to remain entirely detached. Indifference in the face of these pathetic and terrifying documents might not be desirable in any case, since any satisfactory interpretation must allow for the affective power of both witchcraft beliefs and the trials in which they climaxed. If many of the accused confessed to imaginary crimes, condemning themselves to be strangled and burned at the stake, this was scarcely surprising when one appreciates the pressures to which they were subjected. In one of the numerous trials held in the small Lorraine town of Saint-Dié, in the Vosges, twenty-two witnesses appeared in 1592 to denounce Mengette Estienne of Le Paire d'Avould, alleging that her malevolence had caused deaths and sicknesses among men and animals.[1] Mengette was made of sterner stuff than most; despite this concerted hostility from her neighbours, and the denial of legal assistance, she defended herself with spirit and intelligence. The judges nevertheless saw fit not only to have her tortured on the rack, but to precede this with a series of threats whose impropriety never apparently struck them, since they were preserved in the official records. She was exhorted:

> de nous dire et declairer verité, autrement que luy ferions sentir la rigueur de justice affin de l'y induyre. Et que cela luy pourroit causer une perclusion de membre pour estre miserable toute sa vie Oultre que surcellans la verité elle seroit damnée avec les malheureux.

[1] AD M.-et-M., B 8667, No. 4.

[to tell and declare the truth to us, otherwise we would make her feel the torture in order to induce her to do so. And that this could cause her to be crippled in a limb so that she would be miserable for the rest of her life, while if she concealed the truth she would be damned with the wretched.]

During the torture—which seems only to have ended when she lost consciousness—she quite justifiably cried out: 'Vray Jhesus on luy faisoit grand tort Crians alarme alarme y at il plus de pitié au monde et qui nourriroit ses enfans' ['True Jesus they did her great wrong, crying help, help, was there no more mercy in the world, and who would feed her children'].

In fact Mengette's exceptional courage allowed her to endure three sessions of torture without confessing, so that her judges, who had obviously considered her guilty from the start, were finally compelled to release her. She was luckier than Jacquotte Tixerand of Le Bourget d'Amance, tried in 1616, who persisted in her denials through applications of the thumbscrews, the rack, and the *tortillons* (a kind of tourniquet).[2] Here local practice included a fourth technique, the strappado, in which the victim was raised in the air by a rope attached to her hands, themselves tied behind her back, and passed over a pulley. When the executioner added the final touch, tying a fifty pound weight to her feet, Jacquotte's resistance gave out. She now told her judges what they wanted to hear; how she had been seduced by the devil, had attended the sabbat, and had committed various *maléfices*. Such appalling abuses of judicial torture were much rarer in the areas ruled by the king of France than in the lands of the duke of Lorraine, but here too one suspects that the local courts of first instance were far more anxious to convict the accused than to give them a fair hearing. It is easy to see why many earlier historians of European witchcraft described the whole phenomenon in terms of ignorance, brutality, and superstition, the darkest expression of unenlightened age.

Over the past twenty years the historiography of witchcraft has been transformed by a series of careful scholarly studies, many of them drawing on the techniques of social history and social anthropology. This essay examines witchcraft in

[2] AD M.-et-M., B 2199, No. 1.

French-speaking Europe in the light of this work, not forgetting the kind of judicial atrocities just evoked, but trying to understand how they could occur in an age which did have aspirations towards rationality and humanity. This must lead us to consider witchcraft as a nexus of beliefs and practices rooted in local society, so that the persecution of witches is seen to derive from social tensions at local level at least as much as from judicial activity inspired by the ruling classes. In seeking to explain the phenomenon, the historian can also hope to uncover important information about more general aspects of the social structure and of popular belief. To pursue these elusive topics the geographical area covered here extends beyond the borders of the kingdom of France, as these stood until the late seventeenth century, in order to incorporate other French-speaking regions for which important evidence survives. This has the obvious advantage of licensing the use of my own detailed work on Lorraine, but in fact it would be difficult to treat the subject in any other way. For reasons that are not entirely clear, but most plausibly because of the great difficulty of finding primary source material, no major study has yet appeared on the social aspects of witchcraft within the kingdom itself. Generalizations on the subject commonly call on the work of Delcambre, Bavoux, Monter, Muchembled, and Dupont-Bouchat, without necessarily making it clear that all these authors were writing about the borderlands.[3] Some historians, such as Chaunu and Muchembled, have taken it for granted that these were the main areas of persecution, and constructed their interpretations accordingly.[4] Although in broad terms they are probably correct, the situation was rather more complicated

[3] E. Delcambre, *Le Concept de la sorcellerie dans le duché de Lorraine au 16ᵉ et au 17ᵉ siècle (*3 vols., Nancy, 1948–51); F. Bavoux, *La Sorcellerie au pays de Quingey* (Besançon, 1947), *Hantises et diableries dans la terre abbatiale de Luxeuil* (Monaco, 1956), and *Les Procès inédits de Boguet en matière de sorcellerie dans la grande judicature de Saint-Claude* (Dijon, 1958); E. W. Monter, *Witchcraft in France and Switzerland* (Ithaca, NY, 1976); M.-S. Dupont-Bouchat, W. Frijhoff, and R. Muchembled, eds., *Prophètes et sorciers dans les Pays-Bas 16ᵉ–17ᵉ siècles)* (Paris, 1978); R. Muchembled, *La Sorcière au village* (Paris, 1979); *Les Derniers Bûchers* (Paris, 1981).

[4] P. Chaunu, 'Sur la fin des sorciers au 17ᵉ siècle', *Annales: économies, sociétés, civilisations,* 24 (1969), 895–911; Muchembled, *La Sorcière.*

than this. Soman's work on the records of the *parlement* of Paris indicates that the largest body of appeals came from the Ardennes, Champagne, Nivernais, and Berry, all admittedly in the eastern half of the court's *ressort*, but by no means all classifiable as borderlands.[5] His 1,288 appellants before the *parlement* are complemented by 554 other cases which never reached this stage, mostly known by indirect evidence, rather than the actual records of the trial.[6] It is clear that there must have been significantly more of these local episodes, although probably not enough to alter the general picture of either the distribution or the intensity of persecution. The *parlement* of Paris itself ordered the execution of 104 witches, and Soman estimates that between 250 and 350 further executions were carried out locally, many of them illegally.[7] For a region with a population of around eight to ten million, over a period of a century, this is a comparatively modest level. Virtually nothing is known about the statistics of cases before the other provincial *parlements*, let alone from the lower courts they supervised; for the relevant period hardly any records from the courts of first instance appear to survive anywhere in France.

This deficiency of records is particularly serious for the period before 1640, which is likely to have seen the great majority of trials. An exceptional cache of documents reveals a group of trials at Seix in the central Pyrenees in 1562, while Pierre de Lancre has left us an account of his mission to the Labourd in 1609.[8] De Lancre avoided giving any figures, and

[5] A. Soman, 'Les Procès de sorcellerie au Parlement de Paris (1565–1640)'. *Annales ESC* 32 (1977), 790–814; 'La Sorcellerie vue du Parlement de Paris au début du 17e siècle', *Actes du 104e Congrès nationale des Sociétés savantes (Bordeaux 1979): Section d'histoire moderne et contemporaine* (Paris, 1981), ii. 393–405; 'La Décriminalisation de la sorcellerie en France', *Histoire, économie et société*, 4 (1985), 179–203.

[6] These slightly revised figures are given in a study which also considers a regional example; A. Soman, 'Trente procès de sorcellerie dans le Perche (1566–1624)', *L'Orne littéraire*, 8 (1986), 42–57.

[7] Figures given during a seminar at the Institute of Historical Research in April 1986.

[8] J.-F. Le Nail, 'Procedure contre des sorcières de Seix en 1562', *Société ariègoise, sciences, lettres et arts*, 31 (1976), 155–232 (Seix is in the modern department of the Ariège); P. de Lancre, *Tableau de l'inconstance des mauvais anges et demons* (Paris, 1613).

the fantastic total of 600 victims often associated with his name comes from a late seventeenth-century source; the true number could well have been as low as thirty.[9] There is relatively good information on the panics of 1643–4 in Gascony, the Toulousain, Burgundy, and Champagne, not least because by this stage the authorities were becoming generally hostile to such manifestations of what was increasingly seen as popular superstition and emotionality.[10] By this date it seems unlikely that many isolated cases were reaching the courts; when the Rouen *parlement* had to be restrained from proceeding against suspects in 1670 there were implications that there had been little recent activity of this kind.[11] The major episodes involving demonic possession, the object of valuable modern studies by Mandrou and Walker, form a special category on their own.[12] While they certainly relate to the same belief structure, and indeed helped to bring it into doubt, they cannot be equated with the accusations peasants brought against one another. It would therefore be quite misleading to use them to fill up a list of known French cases, perhaps lumping in the 'affaire des poisons' from the court of Louis XIV into the bargain. Such a presentation tends to create a misleading impression of a sustained and intensive witch-hunt, masking the vital point that such activity was very sporadic and patchy, in both time and space. We do remain uncertain about the total numbers involved, since a steady trickle of cases in local courts could have escaped attention, left no records, yet been more important in aggregate than the cases which we know of. Quite apart from the general shortage of judicial records, both Mandrou and Muchembled suggest that it may have been common practice to burn the papers with the witch, obliterating the memory as well as the

[9] The figure of 600 comes from *Factum et arrest du Parlement de Paris contre les bergers sorciers executez depuis peu dans la province de Brie* (Paris, 1695), 63–4.

[10] There is a good general account in R. Mandrou, *Magistrats et sorciers en France au 17e siècle* (Paris, 1968), 372–94. See also Soman, 'La Décriminalisation', for additional information and a discussion of the attitude of the *parlement* of Paris.

[11] See Mandrou, *Magistrats*, 444–58, esp. 451.

[12] Ibid., 195–341, and *Possession et sorcellerie au 17e siècle* (Paris, 1979); D. P. Walker, *Unclean Spirits* (London, 1981). For Loudun, see also M. de Certeau, *La Possession de Loudun* (Paris, 1970).

person of the deviant.[13] However widely this was done, it could hardly have prevented some evidence surviving from severe witch-hunting on a national scale, as indeed it does for certain years marked by serious panics; pamphlets, memoirs, collections of legal decisions, and the deliberations of local authorities do between them give us a reasonable check on more exceptional or sensational events.[14] Some negative evidence may also be gleaned from the writings of the demonologists themselves, for they were inclined to bemoan the laxity of their fellow-jurists, and could only cite a relatively small range of examples.[15]

There is nothing, then, to suggest that France as a whole suffered from a significantly higher level of persecution than England; the ratio of those executed to total population could even have been smaller. As in England there seem to have been marked regional variations, with the eastern and north-eastern frontier areas as one centre of activity, the extreme south-west as another. Since the former fell within the *ressort* of the *parlement* of Paris, which as Soman and Mandrou have shown was always inclined to be sceptical about witchcraft accusations, an important conclusion might be that the demand for persecution depended primarily on local factors, rather than on the jurisprudence applied by the legal élite. The authorities were interested in controlling the process, not in instigating it; by the 1640s they had effectively brought it to

[13] Mandrou, *Magistrats*, 108, and *Possession*, 14. Muchembled, *La Sorcière*, 123.

[14] For an excellent example of how this can be done on a regional scale, see X. Martin, 'Aspects de la sorcellerie en Anjou', in *Histoire des faits de sorcellerie*, ed. J. de Viguerie (Angers, 1985), 71–109. Martin concludes that a priest was probably executed in 1571, followed by three witches and a magician in 1593, and that a suspect was murdered in 1611; a wide range of local sources give no indication that there can have been a significantly larger number of cases, despite the claim by Mandrou (*Magistrats*, p. 551) that there were numerous trials in Anjou.

[15] De Lancre, provided a valuable appendix to his *L'incrédulité et mescreance du sortilège plainement convaincue* (Paris, 1622), with details and extracts from some eighteen arrêts (pp. 659–830); he had clearly worked hard to find as many. The defensive nature of the first chapter of this work (pp. 1–68) is also striking. Bodin, in *De la demonomanie des sorciers* (Paris, 1580) cites only a minuscule number of French cases; his preface complains of incredulity among the *parlementaires*.

a stop. This conclusion does need to be modified to the extent of noting a rather different relationship to official policy and declarations in the Spanish Netherlands, Luxembourg, and Lorraine, which could well have influenced neighbouring regions of France.[16] One may argue that in these smaller territories the rulers were more responsive to popular calls for witch-hunting, but whatever their motivation one must assume that their intervention did tend to make matters worse. This was much more likely because these territories had not developed a formal appeal system on the French model, so that local courts had a far freer hand in the conduct of trials. One crucial feature in the kingdom of France was that the *parlements* could challenge the abusive practice of witch-hunting without ever formally repudiating the demonological arguments, by insisting that certain rules of procedure be followed, and then by trying all the cases themselves. Formal pronouncements might stress the abhorrent nature of the crime, as an exercise in public relations, while the everyday jurisprudence of the *tournelle* made it virtually impossible to bring a successful prosecution. To this one must add Soman's startling findings about the inefficacy of torture as applied by the *parlement*; precisely one of the 185 suspects who entered the torture chamber confessed, and the woman in question was merely shown the instruments.[17] One can only conclude that the judges thought of torture as a useful threat, and perhaps a mild punishment for people who had annoyed their neighbours, but were determined to exclude it from their system of proofs. Soman has extended his argument to argue both powerfully and convincingly that *ancien régime* justice has been totally misunderstood, with a legend of ferocious repression obscuring its real nature. He portrays a system in which the main objectives were to promote reconciliation through informal settlements, to

[16] The series of ordinances affecting Luxembourg, beginning with those of 1563 and 1573, concerned with judicial abuses, but then proceeding to the famous 1592 ordinance of Philip II, is discussed by Dupont-Bouchat in *Prophètes et sorciers*, 86–90. She notes that although the ordinance of 1595 did sound a more cautious note, the archdukes recirculated the 1592 text in 1606. See also Muchembled in the same volume, pp. 165–6; he prints the essential section of the 1592 ordinance in *La Sorcière*, 78–82.

[17] Soman, 'Trente procès', 44–5.

prevent the courts being used to frame enemies, and to ensure that the innocent could be effectively cleared.[18] Shrewd judges knew very well that these were the best ways to maintain both the public peace and the royal authority. Even where the controls were less efficient, as in the Netherlands and Lorraine, everyday criminal justice followed much the same pattern. If witches came off worse here, this was because the local community found it easier to use the legal system for its own purposes, and it is to those communities that we should look for the driving forces behind the persecution.

An approach which concentrates on the local and particular nature of accusations must be combined with an awareness that better legal records, even if we possessed them, would not give us a total picture of witchcraft beliefs. Given the occasional scraps of medieval evidence, and the knowledge that contemporary researchers can still find numerous instances of such beliefs in parts of rural France, we are dealing with a phenomenon which has a continuous (although not necessarily unchanging) existence as far back as we can trace. Intermittent legal persecution over a couple of centuries, however important and puzzling in its own right, must always be considered within this context. In parts of western France it is possible to identify many more supposed witches today than are known to have been tried throughout the period of persecution.[19] Such comparisons need to be handled with caution, since the modern notion of the witch has lost nearly all the religious (and diabolical) charge it once possessed; those who think themselves victims may still hope for relief from ecclesiastical counter-magic—holy water or exorcism—but they normally explain the motives of the 'witch' in terms of straightforward malevolence or rivalry. Yet I think that this really represents a continuity, and that the

[18] Soman, references given in nn. 5 and 6; also 'Deviance and criminal justice in western Europe, 1300–1800: an essay in structure', *Criminal Justice History*, 1 (1980) 3–28; 'Criminal jurisprudence in *ancien-régime* France: the Parlement of Paris in the sixteenth and seventeenth centuries', in *Crime and Criminal Justice in Europe and Canada*, ed. L. A. Knafla (Waterloo, Ontario, 1981), 43–75.

[19] For modern witchcraft see J. Favret-Saada, *Deadly Words* (Cambridge 1984), and P. Gaboriau, 'La sorcellerie actuelle dans le Choletais', in *Histoire des faits de sorcellerie*, ed. J. de Viguerie, 177–86.

diabolism was never more than a secondary, even an imposed element in peasant belief. If this is correct, then it can be argued that the essential features of the popular conception of witchcraft have not changed for at least a millenium, and probably much longer. In making such a claim one is implicitly defining the term in a way which should be made explicit. Witchcraft in this sense is the supposed power to harm other people, animals, or the community at large, ascribed to specific individuals, and carried out through occult or non-physical means.

While no clear line can be drawn, it is possible to identify an allied set of beliefs which may be classified as sorcery. These include image magic directed against enemies (including hostile witches), love or fertility charms, various means of finding hidden treasure, sometimes by invoking spirits, and techniques that shade off into various forms of magical healing. The distinction between witchcraft and sorcery, in a more sophisticated formulation, was a major feature of the pioneering work by Evans-Pritchard on the Azande, and in some African societies appears to be fundamental.[20] Elsewhere in Africa, and generally in Europe, this distinction seems to have only a secondary importance; this is essentially because the concept of involuntary maleficence is very rare, with evil effects seen rather as proceeding from the conscious will of an active agent. Europeans did of course distinguish between black and white witchcraft, the latter claimed by its practitioners and their clients to be free of any diabolical element, but regularly denounced by the church as involving implicit compacts with the devil. For ordinary people the distinction was essentially one of effects rather than of means; supernatural power was an attribute of individuals, as ambivalent as the relationships in which they were engaged. For them the real division was that between the malevolence of evil witches and the benign activities of village wizards or cunning folk. One must be very wary of imposing taut, logical structures on beliefs which by their very nature were and are fluid, double-edged and imprecise. It is striking that although the French language has a specific term for the cunning folk (*devin*), the

[20] E. E. Evans-Pritchard, *Witchcraft, Oracles and Magic among the Azande* (Oxford, 1937).

word *sorcier* has commonly been used to describe white witches as well as black.[21] An analysis of language cannot therefore take us very far, for many terms seem to have been used virtually interchangeably in this way, in line with a fundamental characteristic of the popular conception of witchcraft. This is the confusion or mingling of the natural and supernatural worlds, and the identification of certain groups or individuals as possessing special contacts with the latter, both classic features of an animistic world-view. In such a context the power to injure and the power to heal are virtually inseparable, and are commonly found in the same person, whose role is liable to vary with circumstances. It is no surprise to find a significant number of magical healers among those accused of witchcraft, although the very high numbers apparently present in Soman's sample may be unrepresentative.[22]

THE HISTORICAL AND PSYCHOLOGICAL ROOTS OF PERSECUTION

Persecution of the cunning folk must have resulted in part from the hostility they naturally aroused among powerful groups and distinctly monopolistic tendencies: doctors, lawyers, and above all the church. It was the theologians, with some help from the lawyers, who had elaborated a distinctly Christian theory of witchcraft over a period of centuries, to provide the indispensable theoretical basis for legal action against witches. They assimilated witchcraft to heresy by concentrating on the supposed pact between witch and devil, then added a series of lurid fantasies about the sabbat and sexual relations with demons, largely derived from the accusations already made against groups of heretics.[23] The scheme also included the popular idea of witches as people who

[21] R.-L. Wagner, *'Sorcier' et 'magicien': contribution à l'histoire du vocabulaire de la magie* (Paris, 1939).

[22] Soman, 'Les Procès de sorcellerie', 803. Delcambre, *Le Concept de la sorcellerie*, vol. iii, gives much information on healers accused in Lorraine.

[23] The most reliable guides on this question are N. Cohn, *Europe's Inner Demons* (London, 1975) and R. Kieckhefer, *European Witch Trials: Their Foundations in Popular and Learned Culture*, 1300–1500 (London, 1976). The most convenient collection of information is in J. B. Russell's *Witchcraft in the Middle Ages* (Ithaca, NY, 1972), but his interpretation has been widely criticized for its facile equation of witchcraft and heresy.

harmed others through *maleficium*, using some combination of diabolical power and special poisons; this element was often treated as an illusion of the devil, who acted directly but tricked the witches into believing themselves responsible.[24] Élite and popular beliefs never merged fully, however; in the latter the sources of a witch's power were far less explicit, while even the diabolical pact was rather a marginal element. There is a striking contrast between the stability of the elementary ideas at village level and the shifting opinions and arguments among the educated minority who ruled. The latter tried to distance themselves from the more doubtful elements, with even the great demonologists emphasizing their contempt for popular superstitions. How far this was a real distinction must nevertheless remain a very open question, particularly from the fifteenth to the seventeenth centuries.

The articulate leaders of the early church seem to have regarded witchcraft and sorcery as part of the range of paganism and superstition which they were determined to eradicate; to believe in such powers was almost as sinful as to attempt to use them.[25] How far such attitudes travelled down to parish level one may doubt, when the general impact of Christianity on the rural world seems to have been so modest, characterized by adaptation to local practices rather than their wholesale replacement. The church was attempting a take-over of supernatural power, which in many ways encouraged older modes of thought. The hierarchy, splendour, and ritual of the medieval church may well have enhanced its magical prestige in the eyes of the populace, actually helping

[24] This was a central argument of the sceptical Johann Weyer in his *De praestigiis demonum* (Basle, 1563), but his opponents were quick to argue that although the witches were deceived, they still helped the devil and offended against God. See for example Bodin, *Demonomanie*, Bk. II, ch. VIII (fos. 109–118^r), in the course of which he writes: 'Ce n'est donc pas la poison, ny les os, ny les pouldres enterrées qui font mourir: mais Satan à la priere des Sorcieres par la juste permission de Dieu' (fo. 115^r).

[25] This is the position taken by the Canon *Episcopi* (almost certainly a Carolingian text) and by Burchard of Worms in his *Corrector*. See Cohn, *Europe's Inner Demons*, 211–19; Kieckhefer, *European Witch Trials*, 38–9. As already explained, later demonologists found it possible to deny real powers to witches while still calling for persecution.

to perpetuate the whole structure of animistic ideas which dominated their world picture. Moreover, the church offered a wide range of magical services which suggested that there were evil powers to be combated, from blessing the harvest to anathematizing marauding beasts. There is a considerable temptation to treat this as a crudely Manichean view, with God and the devil as the embodiments of good and evil, and the saints and the clergy as special intercessors to whom one applied for help. Yet this would be gravely misleading, for a very crucial reason. Ordinary people seem to have attached surprisingly little importance to the devil in most circumstances; they were more inclined to think that God himself was responsible for evil, perhaps as a punishment or as a trial for the faithful.[26] In European folklore the devil rarely has the best tunes, being far more likely to appear as the dupe of some cunning peasant, or at best being allowed to punish the egregiously wicked. As Cohn and Kieckhefer have pointed out, those who were convicted of *maleficium* in early witchcraft trials were virtually never accused of diabolism by the witnesses, although they might be tortured into confessing it by judges who could see no other way of making sense of their supposed crimes.[27] While it is plain that the persecution of the early modern period did help to spread the official doctrines concerning diabolism, they always remained marginal in the popular mind, to recede rapidly once the outside influences were withdrawn.

In this area, then, the development of an elaborated vision of the devil and his powers seems to have been the work of the educated clergy, more Manichean than their flock. As the chosen holy men of society, expected to renounce many sensual pleasures and keep higher standards than the laity, they were plainly required to build up a system of internal repression, based on the super-ego, which would impose a high psychic cost. The natural consequence was the projection of forbidden desires onto a personification of evil, which lay

[26] On the use of this argument by the demonologists themselves, in the form 'with the permission of God', there are some justly scathing comments by S. Anglo in 'Evident authority and authoritative evidence: the Malleus Maleficarum', in *The Damned Art*, ed. S. Anglo (London, 1977), 21. See n. 24 above for a typical formulation by Bodin.

[27] Cohn, *Europe's Inner Demons*, pp. 239–55; Kieckhefer, *European Witch Trials*, 31–6, 73–92.

ready to their hands in the intellectual tradition of which they were the guardians. As the reform movements of the early modern period, both protestant and catholic, sought to extend similar high standards of personal morality to the laity, significant number among the élites must have undergone analogous experiences, and come to see the devil as a pervasive presence in the world. The persistence of witchcraft beliefs based on personal rather than diabolical power can therefore be seen as a sign of the churches' failure to impose the new faith effectively on the rural world; here repression was external or social, rather than internalized. Earlier witch-hunting episodes had been inspired by such exceptional ascetic figures as Conrad of Marburg and John of Capestrano, clearly driven by enormous inner tensions, sadistic saints who punished the bodies of deviants even more harshly than they did their own.[28] The fantasies of unbridled sexuality with which such men overlaid commonplace heresy and folk magic were of course exceptionally dangerous. Rooted as they were in the most central part of the psyche, it was all too easy to extract confessions, which then became part of the received knowledge about heresy and witchcraft, and held an appalling fascination for all who encountered them. As 'true Christianity' became ever more synonymous with the denial of libidinal impulses, so the persecution of witches was likely to attract growing support from the ruling classes. Among the leading persecutors one therefore finds a higher proportion of men who seem merely typical of their age, although such appearances are often deceptive. In very different ways both Jean Bodin and Pierre de Lancre, whose writings reveal a great deal about their personalities, exemplify the connections between internal conflict and external aggression.[29] Much of the modern occultist literature, not to mention the groups

[28] For these cases, see Cohn, *Europe's Inner Demons*, 24–31 and 50–3.

[29] There are interesting essays on both men by C. Baxter and M. M. McGowan respectively in *The Damned Art*, ed. Anglo. Baxter emphasizes Bodin's secret Judaism, which left him in a very odd position as a concealed heretic himself. McGowan brings out the way in which de Lancre used literary effects to heighten his portrayal of the sabbat, and the strange quality of excitement this conveys. Alfred Soman has pointed out to me that Bodin was a man of relatively humble birth, who failed to obtain advancement commensurate with his talents, and wrote a brilliant book about his fears.

which practise forms of pseudo-witchcraft, can stand witness to the lasting appeal of the fantasies, as does the bizarre career of the late Montague Summers, cleric-scholar-pornographer and firm believer in the reality of demonic witchcraft down the centuries.[30]

This is not the kind of analysis which historians find most congenial, even though it can only be avoided, in general discussion of witchcraft, by something like a conscious act of *suppressio veri*. In this subject fastidiousness quickly becomes intellectually disabling. No one can enter the minds of the witch-hunters without allowing full value to their libidinal drives or their tendency to prurience, and therefore the persecution itself can only be understood if these elements are allowed for. Yet there are enormous difficulties inherent in such an approach. Psychoanalysis often seems the reverse of an exact science even in the contemporary world; attempts to apply it to long-dead individuals have rarely carried conviction. In the peculiar case of witchcraft we can claim to be dealing specifically with fantasies, but their collective and syncretist nature forces us into the particularly murky waters of group psychology. Above all, like nearly all the general explanations ever advanced in the field, this one constantly threatens to explain too much. If the persecution fits so beautifully into an interpretation of the *Zeitgeist*, based on the repression of libidinal instinct, why then was it so limited in time and space, why were the numbers involved so relatively small, and why were there such divisions among the élites about the reasonableness of the whole enterprise? To produce a more satisfactory interpretation we must invert the normal approach, regard the whole phenomenon as a natural result of powerful tendencies within both Christianity and the development of a higher civilization, and concentrate on explaining why things were not a great deal worse, rather than why they happened at all. Such a reorientation also has the advantage of getting away, at least to some extent, from the crude reductionism which commonly lurks within broad psychological generalizations. Working with a simple basic model, founded on rather banal assumptions about human personality, one

[30] On Summers see Cohn, *Europe's Inner Demons*, 120–1.

can seek to reintroduce the complexities of the real world, as they controlled and inflected its outward manifestations in a specific society and period.

POPULAR CONCEPTIONS AND LOCAL REMEDIES

Perhaps the most fundamental point about French society during the sixteenth and seventeenth centuries, in this respect, was the growing divide between the educated minority and the mass of the population. While this should not be thought of in terms of polar opposition, rather of a steady divergence, the tendency is unmistakable in the long run. It can be associated with the growing dominance of towns, in political, economic, and social terms, and therefore involved a physical separation between the ruling élites and rural society. Most students of the catholic reform movement have seen this in related terms, as carrying alien urban values into a traditional countryside, with predictably mixed results.[31] The persecution of witches must also be placed within this context, as part of a great extension of regulatory criminal justice, which sought to impose higher standards of order and decorum throughout society. It has even been argued, by Christina Larner, that the persecution was an integral part of an 'age of faith', in which secular nation states sought legitimation through religious allegiance, and were drawn into identifying and punishing deviants.[32] Such patterns can certainly be discerned, and help to explain why rational educated men of the period took witchcraft seriously; they must form part of any general explanation. Yet they are often two-edged in their implications. As we have already seen, within the kingdom of France the great royal courts constantly sought to impose their control and their standards on the lower courts, a process which almost from the start led most of them to suspect undue credulity in these inferior judges, so that they first restrained

[31] The bibliography on this subject is immense, but this point of view is represented by J. Delumeau, *Catholicism between Luther and Voltaire* (London, 1977), and R. Muchembled, *Culture populaire et culture des élites dans la France moderne* (Paris, 1978). For a rather different approach see J. Bossy, *Christianity in the West*, 1400–1700 (Oxford, 1985). For a critique of all these approaches, see the Conclusion, pp. 381–413 below.

[32] This idea, which first appeared in *Enemies of God* (London, 1981) was further developed in *Witchcraft and Religion* (Oxford, 1984).

and then eliminated the legal persecution of witches.[33] Equally important, the differences in attitudes meant that there was never a good match between popular conceptions of witchcraft and the elaborated theories of the demonologists. This misfit was worsened by the dearth of effective intermediaries at village level; rural priests (in the days before the seminaries), local notaries and lawyers, medical practitioners of dubious competence, were all more at home with popular than with élite culture. Crucial though these men were in conducting the trials which actually took place, their interests were generally restricted to their immediate neighbourhood, while it would be many decades before most of them were affected by any real degree of internalized repression, if indeed (the clergy apart) they ever were. The legal mechanism for a great witch-hunt was put in place, dangerous ideas circulated in the intellectual world, yet the response at local level seems to have been slow and hesitant.

In principle there should have been an enormous supply of ready-made victims among the population, given the power and durability of the popular beliefs; we therefore need to explore the reasons why the apparently formidable new legal machinery did not evoke a more sweeping response.[34] Peasants must have been used to coping with witches through less formal agencies, which did not become suddenly obsolete. A range of protective folklore was available, much of it based on the power associated with the church; sacramentals such as sacred hosts and holy water were important in this context. Better-educated priests may have put up some resistance over the more dubious rites they were asked to perform, but it was difficult to stand out against persistent popular demands. The *curé*'s very reluctance might encourage the idea that he had exceptional power at his disposal, while if he proved obstinate people could always turn to the *devin* for help in identifying the source of their problems and for magical-cum-religious remedies. There is abundant evidence to show that down to

[33] For this process, see the works by Mandrou and Soman cited above, nn. 5–6, 10, and 18.

[34] There is an obvious danger of exaggerating the capacities of the criminal justice system, especially in the light of Soman's observations (see n. 18 above).

the present century priests have been widely credited with supernatural powers against witches, beliefs supported by a mass of folklore. Even today demands for blessings or even exorcisms do not always go unanswered, and in earlier times many clerics offered services little different from those of the *devins*. Some secret visitation articles of 1615, for the area immediately around Toulouse, reveal the presence of magical operators in nine villages; in two cases the number is unspecified, for the remaining seven it includes eight women, one man, and three priests.[35] One may compare this with an enquiry made by a modern ethnologist over a rather larger area in Anjou in 1960–1, which turned up a total of 144 similar healers known to have practised over the previous decade; seventy-seven were still active, and the grand total included six priests, two members of female religious orders, and two priests' housekeepers.[36] With few exceptions these people used techniques and books dating back to the seventeenth and eighteenth centuries, if not earlier; some of them could also claim a family tradition of healing stretching back as long. In many areas these practices have proved more durable than Christianity; one *curé* claimed that his parishioners, anti-clerical in every other respect, insisted on baptizing their children because otherwise the remedies offered by the *devin* would not work.[37] The most substantial recent study, by Jeanne Favret-Saada, shows how tenaciously the peasants of the bocage country around Mayenne have clung to a belief in malevolent witchcraft, defying the disbelief and contempt of the educated, and providing employment for specialists in counter-magic.[38]

[35] Archives départementales de la Haute-Garonne, 1 G 489.

[36] M. Bouteiller, *Médecine populaire d'hier et d'aujourd'hui* (Paris, 1966), 150–6. The author adds the necessary caution that the clerics in question had a reputation for special powers, and need not have engaged in overtly magical practices. For earlier priests as healers, see ibid., 28–32.

[37] For this *curé*, see ibid., 134.

[38] Favret-Saada, *Deadly Words, passim.* For the widespread witchcraft belief of the late 19th century, see E. Weber, *Peasants into Frenchmen: the Modernization of Rural France, 1870–1914* (London, 1977), 23–9, and J. Devlin, *The Superstitious Mind* (New Haven, Conn., 1987), esp. 100–19, 165–84. There is an excellent example of a late 18th-century witchcraft case in the south-west, which led to the accuser being fined by the courts, in E. Le Roy Ladurie, *La Sorcière de Jasmin* (Paris, 1983), 35–69.

Evidence from other regions of France tends to confirm the view that peasants have always had ready access to a network of supernaturally gifted healers, many of them within easy reach on foot.[39] Valuable members of the community, we might think; quite often giving their services free, curing a large number of people through psychosomatic suggestion, many of their remedies probably less dangerous that the more active intervention of the doctors. Yet it is plain that many of them are also feared, and that few peasants would care to quarrel with such potent adversaries, given the general assumption that the power to heal is inseparable from that to harm. This belief has been widely applied to God and the saints, with a whole range of illnesses traditionally described as 'maux des saints' supposedly sent by an offended saint, and curable by undertaking a pilgrimage to his or her shrine. The ambivalence of the popular attitude to saints was beautifully illustrated, as recently as 1947, when a woman taking part in a saint's day procession remarked: 'L'Bon Dieu, on peut encore le déranger tous les jours, mais ces ch'ti saints, c'est si vengeantieux, qu'il ne faut se risquer à les déranger que le jour qu'ils permettent'. ['You can still bother the good Lord every day, but these miserable saints are so vindictive that you must only risk disturbing them on the day they permit'].[40] Just as individual saints had their specific illnesses, so the healers were specialized; *rebouteurs* dealing with strains and fractures, *tireurs des saints* identifying suitable intercessors and undertaking pilgrimages by proxy, *conjureurs* employing magical prayers and empirical remedies, *leveurs de sorts* revealing the source of bewitchment and operating counter-magic, and so on. While precise classification is impossible, each community would know the personal competence of the local healers, and also be aware of more distant, supposedly more powerful figures to be approached in the most serious cases. This could be on account of their special repute, as with the sinister figure of Dom Jean de Niderhoff in Lorraine, a monk who identified bewitchment; a cleric of this kind naturally attracted a clien-

[39] For this subject in general, see F. Lebrun, *Médecins, saints et sorciers aux 17e et 18e siècles* (Paris, 1983).

[40] This was near Châteauroux; M. Bouteiller, *Sorciers et jeteurs de sort* (Paris, 1958), 204–5.

tèle from far and wide.[41] More generally, however, one can detect the idea that a displacement of ten or fifteen miles gave added force to the whole operation, doubtless aided by the fact that these healers would in general be personally unknown to their clients, in their everyday personae as well as their, magical ones.

Witchcraft has always operated within such an ambience, and still does for some people. Although on one level the belief appears fundamentally negative, concerned with power relations and the infliction of harm, it can be turned round to display some unexpectedly positive features. A diagnosis of witchcraft was also an attempt to find a cure for the evils involved; given the current state of medical knowledge, it often appeared to hold out more hope of effective action than the 'correct' identification of a disease. Before we dismiss magic as simply inadequate technology, we should remember that official medicine was in many ways more dangerous, often being positively harmful. Progress in this field would be exceedingly slow; only towards the end of the nineteenth century did doctors come to recognize their own limitations, while most of the crucial developments towards an effective scientific medicine have taken place well within living memory. The doctors who wrote numerous theses around the turn of the century denouncing superstition and illicit medicine were probably better diagnosticians than the cunning folk, but did not necessarily have better remedies.[42] To the present day some people regard healing as proceeding

[41] Dom Jean makes several appearances in the testimonies given against suspects in Lorraine. In the case of Barbelline Goudot (AD. M.-et-M., B 3327, No. 5), for example, it was reported that he had offered a classic explanation for the loss of two fine horses. He told his client that a woman of the village was responsible, out of envy over some pasture she had wanted for herself, and that she would be a frequent visitor offering comfort to his wife during his absence. In this instance the accused's village of Domjevin was some 22 km from Dom Jean at Xanrey.

[42] There is quite a group of these medical theses published at this time, mostly of rather low quality, and marked by a transparent professional hostility. Among them are P. Bidault, *Les Superstitions médicales du Morvan* (Paris, 1899); A. J. Darmezin, *Superstitions et remèdes populaires en Touraine* (Bordeaux, 1905); G. Foucart, *Des erreurs et des préjugés popularies en médecine* (Paris, 1893); and F. Kerambrun, *Les Rebouteux et les guérisseurs: croyances popularies* (Bordeaux, 1898).

as much from faith and special personal powers as from applied science. Two groups of people possessed such powers, the *devins* or cunning folk, and the witches, the latter of course only by undoing the effects of their own malice. While in principle they were clearly distinguished, individuals could easily cross the boundary. A suspected witch might purchase toleration by the regular offer of healing, just as the *devin* who failed to effect a cure could fall under suspicion. In the period of persecution such changes of status were evidently quite common in France, and it is highly unlikely that this resulted from the attempts of lawyers and clerics to equate all magical practices with the power of the devil. The rich evidence from folklore and modern ethnography demonstrates the total failure of this campaign to change popular belief; not only have healing powers continued to be treated as essentially personal, but *curés* who deplored popular superstitions have found themselves incorporated into the system against their will, cast as magical protectors of their flock.[43]

One striking similarity between *devins* and witches has been the emphasis on heredity; the healing gift is passed down within a family, just as the contamination of witchcraft might be. Some modern healers can trace their antecedents back into the eighteenth century, while in a remarkable case near Sancerre a family was pursued until at least the first world war by a reputation for witchcraft antedating trials and executions that took place in the 1580s. There are other recent instances of the children of suspect families being ostracized at school, a reminder of the greater dangers they would have run in the past.[44] One of the common allegations in Lorraine was that bewitchment had followed a children's quarrel in which terms such as 'fils de sorcière' had been used, while witnesses were quick to refer to family reputation whenever possible. Although the belief in inherited qualities was very general in

[43] Bouteiller has numerous examples in both *La Médecine populaire* and *Sorciers;* in the latter, p. 189, there is a typical story of a *curé* persuaded to look at a sick calf, then held to have cured it. For enduring belief in the magical powers of priests in the Nivernais, see J. Drouillet, *Folklore du Nivernais et du Morvan* (La Charité-sur-Loire, 1957–) iv. 139–41; in the Sologne, C. Seignolle, *En Sologne* (Paris, 1967), 190–1. See also Weber, *Peasants into Frenchmen,* 26–7, and Devlin, *The Superstitious Mind,* 20–1, 206–9.

[44] For these cases, see Bouteiller, *Sorciers*, 99, 159.

early modern Europe, better-educated experts were cautious on the subject, with Boguet considering convicted relatives as a comparatively minor *indice* against a suspect.[45] Here again the notion of the pact did not mesh very well with popular conceptions, and could give rise to awkward debates about the minimum age at which children became liable to the full rigour of the law. Direct accusations against children were very uncommon within peasant society, since the power was supposed to be concentrated in the hands of their mother or father, like a material possession, to be passed on only at the time of death, or perhaps on the child reaching maturity, so that it was normally the parent who was alleged to have exacted vengeance for insults to the family as a whole.

Another important component of popular belief was the idea that witchcraft allowed the intrusion of hostile external power into the normally protected space of the village, as in the beliefs about storm-raising and the protection that might be obtained by ringing the church bells. This emerges in the folklore as a tendency to identify witches as outsiders; migrants such as building workers or shepherds, forest dwellers, pedlars, and of course wandering beggars.[46] In this way the antisocial behaviour of the witch could be explained as the reaction of an individual who was excluded from normal social relationships, a description which could perhaps be extended to solitary old women who had become a burden on the community. It is clear, however, that the folklore is not a good guide to reality on this point, whatever period is in question. In practice the great majority of suspects have always been neighbours of those who accused them, and one may perhaps interpret the emphasis on literal outsiders in the folklore as expressing the communal reluctance to see the evil as internal. Village story-tellers must have developed and preserved such notions, particularly at the winter evening meetings known as the *veillées*, where it is generally agreed that much of the talk and stories centred on the supernatural, notably on the doings of witches and werewolves. One common theme was apparently the evil power of beggars,

[45] H. Boguet, *Instruction pour un juge en faict de sorcelerie*, appended to his *Discours des sorciers* (Lyon, 3rd edn., 1610), pp. 14–15, art. XIV.

[46] Bouteiller, *Sorciers*, 157–84.

always ready to avenge a poor reception by bewitching men, children, or animals, when they did not do so by outright theft, violence, or arson.[47]

This last example again illustrates the pervasive tendency for distinctions between natural and supernatural to become blurred, if they were made at all. Although the theory which explained a witch's powers might invoke menacing external forces, in practice the holder did not carry the kind of threatening supernatural 'charge' we might expect. The witch was a dangerous individual, and not a person to quarrel with, just as a strong and ill-tempered man might be. In the same way that anthropologists have observed in other societies, French villagers did not find it necessary to go around worrying constantly about the presence of witches and magical powers; they treated them with a certain fatalism, as risks of ordinary life. A shrewd person knew how to manage these aspects of the environment, which had nothing very strange or shocking about them most of the time. Given the conception of the power as inherent in certain individuals, and its link to commonplace feelings of ill will, there was a distinctly static or repetitive character attached to village belief. It was left to the demonologists, members of the educated, town-dwelling élite, to turn this accepted state of affairs into a vast diabolical conspiracy which threatened to ruin 'le genre humain', a cancer which needed to be burned out. The temperature of feeling in the village could of course rise suddenly, with the community uniting against the witches under the stress of some immediate crisis, whether general or personal. This had probably always been the case, and from the early middle ages to very recent times there are examples of direct action against suspects, resulting in their death or expulsion. From the *lettres de rémission* issued by the ducal authorities in Lorraine we know of three such cases of murder between 1485 and 1500, one in 1564, and a further ten between 1578 and 1603.[48] It is

[47] On the *veillée* in general, see Weber, *Peasants into Frenchmen*, 413–18. For the kind of stories told, see Drouillet, *Folklore du Nivernais et du Morvan*, i. 221, and for a large selection of recent examples, C. Seignolle, *Les Évangiles du diable* (Paris, 1964).

[48] *Archives départementales de la Meurthe-et-Moselle: inventaire sommaire, série B*, ed. E. Delcambre (Nancy, 1949–53), i. 95, 208, 227, 271; iv. 2, 30, 136, 162,

impossible to know whether these statistics bear any real relation to the frequency of such events, but it is striking how the two major groups coincide with the first outburst of witch trials, and then the major persecution in the time of Nicolas Rémy. For the *ressort* of the Paris *parlement* Soman has found twenty lynchings from the equivalent period, making the further point that no other crime gave rise to popular justice of this kind.[49] There have certainly been sporadic cases in France from the eighteenth century to the 1970s in which identification of a witch by a *devin* has led to murder.[50] In their sudden, brutal unpredictability these events fit into the general pattern of village crime, in which tensions that have apparently been kept under control flare into brief moments of violence. One of the Lorraine cases illustrated the difficulties facing the authorities if they wanted to be lenient; fifty inhabitants of Stainville killed two women accused of *maléfice* and condemned only to banishment.[51] Such possibilities concerned Bodin greatly, for he thought that if the courts did not respond to popular pressure, the people would turn against the judges, stoning them as they had done the witches.[52]

Despite the attempts to reassign them *en bloc* to the devil's camp, the *devins* regularly acted in the role of witch-doctors, confirming the opinions of their clients about the nature of their problems and the identity of the witches. Unlike their African counterparts, they could not use the highly charged atmosphere of public meetings as part of the process, but this may not have made much difference to their effectiveness. They reinforced their clients' wish to explain a special misfortune, the normal reason for a consultation, by a specific human cause, about which they could take some action. While

166; v. 47, 65, 75, 96, 99, 176, 180, 185, 197, 313. R. des Godins de Souhesmes, *Étude sur la criminalité en Lorraine, d'après les lettres de rémission (1473–1737)* (Paris, 1903), cites two more cases of killings by individuals in 1685 and 1689.

[49] Soman, 'La Décriminalisation', 179–83.

[50] For two 19th-century cases, see Weber, *Peasants into Frenchmen*, 25, and for the problem more generally see Devlin, *The Superstitious Mind*, 113–16. For the murder of a supposed witch in the Orne in 1976, see Muchembled, *La Sorcière*, 226–8.

[51] Delcambre, *Inventaire sommaire*, v. 313.

[52] Bodin, *Demonomanie*, fo. 166r.

just about every form of misfortune could in theory be ascribed to witchcraft, in practice people were generally more discriminating, so far as the very limited evidence suggests. Any illness which was not immediately recognizable was an obvious candidate, whether it afflicted persons or animals; the local surgeon, blacksmith, or butcher might encourage such ideas by pronouncing the sickness unnatural. Slow lingering illnesses which failed to respond to treatment inevitably aroused suspicions, as did sudden deaths of previously healthy animals, especially if a particular individual experienced several such misfortunes. Mental illness was frequently attributed to witchcraft, sometimes through the more complex agency of demonic possession. Witches were generally believed to be able to produce hailstorms by beating water, either communally at the sabbat or individually; perhaps this notion reflected not only the devastating nature of such storms, but also their unpredictability and tendency to affect only very local areas. Other communal misfortunes sometimes credited to witchcraft were sudden late frosts and animal epidemics. On a more individual basis there was a widespread fear that impotence could be induced by the 'nouement d'aiguillette' which involved tying a knot in a lace at the moment of marriage, although this seems to belong to sorcery rather than witchcraft, and did not usually appear in prosecutions.[53] Everyday activities such as butter-churning and milling might be interrupted mysteriously after a refusal to give milk or a dispute over the quality of flour. The idea of witches using special powers to avenge themselves naturally gave rise to the notions of making the punishment fit the crime which are apparent here, with the person who had failed to give being magically deprived of the commodity with which they had been so stingy. Taking all the cases together, however, only a minority appear to involve misfortunes which were blamed on witchcraft because their specific nature fitted some general popular conviction. More commonly the victim thought of the possibility because there was a suspect ready to hand, in the form of someone already reputed a witch, with whom some altercation or quarrel had recently taken place. Time and

[53] E. Le Roy Ladurie, 'L'Aiguillette', in his *Le Territoire de l'historien* (Paris, 1978), ii. 136–49.

again the cases reveal this pattern, as when a father and son, Alain and Charles Pelé from Mongues in the Berry, were condemned for bewitching Françoise Plansson in 1608. She denounced the son,

disant que ledict Charles Pelé luy avoit donne un sort, & l'avoit faict venir malade en telle sorte qu'il luy sembloit que les fourmis la mangeoient depuis la ceinture jusqu'aux extremités des orteils, diffamé d'estre sorcier, & luy vouloit mal, parce que ledict sorcier avoit opinion, que ladicte Planson luy avoit faict mourir sa chatte.

[saying that the said Charles Pelé had put a spell on her, and had made her ill in such a way that she felt ants were eating her from the waist to the ends of her toenails, he was reputed a witch, and wished her ill, because the said witch believed that the said Planson had caused the death of his cat.][54]

DISPUTES, GUILT, AND PROJECTION

Confronted with an apparently similar situation in England, Keith Thomas and Alan Macfarlane have been able to expose a further layer of meaning in such histories of quarrels and menaces followed by witchcraft accusations. The supposed victim had usually refused some service which formed part of the normal obligations towards neighbours; the loan of an object, or a small gift of food or drink. The departure of the angry witch had been followed by some abnormal misfortune, with the guilt felt at refusing her turned into anger at her disproportionate reaction. It can even be argued that this pattern, whereby hostile feelings were projected back on to the person originally wronged, was a functional and constructive one, helping to shift communal values in the direction of a more individualistic society, and breaking down medieval notions about charity.[55] It is plain that French cases do possess many common features with their English counterparts, and that there is a basic similarity in the psychological relationship between accuser and accused. The theme of begging is a pervasive one, leading to confrontations such as that between the suspect Clauda Gaillard and one of her neighbours:

[54] De Lancre, *L'Incrédulité*, 794.

[55] K. V. Thomas, *Religion and the Decline of Magic* (London, 1971), 435–583; A. Macfarlane, *Witchcraft in Tudor and Stuart England* (London, 1970).

comme Marie Perrier luy eut une fois refusé l'aumosne, elle luy souffla fort rudement contre, de facon que Marie tomba par terre, & s'estant relevée avec peine, elle demeura malade quelques jours, & jusques à tant que Pierre Perrier son neveu eut menacé la Sorcière.

[since Marie Perrier had on one occasion refused her alms, she breathed very heavily on her, in such a way that Marie fell to the ground, and after getting up with difficulty she remained ill for several days, until her nephew Pierre Perrier had threatened the witch.][56]

The demonologists themselves emphasized the frequency of such occurrences, as when Rémy suggested that most witches were beggars who hoped to make money by curing the illnesses they had caused.[57] He also drew attention to the curses which came 'to the lips of witches in our time when they have been begging and someone has refused them; for nothing is so common as for them to utter a wish that all his family may die of starvation, that his wife may give birth to monsters, and his whole house be infected with prodigies and portents'.[58] All the authorities agreed that witches were poor and miserable, often driven to make a pact with Satan out of despair. Bodin thought that the charitable enjoyed special protection, and in a remarkable passage claimed that:

les sorcieres confessent que celuy qui est aumosnier, ne peut estre offensé des sortileges, encores que d'ailleurs il soit vicieux . . . C'est pourquoy les sorciers qui sont contraincts par Satan de mal faire, tuer, empoisonner hommes & bestes, ou bien estre tourmentez sans relache, quand ils n'ont point d'ennemis, desquels ils se puissent venger, ils vont demander l'ausmone, & celuy qui les refuse, ayant dequoy donner, sera en danger, pourveu qu'il ne sache qu'ils soient sorciers. Car le Sorcier n'a point plus de puissance que sur celuy qui luy donne l'ausmone, s'il sçait qu'il soit Sorcier. Et se faut bien garder mesmes de donner l'ausmone à celles qui en ont le bruict: mais celuy qui ne leur donnera l'ausmone, ne sçachant qu'ils soyent sorciers, à grand peine eschapera il qu'il ne soit offensé, comme il s'est verifié souvent. Et de fait i'ay sçeu, estant à Poictiers aux Grands iours l'an mil cinq cens soixante sept, entre les substituts du Procureur general, qu'il y eut deux sorciers fort piteux & pauvres,

[56] Boguet, *Discours*, 175.

[57] N. Rémy, *Daemonolatreiae libri tres* (Lyon, 1595), Bk. III, ch. v.; Eng. trans. by E. A. Ashwin (London, 1930), p. 159.

[58] Ibid., Bk. II, ch. ix. T. Ashwin, p. 125.

qui demanderent l'ausmone en une riche maison: On les refusa: ils ietterent là leur sort, & tous ceux de la maison furent enragez, & moururent furieux: non pas que ce fust la cause pour quoy Dieu les livra en la puissance de Satan & des sorciers ses ministres, mais que d'ailleurs estans meschans, & n'ayans pitié des pauvres, Dieu n'eut point pitié d'eux.

[the witches confess that the charitable person cannot be harmed by their spells, even if he is otherwise wicked . . . This is why the witches who are compelled by Satan to do evil, to kill and poison men and animals on pain of perpetual torment, when they have no enemies on whom they can take vengeance, go begging, and anyone who has anything to give and refuses them will be in danger, as long as he does not know they are witches. For the witch has no greater power than that over the person who gives alms knowing that he gives to a witch. And one must be very careful not even to give to those who are reputed so; but any who refuse them alms, not knowing them to be witches, will have great difficulty escaping unscathed, as has often been proved. And in truth when I was at Poitiers for the Grands Jours of 1567, as one of the *substituts* of the *procureur général*, I heard of two pitiful and poor witches, who asked for charity at a rich house. When they were refused they bewitched the house, and all the inhabitants contracted rabies and died mad. It was not the cause for God to deliver them into the power of Satan and his agents the witches, but since they were wicked in other ways, and had no pity for the poor, God had no mercy for them.][59]

Despite this striking piece of evidence, the strict refusal–guilt syndrome found in England cannot be exported unchanged to French-speaking Europe. Too many of the specific accusations do not quite fit within it. What becomes clear is that the English cases form a sub-set within a more general aggression–guilt model, which does encompass the majority of the French evidence. Most of the quarrels between witch and victim involved overt hostility, with the denial of neighbourliness extending over a much wider field than the straightforward refusal of alms, so that guilt would not necessarily have been induced by the consciousness of failure to perform a socially expected act of charity. In practice this is unlikely to have made very much difference; it is a matter of common human experience that quarrels and displays of aggression leave the participants shaken and uneasy, guilty

[59] Bodin, *Demonomanie*, fo. 124v.

over their loss of control even when convinced it was justified, and anxious about the hostility they may have aroused in others. Demonologists, judges, accusers, and witches making confessions gave overwhelming support to the idea that bewitchment was the result of spite and the desire for revenge, as in the passage from Bodin just quoted, where it is strongly implied that the revenge motive was actually necessary. In a case cited by Boguet:

Jacques Bocquet ayant este battu par l'hoste de Myjoux, proposa de se venger du tort, qu'il reputoit luy avoir esté faict: il met de la poudre sous le sueil de la porte d'une buge, ou l'hoste tenoit sept veaux, cinq desquels luy appartenoyent, & les deux autres à un sien voisin: les sept veaux retournans des champs passent par dessus le sueil, cinq d'iceux, scavoir les veaux de l'hoste moururent aussi tost, les deux autres demeurent sains & entiers.

[Jacques Bocquet, having been beaten by the innkeeper of Myjoux, decided to take vengeance for the wrong he thought had been done him; he put poison powder under the threshold of a stable, where the innkeeper kept seven calves, of which five belonged to him, and the other two to a neighbour: the seven calves passed over the threshold as they returned from the fields, and the five belonging to the innkeeper soon died, while the other two continued in good health.][60]

Where Boguet puzzled himself over how the diabolical powder could be so discriminating, we may note the specificity of the revenge, operating only against the original aggressor.[61] Jacques Bocquet would have been showing himself obedient to the devil, who according to both Bodin and Boguet told the witches at the sabbat: 'Vengez vous, ou vous mourrez'.[62] Boguet explained that the devil recruited the poor and miserable: 'Il offre aux vindicatifs des moyens pour se venger de leurs ennemis, & pour se faire redouter.' ['He offers

[60] Boguet, *Discours*, 162.

[61] Rémy discusses such powders in *Daemonolatreiae*, Bk. II, ch. VIII, trans. Ashwin, pp. 117–20. He concludes that the powder has no effect in itself, but that the demon causes the harm by unknown supernatural means; he too cites a case where five specific cows were made sick. This was also Bodin's position, *Demonomanie*, fos. 113ʳ–116ʳ, but he seems to contradict himself flagrantly over the Jeanne Harvillier case mentioned below, where he was following the actual depositions in the trial.

[62] Bodin, *Demonomanie*, fo. 86ᵛ, Boguet, *Discours*, 140.

the vengeful means to revenge themselves on their enemies, and to make themselves feared'].[63]

Bodin's star witness Jeanne Harvillier was also seeking revenge when she tried to bewitch a man who had beaten her daughter, although on this occasion the wrong person came into contact with the powder, and the devil frustrated her attempts to cure him.[64] The contradiction here, with the powder in one case affecting only the intended target, in another harming the first person to pass over it, is all too typical of demonological theory as a whole, in which 'heads I win, tails you lose' often seemed to be the normal form of argument. A few years later, in 1586, Marie Martin used powder made from dead men's skulls against Jean Bisel, 'parce que la femme dudit Bisel luy avait prins ses glennes dedans les champs'; gleaning was an important resource for the poor, and a constant cause of minor friction, but this sounds like a dispute over customary rights rather than specifically over charity.[65]

The mass of surviving trial documents in Lorraine give a similar picture, with a wide range of village quarrels adduced to explain particular acts of *maléfice*. On this point as on many others Nicolas Rémy provides a very fair series of examples, which seem quite typical of the trials as a whole; close parallels to all the motives he cites can be found in the manuscript material. Benoit Drigie felt he had been overtaxed by the village assessors so he killed 150 cattle belonging to his neighbours, while when Alexée Belheure was not invited to a baptismal feast she caused the death of the unfortunate child. Jana Armacourt hid three sheaves of corn, stolen from a neighbour's field, in Alexée Cabuse's garden; unfortunately the latter saw her and gossiped about the theft. To avenge her for the evil name she thus acquired, the witch's demon produced a whirlwind which broke the gossip's leg as she chased a runaway animal. It was a saying all over Lanfracourt 'that if anyone wanted to keep himself and his possessions safe

[63] Boguet, *Discours*, 43.

[64] Bodin, *Demonomanie*, fo. 129^{r}. This case is also described, with reference to the orginal trial documents, by P. Villette in *La Sorcellerie et sa répression dans le Nord de la France* (Paris, 1976), 95.

[65] Mandrou, *Magistrats*, 148.

and whole, he must avoid being cursed by Jana Ulderique'; when Jean Conard refused to pay what she asked for helping to watch over the village cattle, she suffocated his baby in the cradle. Jacobeta Weher killed a neighbour's daughter out of jealousy for her lover, while Bertrande Barbier hated her son because she suspected him of stealing money from her, and killed him with a poisoned drink.[66] Most of these individuals were held responsible for a whole series of such crimes, supposedly committed by their demons or the administration of the poison powder supplied by the devil. A similar picture emerges from the work of Dupont-Bouchat on Luxembourg, Muchembled on the Cambrésis, and Monter on the Jura region, who all agree that the witnesses were concerned with such acts of *maléfice*, rather than with the question of devil-worship. Very often the accused, who had heard the testimony against them when they were confronted with the witnesses, did no more than take over the motivation others ascribed to them.

The contrast between French and English witch-hunting therefore appears far less than is sometimes suggested; in both areas the majority of accusations came from below, and concentrated on actual harm supposedly caused by *maleficium*. The element of diabolism in the French cases was superimposed by the local judges and clerics, men who stood somewhere between the village community and the educated urban élites, with relatively little influence on popular belief or effect on the process of identifying a witch. The pattern of aggression or denial of neighbourliness followed by guilt is also common, although with important variations. In England the provision of charity to the poor had become a crucial point of strain, coinciding with questions about the very nature of social obligations, at a moment when a new formalized system of relief was being devised. In France and the other regions under discussion, though the poor were becoming a serious social problem, they were creating strains within a system which was never seriously challenged, and within village communities which retained a more traditional

[66] For these cases see Rémy, *Daemonolatreiae*, Bk. I, ch. II, Bk. II, chs. I, VI, VIII, IX, trans. Ashwin, pp. 3, 92, 114–15, 119, 122.

character.[67] The English peasantry seem to have lost their land far more rapidly than their French counterparts, and perhaps in consequence to have become more mobile.[68] Though the process was extremely painful, it must have broken down some of the more claustrophobic features of village life, which seem to play such an important part in French witchcraft traditions. It may also have contributed to the greater efficiency of English agriculture, which helped to prevent major famines, whereas on numerous occasions between the late sixteenth and early eighteenth centuries large numbers of French peasants virtually starved to death. The highly precarious state of French agriculture put almost everyone at risk, so that the quarrelsomeness so often apparent in the witchcraft cases can plausibly be linked to pervasive fears of a hostile environment. The decline in old-fashioned almsgiving in England had much to do with the reformation and its attack on formalized good works, which was accompanied by the removal of the protective magic associated with the old church. It is not easy to detect the results of this striking change from a comparison of witchcraft persecution in protestant and catholic territories, which is not entirely surprising when so many complex factors are involved. Moreover the contrast was very relative; the Anglican church did not abandon all rituals, nor transform its clergy overnight, while catholic reformers were also seeking to eliminate superstitious practices.

MECHANISMS OF ACCUSATION

Catholic peasants who thought themselves threatened by witchcraft might use sacramentals, such as holy water, consecrated hosts, church candles, or objects blessed by the priest, as counter-magical agencies; often such techniques were advised by the *devins*. They could also ask the priest to come and exorcise buildings, persons, or animals, although there is not a great deal of evidence for this happening in practice except in cases of supposed possession, perhaps

[67] See in particular J.-P. Gutton, *La Société et les pauvres: l'exemple de la généralité de Lyon (1534–1789)* (Paris, 1971) and *La Sociabilité villageoise dans l'ancienne France* (Paris, 1979).

[68] For relative mobility, see P. Laslett, *Family Life and Illicit Love in Earlier Generations* (Cambridge, 1977), 76–86.

because it was rather costly, or because priests were aware that such acts were not really neutral, but came close to counter-magic against the suspect. It is also likely that such generalized practices were seen as protective rather than curative in most cases; once witchcraft was suspected, this was a sign that they had failed. Possession cases were an obvious exception, where the sickness was recognized as more spiritual than physical, and the church itself claimed special powers. In general, however, the priest had failed to establish himself as the only authorized holder of supernatural power, and only if he set himself up in the role of a virtual *devin* would he be the first resort of the afflicted. Pilgrimages and offerings to saints were much more in evidence, associated with beliefs about specific abilities to cure, and implying the propitiation of threatening external powers, but they were normally abandoned once a definite diagnosis of witchcraft had become established. Popular belief seems to have made direct action involving the witch the only really effective counter-measure, either by forcing her to visit and offer healing, or by taking substances from her house (in Lorraine bread, salt, and ashes). If this analysis is correct, then the reformation might have increased general anxiety by removing protective rituals, but would not have made so much difference once suspicions had begun to crystallize. It is also likely that less formal protective magic largely filled the gap, since the church had never held a monopoly in this area. The remarkable coincidence of major persecutions through most of western Europe in the century 1560–1660 makes it very difficult to ascribe great importance to this particular aspect of protestantism, unless one supposes (most implausibly) that an extreme form of post-Tridentine religiosity spread almost instantaneously across catholic Europe, with similar effects.

This is not to deny all importance to the intense religious conflicts and experiences of the period; there is much force in Christina Larner's characterization of the 'age of faith', during which the labelling and punishment of deviants became a preoccupation of the state.[69] The creation and extension of legal machinery and demonological theory, without

[69] Larner, *Witchcraft and Religion*, 113–26.

which there would have been neither trials nor records, can be related to such trends. A necessary condition does not have to be a sufficient one, however, and the detailed study of actual cases gives little support to the idea that French witch-hunting was primarily inspired by the élites. Even where there is a clear link, as with the proclamations made by the archdukes in the Spanish Netherlands in the 1590s, the evidence is ambiguous; the rhythm of trials over a longer time-span seems to follow an inscrutable pattern of its own, and the rulers soon modified their attitude when they realized that persecution could breed social disorder.[70] The machinery could be set up, but in the end the supply of witches depended on the willingness of their neighbours to denounce them, outside the occasional episodes where enthusiastic judges or professional witch-finders set a real panic going. As already suggested, one of the most interesting questions is why the peasantry showed such restraint in employing this terrible weapon against their enemies, and to make any progress towards understanding this puzzle one must return to an examination of typical cases in their local setting.

Within the steady trickle of cases which made up the routine of persecution one can discern two main groups, in terms of the immediate precipitants for arrest and trial. The first comprised those who were denounced by fellow-villagers, acting either individually or as a group. The second, probably at least as large, was made up of persons identified by convicted witches as having been present at the sabbat. In fact the overwhelming majority of this second group were plainly named in this way because they possessed a local reputation for witchcraft, and the two sets of trials hardly differ at all in terms of the numbers of hostile witnesses and the alleged *maléfices*. To be picked out in this way by a witch did not automatically lead to a trial, although it would tend to build up suspicion; probably the decision on how to proceed depended on the state of local opinion. The general picture is

[70] The incidence of trials in Luxembourg and parts of Flanders and Hainault is discussed by Dupont-Bouchat and Muchembled in *Prophètes et sorciers*, 124–38 and 172–8. The increased sensitivity to abuses in the local courts from 1595 onwards is well documented in Villette, *La Sorcellerie*, 181–8.

therefore that spontaneous denunciations were never very common, but once they occurred there was a danger of a snowball effect generating a little chain of interrelated trials. There is plenty of evidence within the trials themselves to suggest that many prominent suspects had the good fortune to slip through the net, and died unmolested in their own beds. The general impression is that there was no particular desire among the peasantry to take witches to court, as long as other means of countering their supposed evil practices provided alternative solutions. Bodin was very conscious of this problem, and suggested that, since 'les pauvres simples gens craignent les Sorciers plus que Dieu, ny tous les Magistrats, & n'osent se porter pour accusateurs, ny pour decelateurs' ['the poor simple people fear the witches more than God, or all the magistrates, and do not dare present themselves as accusers, or as disclosers'] France should follow what he claimed was the example of Milan and Scotland, with special locked boxes in the churches to allow anonymous denunciations.[71] Even had this insidious proposal been put into effect, one may doubt whether it would have made very much difference. Quite apart from the difficulty of keeping anything secret within village society, particularly if it had to be written down, once the trial began villagers would still have had to declare themselves, so a substantial consensus of local opinion would have been just as necessary. Such consensus was normally required before a denunciation was launched, as is demonstrated by the enormous length of time for which most of the accused were said to have been reputed witches.

A typical sequence of events would begin with an individual suffering misfortune in circumstances which made him suspect a reputed witch with whom he had recently quarrelled. He would then ask her to visit his house, and consult her about the illness affecting himself, his wife, children, or animals. If she proved uncooperative, or failed to effect a cure, he might visit the *devin* (if he had not already done so) to confirm the diagnosis, and possibly engage in the more dubious forms of counter-magic which sought either to turn the spell back on the witch, or force her to declare herself. At

[71] Bodin, *Demonomanie*, fo. 168r. This system of letterboxes for anonymous letters was certainly used in Venice; see P. Burke, *The Historical Anthropology of Early Modern Italy* (Cambridge, 1987), 101, 127.

some stage during this process the sickness might go away, thus confirming the efficacy of the steps taken, or the victim might decide to accept the loss of animals or relatives once this was beyond helping. In the latter case the episode would remain as potential evidence in a future trial, often remembered by relatives or friends as well as by the prime sufferer. Such persons formed an alternative route by which the suspect might learn of the allegations against her, either direct or through the grapevine of village gossip; the latter might also carry the news that someone was consulting the *devin*. This might well prove enough to bring the witch round, anxious to demonstrate goodwill and make her peace, thereby confirming both her guilt and the power of the *devin*'s magic. In response to an open or implied demand that she remove the spell, she might then offer some magical remedy of her own; most peasant women would have known some such cures, even if they did not practice regularly as healers. Bodin's Barbe Doré confessed that she had cured several victims by the well-known technique of cutting open a pigeon and placing it on the sick person's stomach, while saying a special prayer to the Trinity and some saints, then ordering some additional religious observances.[72] Even if the trials did not contain innumerable references to such events, the concern of the demonologists would make us suspect that this kind of enforced healing by the witch was commonplace. Bodin, Rémy, and Boguet all went to considerable lengths to show that such cures were in themselves a form of witchcraft, and involved a tacit pact with the devil.[73] The demonologists also complained of the tolerance generally shown towards the *devins*, themselves obviously servants of the devil. Bodin and de Lancre, in particular, blamed virtually everyone for the lamentable failure to repress witchcraft successfully; lax or sceptical judges, mocking doctors, pliant clerics, and cowardly neighbours were all at fault.[74]

The long pre-history of earlier *maléfices* normally recounted once a witch came to court emphasizes not only the slow

[72] Bodin, *Demonomanie*, fos. 145ᵛ–146ʳ.

[73] Ibid., Bk. III, ch. II, fos. 127ᵛ–132ʳ; Rémy, *Daemonolatreiae*, Bk. III, chs. II–IV, trans. Ashwin pp. 142–58; Boguet, *Discours*, chs. 40–2, pp. 246–86.

[74] Bodin, *Demonomanie*, preface and fos. 76ᵛ–77ʳ, 166ʳ–167ᵛ, 196ᵛ, 206ᵛ–207ᵛ, 211ᵛ–212ʳ; de Lancre, *L'Incrédulité*, 1–68, 361.

process of accretion by which reputations were created, but also the degree to which relief was normally sought through less formal agencies than the law. Bodin was certainly right to claim that people were afraid of retaliation, although it is impossible to determine the exact weight that should be attributed to this factor. He cited a case in which a prosecution failed, and the witch 's'en est si bien vengée, qu'elle a faict mourir des hommes & du bestail sans nombre, comme j'ay sceu des habitans' ['avenged herself so well, that she caused the deaths of innumerable men and animals, as I learned from the local people'].[75] Fears of this kind may well have led people to make physical attacks on defendants who were acquitted, so that a century later the sceptical Richard Simon thought it might be wisest to banish them from Normandy, 'car il eut été dangereux de les absoudre absolument, & de les renvoyer dans leur pays, ou leur vie n'auroit pas été en sûreté, tant le peuple est prévenu sur le fait des Sorciers' ['for it would have been dangerous to acquit them completely, and send them back into their localities, where their lives would not have been safe since the people are so prejudiced on the matter of witches'].[76] Indeed a direct physical attack must always have been attractive; the trials often tell us about witches being beaten, and one suspects that murder was commoner than the sources indicate. Until the late seventeenth century it would have been unlikely to have serious consequences, particularly if it was a communal crime (with stoning as the ideal mode for distributing responsibility widely). Such direct action also eliminated the possibility of revenge, whereas to invoke the courts was effectively to threaten the suspect with death, yet lose control over the process. Counter-magic was a much vaguer threat, leaving room for some arrangement of the original dispute, besides encouraging the witch to lift the spell. To the peasants any courts beyond the immediate neighbourhood appeared uncertain, remote, and potentially unsympathetic, even had the costs of an appeal in the French system not entailed the risk of

[75] Bodin, *Demonomanie*, fo. 166r.

[76] Richard Simon [M. de Sainjore], *Bibliothèque critique* (Amsterdam, 1708), ii. 121.

heavy expenses for individual or community. Formal rules of procedure often involved concepts of proof foreign to them, which might lead to gross miscarriages of justice as they saw it. Indeed, anything less than a death sentence on a witch denounced by tacit communal consent must have seemed inexplicable, and could lead to the kind of incident described by Bodin in the late 1570s:

il est advenu depuis un an à Haguenone pres ceste ville de Laon, que deux Sorcieres qui avoient merité justement la mort, furent condamnées, l'une au fouet, l'autre à y assister: mais le peuple les print, & les lapida & chassa les officiers.

[it happened a year ago at Haguenone near this town of Laon, that two witches who had justly merited death were sentenced, one to a whipping, the other to watch it; but the people seized them and stoned them, chasing the officials away.][77]

This case was the pivot of his crucial argument for the ruthless repression of witchcraft, which would eliminate the likelihood of such disorders. Bodin's view was the exact antithesis of that taken by the *parlementaires*, who much more sensibly recognized that such action would subvert judicial rules and encourage peasants to bring what were essentially uncontrollable assertions against their personal enemies. The known evidence certainly suggests that peaks of official persecution, far from pacifying village feeling, generated higher numbers of illegal killings.[78]

The depositions in a typical trial very often give details of previous episodes in which the presumptions against the witch seem so strong that one must suppose great reluctance to employ legal weapons. When Jeanne Huguet of Anjeux in the Franche-Comté (a region where trials were fairly common) was finally brought to court and banished in 1609, the story was told of her dispute with one Thierry, who asked her to contribute some wheat towards the maintenance of the schoolmaster. In her anger she struck him on the shoulder, before finally agreeing to the demand with the open wish 'qu'autant de jour de mal santé il put avoir que de grains de blé qu'elle

[77] Bodin, *Demonomanie*, fo. 166^r.

[78] See pp. 28–9 above.

lui avait donnés' ['that he might suffer as many days of ill health as she had given him grains of wheat']. Thierry became ill the same day, then lingered miserably for more than three years with some kind of paralysis until he died.[79] There are numerous similar cases in Lorraine, in which it is plain that the sufferer was convinced that a particular witch was to blame, yet during a long and fatal illness no action was taken. In contrast, once a case reached the courts large numbers of villagers were often ready to testify, apparently determined to take this opportunity to rid themselves of a neighbour they had long feared or hated. The merciless use of torture by some courts becomes easier to understand, although not to excuse, when one recognizes that the judges were very aware of the strength of local feelings, and of the selectivity with which legal proceedings were initiated. Once they did begin, however, the 'snowball' process already described gave a chance to settle accounts with other prominent suspects, who might include members of the original witch's family, and could also take in the *devins*. There was a serious danger of manufacturing more and more suspects by this method, as does seem to have occurred in a number of German cities, but within the French-speaking regions there are no obvious examples of this kind.[80] Judges did not extract great lists of participants at the sabbat, then torture them into confessing in their turn and naming yet more victims; they could actually see the dangers of such procedures. In the works of Bodin and Boguet, for example, we find great stress laid on the actual evil deeds of witches, alongside relatively moderate discussions of the extent to which common fame and single uncorroborated testimony could be sufficient to justify the use of torture.[81] Horror at the consequences of their arguments should not blind us to the fact that these judges were making a serious attempt to apply the law as they understood it; they were also conscious that they were pleading a case which did not command automatic agreement among their peers. While

[79] Bavoux, *Hantises et diableries*, 56–7.

[80] The German abuses are very well discussed by H. C. E. Midelfort, *Witch Hunting in Southwestern Germany, 1562–1684* (Stanford, Calif., 1972).

[81] Bodin, *Demonomanie*, esp. Bk. III, ch. II. fos. 172^{v}–180^{r}; Boguet, *Six advis en faict de sorcelerie*, printed as appendix to his *Discours*.

Boguet eventually came down in favour of sending to the stake those who confessed attendance at the sabbat alone, without any *maléfice*, he realized that many scrupulous lawyers disagreed.[82]

THE DOUBTS OF THE ÉLITES

All the demonologists present themselves as arguing a case against reluctant or sceptical colleagues, and this was no mere literary flourish. Mandrou demonstrated how the Paris *parlement* began by prohibiting the dangerous popular practice of the swimming test in 1601, a marginal step in itself, but indicative of a concern with the forms of procedure which led to the institution of an automatic appeal hearing in 1624, so that from that date any sentence involving torture, corporal punishment or execution had to be referred from the local courts to the *parlement* itself.[83] This decision was the prelude to the effective cessation of capital sentences, only three of which were carried out between 1626 and 1639, all on men who had made written pacts with the devil.[84] The *parlementaires* had considered such moves much earlier, however, even in Bodin's lifetime, for the automatic appeal system was proposed in 1588, when it may well have been the political chaos associated with the wars of the League which postponed its implementation.[85] The judges did not deny the possibility of witchcraft, but their own conduct of trials was governed by a more pragmatic awareness of the immense difficulty in obtaining satisfactory proofs of such hidden crimes.[86] They intervened to take all trials into their own hands because they were repeatedly confronted with evidence of grossly abusive procedures at the local level. Their efforts to control the misbehaviour of the lower courts, apparent as early as the 1580s, reached a climax in 1641 when the *lieutenant de justice, procureur fiscal* and *greffier* of Bragelonne, convicted of the murder of a suspected witch, were hung in the Place de

[82] Boguet, *Discours*, 541.
[83] Mandrou, *Magistrats*, 180–4, 340–8.
[84] Soman, 'La Décriminalisation', 197.
[85] Soman, 'Les Procès de sorcellerie', 809–11; 'La Décriminalisation', 189–96.
[86] This emerges very well from the remarks of the councillor Laisné in 1643, cited by Mandrou, *Magistrats*, 360–1.

Grève.[87] In the normal way slightly less flagrant abuses had been dealt with by informal reprimands to the relevant judges.[88] It is noticeable how the panics of 1587–8, 1623, and 1643–4 all produced sharp reactions from the judges, deeply concerned to put a stop to outbreaks of lynch law.[89]

As these facts suggest, the structure of the judicial system, itself a reflection of the power of the crown, was very important in determining the nature of persecution within the kingdom of France. A pyramidal organization, with the great royal courts supervising the local jurisdictions, and rights of appeal to the top in all serious cases, it gave the senior judges in the *parlements* a power and prestige unequalled elsewhere in Europe. These arrangements also ensured that such full-time professional judges built up relevant experience, with little chance of this being lost through long gaps between cases. The effect was certainly to put some kind of check on the system as a whole, discouraging abusive procedures in the lower courts and systematically reducing their sentences on appeal. Provincial *parlements* such as those of Toulouse, Dijon, and Bordeaux followed the lead of Paris in becoming steadily more critical of the details of procedure; only the *parlement* of Rouen stands out for its determination to support witch-hunting within its *ressort*. The one other *parlement* whose attitude appears dubious was the relatively minor court at Pau, whose members quickly backed down before the intervention of the council.[90] The attitude of many *parlementaires*, in the provinces as in Paris, was evidently one of practical scepticism, not denying the possibility of witchcraft, while fearing that persecution led only to miscarriages of justice and social disruption. By the seventeenth century shrewd observers were becoming

[87] Mandrou, *Magistrats*, 354–6.

[88] I owe this point to Alfred Soman; see n. 89 below for some of his numerous references to cases where formal sanctions were taken against local judges.

[89] Soman, 'La Décriminalisation', 189–97. For the reactions of other *parlements* in 1643–4, Mandrou, *Magistrats*, 372–5, 377–9, 383–94.

[90] Mandrou, *Magistrats*, 425–66, for this final panic and reactions to it. There is an interesting study of the end of prosecutions in Alsace by J. Klaits, 'Witchcraft trials and absolute monarchy in Alsace', in *Church, State, and Society under the Bourbon Kings of France*, ed. R. M. Golden (Lawrence, Kan. 1982), 148–72.

aware that there was a relationship between the numbers of suspects coming before the courts and the intensity of local concern; where witches were not actively pursued they somehow tended to fade out of sight.[91] Opinions on the subject among the élites were in fact extraordinarily diverse, which makes it difficult to interpret the persecution primarily as some kind of exercise in social control; it appears more as an unintended side-effect of a general assault on superstition, disorder, and immorality. The first two of these elements were just as likely to provide motives for disapproving of the procedures, since the accusations were not only primarily based on peasant superstitions, but led to judicial irregularities and social disruption. The Paris *parlement* was very conscious of these factors, and its appellate jurisdiction would have been still more effective in curbing witch-hunting if the sheer size of the court's *ressort* had not created a major problem of enforcement. The remarkable figures produced by Alfred Soman demonstrate that rather less than 10 per cent of those whose appeals were heard in Paris ended up at the stake; the rate did fall after 1610, but at its maximum in the 1590s was barely more than 20 per cent.[92]

Even the *parlement* of the Franche-Comté at Dôle, whose magistrates were fairly characterized by Monter as intellectual mediocrities, was surprisingly effective in checking enthusiastic local persecutors. A cyclical pattern is evident here with three short periods of heightened activity, in 1604–11, 1628–31, and 1657–61; each time the *parlement* began by confirming most of the sentences passed by lower courts, then became more critical as the defects and abuses in the trials made themselves apparent.[93] One leading figure in the witch-hunt after 1628, the *bailli* of Luxeuil Jean Clerc, seems to have been discouraged by a series of appeal decisions of which roughly two-thirds went against him. Clerc's enthusiasm was probably too evident, and worried others, like

[91] Claude Pellot, premier president of the Rouen *parlement*, wrote to Colbert in 1670 'l'on appréhende que plus l'on en condamnera, et plus l'on en descouvrira, et il en paroistra'. Mandrou, *Magistrats*, 447. Richard Simon made a similar comment, *Bibliothèque critique*, ii. 120–1.

[92] Soman, 'La Décriminalisation', 189, 194.

[93] Monter, *Witchcraft in France and Switzerland*, 72–87.

the *grand procureur* of the *officialité* who feared that depopulation would cause an increased poor relief bill, while the *parlementaires*, although convinced the devil was at work, did not forget that he was the father of lies. They may also have been concerned by the large proportion of children among the accusers; children seem to have been unusually prominent both in this role and as suspects themselves in the Franche-Comté.[94] Potentially the most dangerous panic was the last, inspired by the Inquisitor Father Symard, who was able to generalize the hunt by having *monitoires* read by the *curés* of every parish, calling for anyone with suspicions about witchcraft to reveal these to the inquisitors. The resulting flood of denunciations and arrests did not produce an immediate response from the *parlement*, perhaps because it is only known to have heard three witchcraft appeals between 1634 and 1654, and past experience had largely been forgotten. From the end of 1659, however, it was issuing procedural warnings, and moderating the great majority of sentences; when in 1660 news of Symard's proceedings reached the sceptical Roman Inquisition he was removed from office.[95] The case of Franche-Comté does seem to demonstrate how an appellate jurisdiction on the French model tended to restrain persecution, although the judges never apparently doubted the reality of witchcraft as such. The situation compares interestingly with that in Lorraine, where the ducal court of the *échevins* of Nancy possessed only a much more modest right to advise local judges in the course of the trials, without ever hearing formal appeals. Even this system probably did something to ensure that proper procedures were followed, but the Nancy court, whose attitudes were strongly influenced by the Rémy family in any case, was far from possessing the authority and self-assurance of a *parlement*, and would have found it difficult to pressurize the lower courts if it had tried.[96]

[94] The brief description of this episode by Monter should be supplemented by the fuller account in Bavoux, *Hantises et diableries*, 71–182, esp. for the fears of the *grand procureur*, p. 121, for the devil as the father of lies pp. 129–30, and for the prominence of children pp. 109–10.

[95] Monter, *Witchcraft in France and Switzerland*, 81–7.

[96] The legal situation in Lorraine is outlined in C. Bourgeois, *Pratique civile et criminelle pour les justices inférieures du duché de Lorraine* (Nancy, 1614).

It is important to distinguish between the behaviour of judges hearing individual cases and the formal position adopted by the agencies of government. Any pronouncement concerning witchcraft was virtually bound to assert the reality of the crime, given the weight of case law and the difficulty of proving a negative, but the authorities in both the Spanish Netherlands and Lorraine went rather further in the late sixteenth century, encouraging their subjects to prosecute witches in the courts.[97] It does not appear that these moves had any very powerful impact, since they were general exhortations which offered no new opportunities for litigants, unlike Symard's exploitation of the inquisitorial machinery in the Franche-Comté, which fortunately came late enough to attract rapid and unfavourable notice. Direct action by a determined individual witch-hunter usually had a catalytic effect, generating large numbers of prosecutions or killings, but also tending to attract sharp reactions from judges who saw both their authority and the public peace under threat. There is little doubt that in French-speaking Europe at least it was this 'establishment' attitude which brought an end to persecution, gradually gaining wider support, not least from provincial magistrates anxious not to appear behind the times. The thoroughgoing scepticism to which Reginald Scot gave the most striking expression was less useful in this context than the careful exposure of irregularities in the trials themselves, memorably achieved by the Jesuit Friedrich von Spee. The publication of a French translation of Spee's book by enemies of Symard in 1660 underlines the point.[98] In the Netherlands there was another interesting example of debate within the élite over the outbreak of possession in the Brigittine convent at Lille in 1614. What might have turned into an epidemic was checked, against the wishes of a prominent

[97] See nn. 16 and 70 above. In Lorraine Nicolas Rémy appears to have intervened to stimulate local activity; cf. the letter of 1594 printed by C. Pfister, 'Nicolas Rémy et la sorcellerie en Lorraine', *Revue historique*, 93, (1907), 233; he also intervened personally to follow up the accusations made by a possessed girl at Épinal in 1596 (AD M.-et-M., B 7314).

[98] R. Scot, *Discoverie of Witchcraft* (London, 1584); F. von Spee, *Cautio Criminalis* (Rinteln, 1631). The French translation was by F. B. de Velledor [F. Bouvet], *Advis aux criminalistes sur les abus qui se glissent aux procès de sorcellerie* (Lyon, 1660).

member of the Council, the comte des Estaires, and various clerics, by the efforts of a group led by the abbot of Loos and backed by the papal nuncio.[99] Here too the argument turned on the unverifiable nature of the evidence, as well as the danger of scandal and disruption in the church. In the same area the belated trials at Bouvignies, analysed in great detail by Robert Muchembled, could only take place at all because the court at Douai evaded consulting a higher jurisdiction, as it should have done under the newly established French rule in this region.[100] It remained possible for individual courts to swim against the tide in this way, most notably the *parlement* of Rouen. Here there seems to have been a group of powerful *conseillers*, associated with the *dévot* movement, and jealous of the *premier président* whom Colbert had recently foisted upon them from outside, while a certain resentment towards the Paris *parlement* may also have operated.[101]

The crisis in Normandy resulted from a final wave of witchcraft panics affecting parts of France in 1670. They were a lesser echo of the events of 1643–5, which seem to mark a final turning point for the kingdom as a whole. In three areas these years saw a popular witch-hunt develop, threatening to brush aside the normal legal processes and become an affair of instant popular justice. The first epidemic broke out in Gascony and the Toulouse region around the end of 1643, with 'much talk of putting all the witches to death'. The following months saw various travelling 'connaisseurs de sorciers' passing from village to village, those they identified as witches either being killed on the spot or sent before the *parlement* at Toulouse. A few legal convictions and executions followed, but most of the accused, whose number embarrassed the judges, were acquitted and sent back home. The *parlement* also ordered the arrest of the 'connaisseurs', three of whom were summarily hung in June 1644, and followed this up with threats of severe penalties against the local judges if they proceeded in such an irregular fashion.[102] The spring of 1644

[99] A. Lottin, *Lille, citadelle de la contre-réforme? (1598–1668)* (Dunkerque, 1984), 164–86; 'La "deplorable tragédie" de l'Abbaye du Verger en Artois (1613–1619)', in *Les Faits de sorcellerie*, ed. J. de Viguerie, 111–32.

[100] Muchembled, *Les Derniers Bûchers*, 39–41.

[101] Mandrou, *Magistrats*, 444–58.

[102] Ibid., 377–81, 391–4.

brought similar panics in Burgundy and Champagne, probably linked with severe late frosts, particularly serious in these wine-producing areas.[103] In July 1644 the archbishop of Reims wrote to the Chancellor Séguier, denouncing the local judges in Champagne for conniving in the use of the swimming test and condemning suspects out of hand; he said that up to thirty or forty were being accused in a single village, and asked for a *maître des requêtes* to be sent with a special commission to check these abuses.[104] In Burgundy, where there is particularly good evidence that the witches were blamed for exceptional hail and frost that spring, we find another mobile witch-finder, a young shepherd whom the peasants called *le petit prophète*.[105] Like many of his kind, he based his technique on observing small blemishes in the suspect's eye, diabolical marks only he could discern. The *parlementaires* of Dijon had already adopted the same formal position as those of Paris, by an *arrêt* of 1635; they now reacted by systematically commuting the death sentences passed by lower courts, and ordering harsh action against those responsible for the witch-hunts.[106] One can detect in these events the traditional beliefs that witches sought to damage the fruits of the earth through hailstorms and the like, while the hardships inflicted by the war policy of Richelieu and Mazarin may have exacerbated local tensions. It is also arguable that they stood at the end of a long evolutionary process, in which the judicial pursuit of witches had encouraged both popular fears and the desire for a full-scale purge of these polluters of the Christian commonwealth. It would certainly be ironic if this were so, with the limited earlier acceptance of persecution by the authorities helping to bring fears of witchcraft to a peak at the very moment when the *parlements* were revising their position. Nothing was more likely to make that revision permanent

[103] Ibid., 372–5, 385–90.

[104] R. Mousnier, *Lettres et mémoires adressés au Chancelier Séguier (1633–1649)* (Paris, 1964), 636–7.

[105] J. d'Autun, *L'Incrédulité savante et la crédulité ignorante au suject des magiciens et sorciers* (Lyon, 1671), preface; for the devastating frost of 23 June 1644, and the popular belief that the witches had caused it, see also the statement by the *procureur général* of the Dijon *parlement*, cited by Mandrou, *Magistrats*, 420.

[106] Mandrou, *Magistrats*, 385–90.

than the spectacle of uncontrolled popular justice at work, which helps to explain why these panics mark the point at which a decisive break took place between popular belief and educated opinion about witchcraft.

In his study of élite attitudes Robert Mandrou laid great emphasis on the notorious cases of demonic possession which occurred at intervals throughout the period of persecution. His argument was essentially that these had far greater impact than the commonplace pursuit of village witches, since they involved members of the ruling classes, and had significant political, social, and religious overtones. They generated widespread debate, which can still be traced through an abundant polemical literature. Although Mandrou recognized that these debates were inconclusive, with opinion remaining divided, he believed that they drew attention to the crucial problems of what evidence was admissible, and underlay the increasing practical scepticism of the *parlementaires*. The theory is both elegant and plausible; there is every reason to suppose that these great scandals made a deep impact of just this kind. The blatantly political motivation evident in the most famous case of all was not likely to escape contemporaries. At Loudun in 1633–4 Richelieu's sinister agent Laubardemont was clearly intent on destroying a local opponent, manipulated the possessed nuns to that end, and employed special commissions to bypass the ordinary judicial machinery. The extraordinary courage of Urbain Grandier, who refused to confess despite being subjected to atrocious tortures after his conviction, must have left profound doubts hanging over the whole affair. These could only be reinforced when the nuns continued their theatrical displays, while successive exorcists became deranged themselves.[107] Yet there are difficulties in attributing crucial importance even to the devils of Loudun. Since the *parlement* of Paris had virtually ceased to impose the death sentence for witchcraft a decade earlier, at a time free of major possession cases, reaction to Loudun can only have confirmed sceptical views, not created them. Like many other interpretations, Mandrou's depends on chronological links which fail to withstand detailed investi-

[107] Mandrou, *Magistrats*, 195–363. For Loudun see also de Certeau, *La Possession de Loudun*.

gation. His analysis remains valid and pertinent in many respects, but it exaggerates the significance of what were really secondary factors, to impose an artificially neat pattern on a complex situation. Nor can possession cases be exactly equated with witchcraft ones, given the differences in the way accusations began, the nature of the evidence, and the special elements of fraud and hysteria involved. Just as with witchcraft, there were divided opinions from the start; Marthe Brossier, whose supposed possession carried enough political dangers to worry Henri IV himself in 1599, soon came to be widely regarded as an impostor.[108] Scepticism over both forms of diabolic activity would clearly have been mutually reinforcing, so the story told by Mandrou remains highly relevant, but as part of a more elaborate set of arguments.

THE ACCULTURATION OF THE VILLAGE?

Some recent scholars, notably Robert Muchembled, Alfred Soman, and Yves Castan, have done a good deal to build up the necessary layers of explanation, stretching from the village community to the rulers of society. Muchembled has made an important contribution, in a series of books based largely on the archival material from the Cambrai region.[109] These are unusually rich sources, allowing the kind of reconstruction of village beliefs which has generally proved very difficult for France itself (to which this area only passed in 1668, shortly before the final outbreaks of witch-hunting). They have allowed the kind of 'thick description' to which the anthropologically-minded historian always aspires, and which is naturally far too complex to be summarized adequately here. The analyses of peasant conceptions of the supernatural, of the role of *devins*, of the female world of magical healing, sociability, and gossip, are all subtle and convincing. There are many keen insights into the meaning of witchcraft for the peasantry, together with an awareness that it was inextricably involved with the inner tensions of village society. Muchembled sees the period as one during which power was

[108] For this case, Mandrou, *Magistrats*, pp. 63–79; Walker, *Unclean Spirits*, 33–42.

[109] *Prophètes et sorciers; La Sorcière au village; Les Derniers Bûchers*. See n. 3 above.

being concentrated right down the social ladder, with church and state combining to promote the authority of natural rulers at every level, ending with the control of the father within the family. The ideology of counter-reformation catholicism was the most powerful expression of this assertion of hierarchical values, and found one of its main expressions in a campaign against the 'pagan' superstitions of the villagers. The small groups of substantial tenant farmers who were tending to monopolize power in the village can also be seen as a bridging element between the old culture and the new, increasingly literate and self-conscious about their position. When such men orchestrated persecutions, or appeared as witnesses against their social inferiors, they were demonstrating that they belonged to the superior world of the ruling classes, repressing superstitions which they now felt guilty for not having completely destroyed inside themselves. This is an attractive hypothesis, whose component parts are skilfully argued, supported by convincing examples, and all recognized as trends within early modern society by most of its historians.[110] There is certainly much truth in it as an explanation for the behaviour of certain individuals during the process of bringing witches to trial. It is also highly probable that those who lost out in the economic changes of the period were resentful, and that their richer neighbours were well aware of such feelings. Why then does this interpretation, so carefully based on primary source material, and rich in genuine insights, ultimately fail to convince?

The first difficulty is a methodological one. The number of full dossiers surviving in this region appears to be very limited, for Muchembled employs only a small range of examples, which appear repeatedly in his books. Even this sample suggests that there were several different scenarios, but there is no way of deciding, on such a tenuous basis, which were really typical. Working with a much larger group of trials in Lorraine, I have found analogues for all the patterns he identifies, but only as small proportions of the whole. Generalizations in social history tend to have hidden statistical assumptions, and the very least one would require of

[110] This is not to say that I agree with it; see the Conclusion pp. 384–9 for criticism of the wider thesis.

Muchembled's model is that 50 per cent of the cases could be assimilated to it; there is no sign that anything like this figure will emerge from the Lorraine material. The natural inference would be that we have here a most accomplished reconstruction of one type of trial, which includes many features common to the phenomenon as a whole, but is in other respects exceptional. This view is strengthened by an examination of the second set of difficulties, those internal to the actual examples used. These are taken from a very long period of time, stretching from the middle of the fifteenth century to the end of the seventeenth, but the most detailed ones not surprisingly come from the later decades. The most ambitious reconstruction, that of the trials at Bouvignies in 1679, comes from the very end of the period, when the pattern ought to be unusually clear.[111] Although the dominant group of richer peasants can certainly be seen managing the legal proceedings, this is in itself no proof of anything beyond their natural predominance in local society. Given the networks of marriage alliance, protection, and neighbourhood which held communities together, providing links across the different levels of prosperity and influence, the real surprise would be to find anything else happening. Since it was the prosperous tenant farmers who sat on the local court, only in peculiar circumstances would their inferiors bring cases without support among them. There is no clear evidence for any difference of attitude between social groups, nor for the influence of élite culture and religion; the illiteracy of a number of the judges must have been a hindrance here. Nor is it easy to see that disputes and jealousies between groups had any special significance, although they were bound to come into effect once the accusations gained momentum. Muchembled builds up a complex picture of such possible divisions within the community, interpreting them as evidence of a struggle for power which was in some way expressed and furthered by the persecution.

The cumulative effect of listing numerous tensions and disputes is considerable, tending to mask the unfortunate fact that there is no proof of any *significant* links between these social factors and the beginning of specific prosecutions.

[111] *Les Derniers Bûchers*.

Villages must have been full of such minor rifts, at almost any time, and it is certain that they help to explain why certain persons testified against others. It is a very different matter to demonstrate that they were the mainspring of the whole process, with the trials serving a functional purpose in affirming the patterns of domination within local society. The attempt to do this for Bouvignies depends on juxtaposing a series of analyses which never actually meet at any crucial point, while there is also a certain psychological implausibility in reducing the intense and genuine fears aroused by witchcraft beliefs to an occasion for the settling of scores. Muchembled's determination to impose a preconceived pattern by the use of much circumstantial evidence rather obscures the fact that the central actors in the drama seem to have been performing to a different script. The depositions against the prime suspects suggest personal rather than social conflicts, with hardly a hint that witchcraft was being used to take revenge against the dominant families. By far the most plausible explanation for the animosities behind the first trials, that of Peronne Goguillon, appears to lie in disputes within her own family group, which are only briefly evoked.[112] As far as the legal action itself is concerned, there is no need for any complicated explanation; this was one of those cases which began with an unwise suit for defamation, brought by the victim's husband. It is possible to believe that the tavern-keeper Gilles Fauveau played a more important part in setting up the affair than the formal record suggests, and that he was ambitious to raise his status and power in the village, although both these hypotheses (Muchembled's own word) are unverifiable.[113] If this is so his illiteracy hardly fits with the general argument about acculturation of the village élite, while it is hard to see what solid advantages he would have gained from behind-the-scenes management. The clerics who show up in the record do not seem too consistent in their attachment to the new *mentalité* either, with the *curé* of Raimbeaucourt and the Capucins of Orchies both following the traditional practices of the *devins* to identify witches.[114] At

[112] *Les Derniers Bûchers*, 157–61.

[113] Ibid., 154–8; the phrase 'à titre de simple hypothèse' is on p. 157. For the legal steps leading to the trial, see p. 18.

[114] Ibid., 226–7.

every point the evidence is ambivalent and hard to assess; one might have expected that trials at this late date would have given far more decisive support to an interpretation of this kind.

CHECKS AND BALANCES: TOWARDS A GENERAL INTERPRETATION

The ambiguities and complexities of the problem are very well conveyed in the excellent book by Yves Castan, *Magie et sorcellerie à l'époque moderne*, which has received far less attention than it deserves.[115] There is a certain obliquity about his approach, with its concern to explore the multiple links between witchcraft and the peasant world, and its refusal to offer sweeping general explanations, which has no doubt limited its appeal. Yet I would suggest that this is the best way to make sense of the whole phenomenon, and that Castan's example deserves to be more widely followed. Most functionalist interpretations of witchcraft tend to treat the persecution as somehow directed to ends outside itself, serving the interests of particular groups, and perhaps reinforcing changes in the structure of society. Although almost any set of conflicts is bound to have overtones of this kind there is no reason why they should be regarded as primary causes. We are not dealing with rituals, whose very purpose is to express some aspects of reality or belief, but with something which is a social fact in its own right. There is an immense attraction about ingenious theories which tell us that something is quite other than it at first seems, purporting to uncover the hidden meaning beneath; those who feel they have been let into the secret are liable to suffer a permanent suspension of critical judgement. Witchcraft has already suffered from an extreme example of this with the wild fictions perpetrated by Margaret Murray, based on gross misrepresentation of the evidence, which enjoyed an amazingly long innings.[116] There is some danger now that far more sophisticated arguments will take their place, obscuring the banal realities beneath a redundant superstructure of interpretation. These theories have all the more chance of success because they generally contain a good

[115] Paris, 1979.

[116] There is a definitive refutation in Cohn, *Europe's Inner Demons*, 107–20.

deal of truth (even Murray had a point in stressing that there were pagan survivals involved). The fluid and adaptable nature of the whole system of belief and persecution made it a vehicle for many external forces, which influenced its application in particular areas or particular cases. For this reason witchcraft can provide marvellous insights into social conflicts, but it is essential not to slip into a facile inversion at this stage, and treat such conflicts as the 'fundamental' causes of the whole long tragedy. It is perhaps a pity that Castan did not spell out his own position more clearly, for one of the greatest merits of his book is that it keeps witchcraft at the centre, while maintaining an admirable sense of proportion about the importance of external factors.

In the present state of witchcraft studies there is much to be said for reiterating certain obvious points which are in danger of being lost from sight. The persecution really did stem from the universal peasant belief in occult personal powers, which led to certain individuals being identified as witches. The combination of this perennial pattern with the development of demonological theory, and the extension of legal systems to provide greater social control, produced a wave of witch-hunting through the courts. The legal mechanisms were still relatively weak, as both Castan and Soman make clear; with relatively few exceptions they only operated where denunciations forced them into action. It is possible to construct a scenario in which the timing of the various processes differed, so that large-scale persecution was averted until élite opinion had evolved so far as to make it impossible; Ireland and southern Italy might be areas which escaped in just such a way. On the European scale, however, it would have needed exceptional good fortune for this to happen generally, so that the phenomenon looks to have been nearly inevitable. Not least because the scepticism which spread in the seventeenth century owed so much to the experience of the trials themselves; Stuart Clark's work on the scientific status of demonology might be developed to suggest that witchcraft debates played an important part in the development of rationalism, and in this sense some good did come out of evil.[117] The

[117] S. Clark, 'The scientific status of demonology', in *Occult and Scientific Mentalities in the Renaissance*, ed. B. Vickers (Cambridge, 1984), 351–74, and the further references given there.

experience of the French-speaking regions is interesting in this respect, with the critical attitudes of the Parisian judges gradually spreading to all other areas, aided by the extension of French political power into the Netherlands and towards the Rhine. A zone of persistent witch-hunting at a moderate level, such as Lorraine, which lacked a sophisticated judiciary, might have kept the process going far longer but for this growth of French influence.

The timing of persecutions, both general and individual, clearly does relate closely to the socio-economic context; it is not likely to be an accident that the decades around the turn of the sixteenth and seventeenth centuries, which saw real incomes drop to their lowest known point in the whole millenium, were the peak period for witchcraft trials. The difficulties of these years must have had a traumatic effect on many people, forced to sell land and possessions in order to survive, and often suffering a drastic loss of status in the process. Village society was permeated by subtle gradations which cannot be fully represented by the mere counting up of landholdings, but which ranged from the possession of one or more cows to a reputation as a hard and effective worker. Relative independence and security could depend on these fragile assets, and the esteem they helped secure. Given that witchcraft was essentially an explanation for misfortune, it was particularly likely to be suspected at a time when minor personal disasters were not only unusually frequent, but also harder than usual to repair. It does seem to make psychological sense to see the typical accuser as someone whose position was endangered, and who was striking back at the suspect in desperation, rather than to posit a consensus (however unconscious) among the local élites directed to affirming their power. One of the most convincing attempts to replace a witch-hunt in its local context, that by Boyer and Nissenbaum in *Salem Possessed*, suggests just such a pattern underlying this famous episode, with a family in difficulties producing the crucial allegations.[118] Questions of power were certainly involved, for the witch wielded occult power to produce the misfortune, motivated by envy or ill will. Such dangerous capacities might be invoked to explain a mysterious rise in

[118] P. Boyer and S. Nissenbaum, *Salem Possessed* (Cambridge, Mass., 1974).

wealth and status, but much more often they were attributed to those who were in a weak position, which left them no other means to assert themselves against their neighbours. It is in this sense that increasing social differentiation might indeed lead the richer peasants to feel threatened by envious inferiors; conscious of the precarious nature of their position, their reactions were surely defensive, and indicated how far they were from renouncing traditional beliefs. The predominance of women among the accused can also be partly explained in similar structural terms, for in many disputes it was only by such occult means that they could have damaged their enemies, particularly if they were widows or spinsters with no husband to back them up. There was also a danger, even for married women, that their husband's family would regard them as outsiders, not entitled to the same degree of support as blood relatives, and better abandoned if their reputation threatened to taint the whole family. Combined with misogynistic views about their sexual waywardness, the ambiguity of their roles as healers and as repositories of folklore, and masculine fears of the whole world of feminine gossip, one must finally recognize the tendency for stereotypes to establish themselves as the trials followed one another. The habit of using 'witch' as a general term of abuse against women reflected this; it was also used to provide evidence against those actually charged, when they had allowed such insults to pass.[119]

It is possible to extend such arguments, and regard large numbers of women (in particular) as possessing marginal reputations as witches. At some points Muchembled does rather argue as though there was a large pool of potential suspects of this kind, from whom occasional victims were chosen for reasons which had very little to do with witch-

[119] One of the most puzzling features of Soman's figures is the almost equal numbers of men and women accused in the *ressort* of the Paris *parlement*; one would be tempted to suspect that this reflected differential use of the appeal procedure, but the same proportions are repeated in his sample from the lower courts. Widespread fear of the male witch may be a distinctively French cultural phenomenon, although it is none too easy to think of an explanation for this.

craft.[120] Again this is a useful concept, but one which must not be taken too far. As individuals passed through their life cycle, in a rural society where privacy was minimal, they must have acquired a complex image which was liable to retrospective reassessment. There was probably also a tendency for them to build up enmities, major or minor, and to become increasingly dependent on family and close friends for support in any disputes. One might envisage the whole process as resembling role-playing games of the 'Dungeons and Dragons' type, in which players amass points under various headings, but with the scores of most characters being negative, in a situation where resources and tolerance were often stretched very thin. While in theory this might have left large numbers of people exposed to possible identification as witches, in practice such risks were heavily concentrated on a few individuals; it needed extreme bad luck for those with only a marginal reputation to get caught up in the judicial machinery. No doubt a sequence of unlucky coincidences might create a reputation as effectively as an abrasive personality could, but whatever the origins of such a dangerous image, it finally became very specific. A small group of prime suspects was usually to be found in a community, although remarkably few of them would ever be brought to account unless there was a major panic. As rumours about their powers spread they were subjected to periodic acts of hostility by their neighbours, and expected to perform cures, but there is little sign of a persistent will to eliminate them totally. Trials might indeed have a cathartic effect once they began, asserting the unity of the village against those who brought in evil from outside, and reaffirming orthodox hierarchies. It does not follow, however, that these incidental benefits explain why persecution occurred; if they had provided the main motivation, then one would have expected the whole process to be much more regular and predictable. At this stage we can revert to the general point made much earlier, that a satisfactory interpretation must allow for the episodic, patchy nature of witch-hunting, rather than invoking sweeping general factors without giving

[120] See particularly the very interesting chapter 'Paroles de femmes' in *Les Derniers Bûchers*, 187–216.

sufficient attention to the specific ways in which they operated. One is repeatedly aware of missing links in such arguments, just at the vital crossover points between general and particular.

This essay has sketched an alternative model which preserves much that is valuable from existing analyses. The core of witchcraft beliefs lay in *maleficium*, the act of damaging others through occult power. This belief allowed almost any misfortune to be attributed to the malice of an enemy, with the more hopeful correlative that those who inflicted such harm might also be able to remove it. The attractions of such a system of thought are obvious, even if we omit the normal human fondness for blaming other people. Once the formal mechanism for persecuting witches had been established, for reasons already discussed, there was a terrifying prospect that accusations would multiply until no one was safe. Set against this possibility, the actual rate of persecution looks remarkably moderate in most of Europe, and certainly in most of French-speaking Europe. The real difficulty is therefore not to explain why some witches ended at the stake, but to understand the various forms of restraint which prevented far more from joining them. Like the pressures for prosecution, these restraints operated at both popular and élite levels. The system can be seen in binary terms, as a series of gates through which an individual or an event had to pass before a categorization as witchcraft occurred, and then before legal steps were taken. This can be represented visually as a flow chart; an excellent example is provided by Christina Larner.[121] Her distinction between natural and unnatural methods of witchcraft control, perfectly valid in its own right, also corresponds to the later part of the series of 'gates', those which determined whether the ultimate weapon of the formal prosecution was brought into play. The earlier stage would comprise the sequence of interpretative decisions by which a particular event might or might not be classified as a product of witchcraft, and a suspect then identified. The model is of course not unilinear; there were several possible starting points, and the options multiplied thereafter. Nor does it assume that there was a constant flow of potential accusations

[121] Larner, *Witchcraft and Religion*, 131.

at a more or less even rate, a high proportion regularly being turned aside before they reached the courts. The level of suspicions clearly fluctuated, responding to the variable fortunes of communities and individuals, and probably being enhanced where there were regular trials with their attendant publicity. The differing results by area and period are just what would be predicted for a phenomenon which represented subjective rather than objective reality, and was mediated through so many variables. Positive causation, such as popular reaction to exceptional weather or epidemic disease, could produce a surge of accusations capable of breaking through many of the restraints, as could the activities of a determined witch-hunter. As already explained, such episodes commonly engendered a further series of reactions, with higher authorities intervening to limit the social disruption they threatened. Under normal circumstances the negative factors operated as a filter at the individual level, allowing only a small proportion of suspects through for the consideration of the judges and ultimately of the historian.

Witchcraft beliefs, then, formed an integral part of village social relationships, with the trial records allowing us access to only a small part of the network of fear and hostility which they represented. This does create some grave difficulties, for ideally we would like to be able to compare cases that came to court with those that did not. The problem can be mitigated by looking at the numerous episodes of past history recounted during the trials, however, and these strengthen the impression that there were many informal crises for each one that directly caused a prosecution. What the records do show beyond doubt is that once the possibility of using the legal system had become well known the great majority of cases were initiated from below; élite attitudes were always uncertain, and within a few decades moved decisively towards scepticism. They might have done so faster if the peasantry had been more reckless in pursuit of witches, thereby creating endemic social problems. In many respects peasants reacted to witchcraft just as they did to other crimes, preferring informal settlements and treating the courts with grave suspicion. Individuals were rarely willing to proceed without the support of kin and neighbours, who often proved reluctant to involve themselves in public and highly divisive actions which might leave a dangerous legacy of bitterness behind them. To make oneself a *partie formelle* against a witch carried risks of this kind, but also liability for costs if the suit failed, and the alarming need to submit one's case before judges one could not entirely trust. This last factor would have been particularly important within France proper, where the royal courts had made such progress towards taking higher justice into their own hands. Even so, it was probably the solidarity of the wider family group, reinforced by the wish to prevent the *lignage* acquiring the taint of proven witchcraft, that offered much the strongest protection. It was far from invulnerable, since quarrels often occurred between relatives, and could themselves generate virulent suspicions, while at a certain point the family might abandon a member who had become so hated in the community as to be a major liability. Those who lacked supportive relatives might still survive, in a distinctly risky way, either by making themselves so feared that no one dared take the lead against them, or by being ready to offer cures whenever necessary. The whole system was so finely balanced at so many points that it is often impossible to give a definite interpretation of any particular episode, even where full documentation survives. This matters a great deal less so long as we place the popular beliefs themselves at the centre of the explanatory model, and do not expect to reify witchcraft into an outgrowth of some specific set of social tensions. It then becomes much easier to allow for such conflicts as powerful secondary factors, which may appear individually in only a minority of trials, and do not provide any magic key to what witchcraft was 'really about'.

Much of the interest of the relatively moderate persecution of witches in French-speaking Europe is precisely that it provides evidence on the social relations within peasant society, and on popular conceptions of religion, morality, and nature. This is the main reason why the subject remains so eminently worth studying within the social and intellectual context of the time. The best answers to the question of why it happened remain the obvious and banal ones; rather than spill more ink in fruitless attempts to supplant them, we might more profitably be seeking to reintegrate witchcraft within the society where it played such a central role. Whatever else it was, it was certainly not some mere aberration, but a fundamental and long-standing feature of peasant *mentalités* throughout Europe. Perhaps historians themselves have spent too much time looking for scapegoats, trying (quite unconsciously) to find special explanations for something they would like to regard as a bizarre departure from normal western standards of rationality and justice. Greatly exaggerated ideas of the number of victims, and a tendency to concentrate on the more sensational trials, have also encouraged the misleading image of a 'witch-craze' sweeping across Europe. The fact that tens of thousands of people were put to death for an imaginary crime is indeed horrifying, however far our own age has surpassed it in terms of institutionalized cruelty. Yet retrospective indignation, however proper, is not much help to historical understanding. Witches should not be assimilated to such alternative targets for attack as Jews, for the basis for most prosecutions was individual and personal rather than general, and those convicted were distinguished only by the general belief that they had harmed their neighbours out of malevolence. It must be recognized that many of them had probably been quarrelsome and difficult members of the community, and that a proportion had resorted to threats and extortion. Curses and spells employed with malevolence are real dangers when everyone believes in them, so that questions of guilt and innocence are not quite so simple as they may seem. Even when all the logical faults which rendered the reasoning worthless are fully borne in mind the depositions in the trials retain an extraordinary plausibility, powerfully reinforced by the confessions of the accused. Familiarity with this material makes one feel that the great surprises are the moderation with which peasants sought legal remedies, and the shrewd scepticism which developed so quickly among the élites. The history of witchcraft in France and on her borders is not quite the straightforward story of human folly one might expect.

2

Witchcraft and popular mentality in Lorraine, 1580–1630

DETAILED records of early criminal trials are scarce, and one of the most extensive collections to survive for the turn of the sixteeth and seventeenth centuries is that of the ancient duchy of Lorraine, now housed in the Archives Départementales of the Meurthe-et-Moselle at Nancy. Among these documents are some three hundred complete dossiers for those tried on charges of witchcraft, nearly all of them for the half-century from 1580 to 1630. Although this probably represents only something between 5 and 10 per cent of Lorraine's witchcraft prosecutions (for the names of many hundreds of others convicted can be recovered from less complete records), it constitutes an admirable working sample; this chapter and the next explore various aspects of these cases. This material is of a kind not normally found in England or France, and only sporadically elsewhere in Europe. It includes full witness depositions, commonly from fifteen to twenty-five witnesses; the interrogation of the accused on the basis of these testimonies; the confrontation of the witnesses and the accused; and normally one or more sessions of interrogation under torture. The nature of the records is very important because they give us an unadulterated view of the first stage of accusations, without any serious likelihood of editing by the lawyers and judges. It is the earlier stages of the trials, rather than the confessions under torture, which enable one to build up a picture of the popular attitudes that had prompted the accusations. The confessions that were eventually extracted from the vast majority of the defendants also have their interest, however; the records generally allow one to distinguish between those admissions made spontaneously and those that resulted from promptings by the judges. This is important, for example, in the case of the sabbat, where the Lorraine material offers us a direct way into popular, as distinct from learned, views about these diabolical festivities.[1]

Lorraine has traditionally been portrayed as the scene of intense witchcraft persecution, and its judges, from the demonologist Nicolas Rémy, *procureur général* of the duchy, down, have acquired an evil name. As so often in the history of witchcraft, there is an element of exaggeration in this. What can be fairly said is that once a suspect reached the courts, his or her chances were poor; the conviction rate generally approached 90 per cent. On the other hand, if one takes the reasonable estimate of around 3,000 trials for the period under consideration, this is around sixty a year in a duchy with a population of at least 400,000. As a per capita rate it is not markedly different from the peak rates achieved in Elizabethan Essex, although the proportion of executions was far greater.[2] The accused were a highly selected group, and there are very few clear examples of people who were pulled in because of a casual denunciation made under torture—the chain of accusation that became infamous in some German cities.[3] The attitude of Rémy and other judges may have encouraged people to use the courts, and it was normal to interrogate those who confessed about their accomplices, but this was done with some caution, and there is no real sign that suspects were manufactured by such means. The typical accused had a long local reputation, twenty years being commonplace. He or she was charged with a range of acts of *maléfice,* causing actual harm to neighbours and their animals, stretching many years back. Suspicious noises and nocturnal comings and goings were sometimes mentioned, but village belief was firmly based on the actual damage caused to community and individuals.

[1] These documents were employed by E. Delcambre in *Le Concept de la sorcellerie dans le duché de Lorraine au 16ᵉ et au 17ᵉ siècle* (3 vols; Nancy, 1948–51), a scholarly but curiously limited work that ignores social factors and generally eschews analysis.

[2] For Essex, see A. Macfarlane, *Witchcraft in Tudor and Stuart England* (London, 1970).

[3] The process is admirably discussed in H. C. E. Midelfort, *Witch Hunting in Southwestern Germany, 1562–1684* (Stanford, Calif., 1972). These chains of accusation differed from the much smaller ones in Lorraine (discussed later) in that they commonly extended to many persons never previously suspected.

A contrast is often drawn between this local belief, founded on specific acts of *maléfice,* and the learned tradition that emphasized the diabolical pact and the sabbat, with witchcraft becoming the most extreme form of heresy. Technically this distinction can certainly be made in Lorraine: the judges sought to obtain confessions to the pact above all, and these were sufficient for a capital sentence even if unaccompanied by admissions of actual evil-doing. Such a bold statement would, however, be misleading. The local commentators—Rémy and the legal writer Claude Bourgeois—were far from disregarding the importance of *maléfice.*[+] Judges continued to press for admissions of this even after they had secured the basic confession. Furthermore, the accused always began their confessions with an account, often in pathetic circumstantial detail, of how they had been tempted by the devil in a moment of distress or weakness and had succumbed. The pact was clearly a part of popular belief; perhaps the accused may have regarded it as less of a social sin than harming their neighbours through active witchcraft, since several of them denied any such acts, despite being beaten and brutalized by the devil. At the least, the pact might allow the displacement of guilt for the harm done to neighbours on to the devil, who had allegedly compelled the performance of such evil. Confessions to attendance at the sabbat, however, often had to be elicited by direct questioning, even though most of them reveal a standard popular image, of a rather unimaginative kind, which must again reflect widely held folk beliefs.

There are other reasons why it would be hard to maintain any real divisions between élite and popular conceptions of witchcraft in Lorraine. The great majority of cases were tried in local courts, some of whose judges were illiterate: the central tribunal of the *échevins* of Nancy reviewed the proceedings, but did not exercise a direct appellate jurisdiction. Rémy's own limitations are interesting here: despite the classical references with which he interspersed his material, the interest of his *Demonolatry* lies exclusively in the discussion of practical details. His view is really more characteristic of the

[+] N. Rémy, *Daemonolatreiae libri tres* (Lyon, 1595), esp. Bk. I, chs, II–III, Bk. II, chs. VII–VIII; C. Bourgeois, *Pratique civile et criminelle pour les justices inférieures du duché de Lorraine* (Nancy, 1614), fo. 43.

popular than of the learned tradition, as in the confused passage in which he fails to resolve the question whether it can be right to force a witch into healing her victims. The book is direct and notably accurate when describing actual trials, only to lapse into verbose futility when it moves to general issues. It is, however, remarkably free from any hysterical or paranoid fears of a grandiose international conspiracy of witches, for Rémy viewed the 'vile rabble of sorcery' with a certain contempt and was serenely confident in his own invulnerability as a judge. It was in line with such attitudes that he remarked: 'For witches make it their chief business to be asked to perform cures so that they may reap some profit, or at least gratitude; since they are for the most part beggars, who support life on the alms they receive.'[5]

This last comment will remind many of the analyses of English witchcraft by Keith Thomas and Alan Macfarlane, with their stress on the refusal of charity and subsequent inversion of guilty feelings by the witch's supposed victim.[6] As an explanation of the internal logic of the accusations this remains the biggest single step yet made towards understanding the reality of European witchcraft persecutions, and it can be extensively confirmed by reference to the evidence for Lorraine. While a single example proves nothing, it will at least give the flavour of the material. In 1584 Catherine la Blanche, a widow in her sixties, was on trial. One of the twenty-five witnesses, Cleron Baltaire, said that five years before, when she and her husband had been fattening a bull, Catherine

vint à sa porte mendier, comme elle faisoit souvent. Elle deposante luy dit Catherine, allez pourchasser et demander vos aulmosnes aultre part, car je ne vous veulx plus rien donner, à rayson que j'ay des enfans pupilz et pauvres enfans de feu le frère de mon marit qui sont sur mes bras et qu'il nous fault nourrir. Pour l'honneur de dieu il vault mieux de les nourrir que vous et pour ce allez vous en.

[came to her door to beg, as she often did. The witness said to her 'Catherine, go to seek and ask for your alms elsewhere, because I do not wish to give anything to you any more, for the reason that I

[5] Rémy, Bk. I, ch. xiv; Bk. I, ch. ii; Bk. III, ch. v.

[6] Macfarlane, *Witchcraft*; K. V. Thomas, *Religion and the Decline of Magic* (London, 1971), 435–583.

have dependent children and the poor children of my husband's late brother on my hands, whom we have to feed. For the honour of God it is more fitting to feed them than you, therefore take yourself off'.]

Although there was apparently no threat or other reaction from Catherine, Cleron nevertheless blamed her for the subsequent death of the bull.[7] Cases that come so close to the English model do, however, pose some awkward problems. If accusations in a thoroughly Catholic and rather traditional area like this follow an almost identical pattern, what happens to those very plausible general explanations in terms of Protestantism and rapid socio-economic change?

Some possible answers do suggest themselves and can be developed to illuminate wider aspects of the topic. First, it is easy to overdo the distinction between Protestantism and Catholicism, both at village and élite levels. The faith of the urban élites in Catholic Europe was showing a distinct tendency towards emphasizing individual responsibility, which the whole pastoral effort of the counter-reformation was to encourage, while the villagers rather illogically yet sensibly combined magical beliefs in the efficacy of the sacraments with a habit of judging individuals by their actual behaviour. The whole business of the diabolical pact was presented as a matter of individual fallibility, even if it was claimed that the devil was too powerful to escape once the fatal step was taken. Accused and judges not infrequently concurred in seeing the trials as a way of reconciling the sinner with God; confession, repentance, and expiation at the stake were saving souls.

Apart from the psychological pressure built up by the legal proceedings themselves, numerous accused witches were probably aware that they had borne their neighbours genuine ill will and may have come to accept responsibility for the ensuing misfortunes. Others remained unconvinced and sometimes tried to revoke their confessions, alleging that to confirm them would risk damning their souls by dying with falsehoods against their name. When Barbelline Goudot was tried in 1604, she revoked her confessions on the grounds that 'ayant demandé à son père confesseur familliairement sy

[7] Archives Départementales, Meurthe-et-Moselle, B 4495. Cited hereafter as AD M.-et-M.

ayant confessé chose non véritable elle en recevroit peine en l'autre monde lequel luy dit qu'il ne falloit dire que la vérité, qu'il fut la cause qu'elle avoit renyé le tout' ['having asked her father confessor privately if after having confessed something that was untrue she would be punished in the other world, he told her one must only tell the truth, which was the reason she had denied it all']. She then confessed again, to the relief of her judges, who urged her to further admissions 'd'aultant que le crime est sy oculte que le Juge n'en peult sainement juger qu'après la pure et simple confession de celuy ou celle qui en est coupable' ['above all because the crime is so secret that the judge can only properly judge it after the pure and simple confession of the person who is guilty of it'].[8]

In terms of ideas of personal responsibility, then, there is little to differentiate Protestant and Catholic positions in practice. More surprisingly, what is absent from these records is any evidence of ecclesiastical counter-magic in operation, apart from pilgrimages to shrines and the burning of the occasional candle. The *curés* are curiously missing from most trials; they never seem to testify and are involved only indirectly. At the trial of Jeannon Poirson, who was 'renvoyée jusqu'au rappel' (the nearest one could get to an acquittal) in 1602, it was alleged that the late *curé* of Leintrey had seen her dancing strangely in the fields, and then said 'qu'il n'avoit jamais voulu croire qu'il fut des sorcières mais qu'à ceste heure là il le croyoit' ['that he had never wished to believe that there were witches, but at that moment he believed it'].[9] Against this expression of relative scepticism one can set the cases of a *curé* who sought magical remedies from the suspect, and another who diagnosed witchcraft from objects found in victims' bedding.[10] It was probably crucial that the *curés* did not take a more active part in instigating the persecution of witches; had they done so, there would have been far more trials than seem actually to have taken place. The position of the *curé* as a local *notable* and a natural arbiter of disputes would have made him the ideal orchestrator of a persecution. Perhaps his role as confessor to his flock was crucial in

[8] Ibid., B 3327, No. 5.
[9] Ibid., B 3323, No. 10.
[10] Ibid., B 8667, No. 8; B 4126, Nos. 2 and 3.

inhibiting him, since any accusation might well suggest that he was breaking the secrecy of the confessional.

A second respect in which the situation may be closer than expected to that in England concerns social and economic changes. It is certainly true that peasant society in Lorraine was not disrupted by the development of a full market economy of the kind that was emerging in England. On the other hand, divisions between rich and poor did widen sharply in Lorraine, and most notably, according to the magisterial study by Guy Cabourdin, in the period between 1580 and 1630. Substantial amounts of land were transferred from peasant ownership to that of the prosperous few, communal rights were eroded, and peasant indebtedness rose very rapidly.[11] While there are many reasons to be suspicious of the 'strain-gauge' explanation of increasing witchcraft tensions, the trials do contain a good deal of circumstantial evidence that would link them to antagonisms between rich and poor. Around 1583 Jean Diez of la Bolle told George Colas that although he was now rich he would become poor, while Jean himself would acquire property; when Jean Diez came to trial in 1592 Colas's widow claimed that the threat had been fulfilled. Despite hard work and a frugal life style they had been reduced to extreme poverty.[12]

Another witch from the same group of trials, Zabel de Sambois, had been unwise enough to get into dispute with the *maire,* Dieudonné Galand, who believed that she had caused him various misfortunes. The *curé* persuaded her to a formal reconciliation and seeking of pardon from the *maire,* on the grounds 'que les pauvres doibvent plier pour les riches' ['that the poor should give way to the rich']; at her trial, however, she objected that the accusations against her were false, 'le tout par envie et malveillance et qu'on faict toujours ainsi contre les pauvres gens, et que sy on scavoit tout le fait dudit maire Galand, qu'on ne teindroit pas beaucoup plus de compte de luy qu'on faict d'elle' ['all of it through envy and ill will, and that poor people are always treated thus, and if they knew all the doings of the said mayor Galand, they would not

[11] G. Cabourdin, *Terre et hommes en Lorraine, 1550–1635* (2 vols; Nancy, 1977).

[12] AD M.-et-M., B 8667, No. 5.

think much better of him than they did of her'].[13] In 1602 Babelon Henri alleged that 'à cause qu'elle est pauvre l'on ne tenoit grande conte au sabat et y avoit bien peu de credit, mais que les riches y ont toujours plus de credit, et sont les plus avant à la besogne' ['because of her poverty she was not much regarded at the sabbat and had little credit there, but the rich always had more status there, and were the most active in all the doings'].[14] To emphasize their predominance, the rich sat higher and had more meat. A similar picture was given the following year by Catherine Charpentier, who added that the rich

> disoient, avoir encore des bledz assez en provision, fussent en volonté, et proposoient, de gresler et gaster les bledz et biens de la terre. Que jamais quant à elle, elle n'y voulut consentir, par la crainte qu'elle avoit d'avoir besoing, cognoissant, comme elle faisoit, la pauvreté de son marit, aussy, elle a esté par plusieurs fois battue, dudit son Mre. Persin, qui enclinoit à la volonté des autres.
>
> [who said, that they still had a sufficiency of grain in store, wished and suggested that they should make hail and destroy the grain and fruits of the earth. As for her, she had never wished to agree, because of her fear of being in want, knowing as she did the poverty of her husband, also, she had several times been beaten by her said master Persin, who supported the wishes of the others.][15]

This theme of social division at the sabbat, with the poor forced to consent to activities which actually harmed them while advantaging the rich, could be illustrated from several other confessions. Despite its imaginary context, it points to the very real social strains found in everyday life. People of even modest wealth are very rare among those accused; the rich witches were as much a fantasy as Maitre Persin and the sabbat itself. The status of Persin and his analogues had a certain ambiguity, for it was rarely clear whether he was the devil in person or merely one of his lieutenants, but he was naturally always found promoting the most destructive course of action.

Having emphasized likenesses, the third point is one of dissimilarity. Although the psychological spur for accusations

[13] Ibid., B 8667, No. 6.

[14] Ibid., B 8691, No. 5.

[15] Ibid., No. 13.

was basically identical—a dispute, in which the accuser was quite often seen in an unfavourable light, followed by a misfortune—the range of disputes seems to have been much wider than in the occasional English trials we can follow in comparable detail. Fewer of them turn on the refusal of recognized neighbourly services or consideration; although these last are naturally common, they are not really predominant. It does seem plausible to suppose that, as the development of the poor-relief system would suggest, obligations to poorer neighbours had become a source of acute tension in England. In Lorraine the stress was perhaps distributed more widely, and it would be difficult to show that witnesses were commonly of a higher social or economic standing than the accused. Muchembled's suggestion, based on a handful of instances from the Cambrésis, that members of the powerful minority were asserting their social control over their inferiors, would be extremely hard to justify from the mass of Lorraine trials, although, as one would expect, a handful do hint at such antagonisms.[16] In truth, the kinds of tensions revealed are those that must always have been part of village life, as were the misfortunes. The accused sometimes pointed out that it was as reasonable to blame chance or the will of God as to name witchcraft as the cause when animals or children were stricken by sudden or unknown illness. Another subtle difference from the English case concerns the 'inverted guilt' pattern; this was very commonly present, but far from being a rule. Judges and witnesses alike plainly assumed that bewitchment would follow a quarrel, and a witness who did not recount such as episode as a prelude to misfortune was likely to be specifically asked if there had been any dispute with the accused. Ill will was not unmotivated, but there is no clear implication that the offences or the aggression should have come from the victim.

The accused cannot have been as surprised as they sometimes claimed to be when they came before the judges. In the great majority of cases, not only did the witnesses allege a reputation stretching back many years, but the evidence revealed that one or more public accusations had been made

[16] R. Muchembled, 'Sorcières du Cambrésis', in *Prophètes et sorciers dans les Pays-Bas (16ᵉ–18ᵉ siècles)*, ed. M.-S. Dupont-Bouchat, W. Frijhoff, and R. Muchembled (Paris, 1978), 210–14.

against the supposed witch. The fact that no reparation had been sought was a major presumption against the suspect, yet there were powerful motives for taking a chance in letting such insults pass, for an attempt to obtain an apology or damages could often turn into a trial on the normal pattern. Every village seems to have contained individuals whom their neighbours believed to be witches. How did such identifications take place, and at what point did a formal prosecution result? At least three quarters of the accused were women; most of these were at least into their late forties and many much older. The great majority were poor, their property commonly insufficient even to meet the modest costs of the trial. Some were beggars, although Rémy certainly exaggerated here. Other categories found quite commonly were individuals who made themselves obnoxious by their quarrelsomeness; those who were of dubious sexual morality; and village herdsmen and women who were often involved in treating the illnesses of animals and who shade into the category of magical healers, often themselves prosecuted as maleficent witches. Above all, however, ill repute was inherited: parents, siblings, or other relatives already accused were a mortal danger. An extreme case was that of Hellenix le Reytre at Blamont in 1606, whom the judges pressed for details about her family. It became clear why they did so when she admitted that her brother had been executed thirty-seven years earlier, while of her four sisters three had also been executed and the fourth accused.[17] In many other cases it was claimed that relatives had been suspected, even if never tried.

Identification might also take place through the white witches or *devins* who specialized in counter-magical healing. Much of their skill lay in persuading the client to articulate his own suspicions, but there could plainly be a random element in the operation. This emerged alongside the theme of inherited witchcraft when Mengette Estienne of Le Paire d'Avould was accused and offered the explanation that the family reputation originated when

ung jour sa mère allant querir du feu chez ung de leur voisin, là où il y avoit ung qui estoit dans ung bain ayans mal en ung jambe, et ne pouvant estre gueri il feit aller au devin laquel devin dict Que ce

[17] AD M.-et-M., B 3335, No. 2.

pourroit avoir faict quelconque de ses voisines, sur ce ladicte bruict fut donné à sa mère parce qu'elle avoit esté querir du feu encore qu'elle n'en eust jamais esté suspitionée.

['one day her mother went to seek fire from the house of a neighbour, where a man who had a bad leg was in a bath, and, unable to find healing, he had sent to consult a *devin,* who had said one of his neighbours might be responsible, which was the reason her mother acquired her reputation, because she went to seek fire, although she had never been suspected'].[18]

In other cases knowledge of a visit to the *devin* seems to have induced the suspect to appear and offer some kind of healing, which would confirm his or her reputation even if it worked. When Mengeon Lausson and his wife Mengeotte were tried in 1620, it emerged that he had prevented her from undertaking a pilgrimage for a neighbour who had lost her milk, on the grounds that she was already suspected of causing similar harm to another woman, and to act as requested would confirm this belief.[19] Suspects usually knew of the graver suspicions against them and had to decide what attitude to adopt; although the situation was horribly dangerous for them, it did at least give them a certain negative power over their potential accusers. The prime mover in the accusation against Georgette Herteman of Brouvelieures in 1615 was the blacksmith Nicolas Mongeot, who believed she had bewitched his wife; she told others that 'elle auroit bien pu fournir quelque chose pour guerir sa femme, mais puis qu'il s'estoit porté sy terrible, elle la laisseroit là' ['she might well have provided something to cure his wife, but because he had been so hostile, she would do nothing for her'].[20]

The villagers were equally conscious of the dangers in crossing those reputed to be witches; many testimonies emphasize how they were feared and humoured. According to the local *tabellion,* Fleuratte Maurice of Docelles was so feared 'que personne du village ne fait banquet de nopces ou autre sans luy envoier quelque present de chair ou autre vivres' ['that nobody in the village held a wedding feast or other celebration without sending her a present of meat or other

[18] AD M.-et-M., B 8667, No. 4.
[19] Ibid., B 3804, No. 3.
[20] Ibid., B 3789, No. 2.

food'].[21] It is striking that in many such cases these individuals were apparently tolerated for many years before a formal accusation was brought; although suspected of this appalling anti-social heresy, they were apparently treated as just one more danger of everyday life, rather than arousing any immediate or panic-stricken reaction. Many must have died without coming to trial at all, given the length of the reputations of those who did. It is almost impossible to understand why at a certain point formal steps were taken, for nothing seems to mark off those *maléfices* that acted as catalysts from those dating back many years. We are almost certainly dealing with a situation in which there was great reluctance to prosecute one's neighbours, in view of the continuing ill will that might result and of the costs that might be incurred if one came forward as a 'partie formelle' to bring the charge. The witch might be removed, but his or her kin still had to be reckoned with.

The troubles of Nicolas Mongeot, mentioned above, did not end with the execution of Georgette Herteman; he appeared again as a witness at the trial of her husband, Nicolas Herteman, to tell how his wife had relapsed after Nicolas had reproached him 'qu'il estoit cause de la mort de sadite femme, et en quoy on luy avoit faict grand tort, mais que cela ne dormoit encore et n'estoit oblié' ['that he was the cause of the death of his said wife, and they had done him great wrong, and that the matter was not yet asleep or forgotten']. Nicolas Herteman was released after withstanding the thumbscrews and the rack, leaving one to imagine the future relations between these neighbouring families.[22] Another witch mentioned earlier, Hellenix le Reytre, deterred a potential accuser by declaring loudly 'qu'elle avoit desja heu cinq proces et les avoit tous gagné. Qu'elle seroit encore bien aysé d'en avoir ung autre pour y faire conformer quelques personnes jusques à leurs chemises' [that she had already had five lawsuits and had won them all. That she would be very happy to have another to make some people toe the line at the risk of losing their shirts'].[23] This was an exceptionally aggressive

[21] Ibid., No. 1.
[22] Ibid., B 3792, No. 1.
[23] Ibid., B 3335, No. 2.

reaction, but several other suspects put the same message across in slightly more veiled terms. Such confrontations emphasize the extent to which witchcraft was a double-edged factor within the complex relationships of village society, allowing a certain status to some of its more rebarbative members.

One way round the dangers of accusing these potentially vindictive neighbours was to seek a direct intervention by the ducal prosecutor or other competent authority. This was difficult to accomplish secretly, however, and involved dealing with a relatively elevated and often distant personage. Much commoner was reliance on the accusations made by the convicted against their accomplices, those with whom they had supposedly gathered at the sabbat. As participants in the world of village gossip, the condemned naturally directed most such nominations at well-known local suspects. Little chains of prosecutions would result, although not all such charges automatically produced further trials without there being any obvious reason for this. Once a trial was under way rumour and tension would commonly spread through the surrounding villages, with talk of taking all the witches. Numerous testimonies expose the fear and agitation of those who knew themselves threatened; they would sometimes make their relief rather too obvious when they heard that they had not been named. They often talked of flight, but few had the courage to cut loose from their local ties and modest property in this way. It is clear from one exceptional case that good repute and the support of one's neighbours could offer some protection. In 1592 Mathieu Blaise of Saint Margarée was separately accused by three convicted witches, but thirty-seven witnesses produced no serious charge against him, while many testified to his good character and generosity to others. Even Nicolas Rémy was compelled to order his immediate release. Yet Mathieu, whose nickname 'le gros' was evidently a reference to his corpulence, did have something of a reputation. One favourable witness told how, talking outside the church of the nearby village of Combrimont a decade earlier, a man had come up to him and said

que l'on parloit bien des sorciers et sorcières et que sy on brusloit Mathieu Blaise il y auroit bien de la gresse. Ce qu'ouy par luy deposant, luy dict que sy ledit Mathieu estoit present et qu'il l'eust ouy, qu'il eust bien reparti en sa reverence, lequel devint tout rouge et s'en alla incontinent.

[that there was much talk of the witches, and that if Mathieu Blaise were burned there would be a lot of grease. When the witness heard this, he told him that if the said Mathieu were there and had heard him, he would have known how to make a good answer, at which the other went all red and left in a hurry.][24]

Despite such incidents, and the failed prosecution, the reputation stuck, so that during a fresh batch of trials in 1603 we find Mathieu being named again by several of those who confessed.[25]

Once gained, it seems, a reputation for witchcraft was almost impossible to lose. For those who neither antagonized their neighbours excessively nor engaged in dubious kinds of healing there were two main ways in which this kind of reputation was acquired. The first arose when a sickness was diagnosed as unnatural, either by the *devin* or by some more orthodox specialist such as the local surgeon, leading to the idea of bewitchment, and inducing the victim to identify a plausible suspect with a grievance against him. The second was through the general awareness of family background, as expressed for example in the investigations of prospective marriage partners by members of the families concerned. As the witchcraft persecutions continued, this latter mode of generating suspicions must have become more and more dangerous, so that an increasing number of those accused owed their reputations to the misfortunes of their relatives.

In theory such identifications might have continued to multiply until a very high proportion of the population was under suspicion. If the number of trials is any guide, this does not seem to have happened; the peak was probably reached in the late sixteenth century, the numbers dropping slightly thereafter until the cataclysm of the Thirty Years War brought an end to virtually all features of normal life, witchcraft among them, in the 1630s. It seems likely that some kind of control mechanism was at work, but its exact nature remains elusive, for this is just where the documents, by their

[24] AD M.-et-M., B 8667, No. 3.
[25] Ibid., B 8691.

own character, are least helpful. It is in fact far easier to understand witchcraft beliefs and persecution synchronically than diachronically. The kinds of disputes and misfortunes that were used in evidence must have been common to all village societies. The use of counter-magical techniques cannot have been a complete answer for European peasants, any more than it was for the Azande in pre-colonial days; if one's child or cow died anyway or, still worse, continued to languish, one would look for some more positive action.[26] To employ the witchcraft explanation in such cases was normally a way of seeking practical relief, which might be provided by extracting a show of goodwill and efforts to cure from the suspect, but with the dangerous side effect of building up evil reputations.

The natural sanction against those who became too obnoxious, or failed to cure their supposed victim, was beating or even lynching. To explain the rise of persecution through the courts one needs to dovetail popular belief and practices with a number of parallel developments. These include the extension of the criminal law and the system of public prosecutors, the spread of demonological theory by the printed book and pamphlet, the general tendency of the social élites and the churches to seek more direct enforcement of social controls, and the rapid socio-economic changes in rural society in the later sixteenth century. The one thing that does seem plain is that no monocausal explanation is likely to be correct. The related problem of the reasons why persecution through the criminal law ceased cannot be illuminated by the experience of Lorraine, where the devastation of war was followed by a lengthy French occupation, bringing with it the more sceptical attitudes already developed by French lawyers and judges.[27]

Another area of great difficulty is the relationship between popular beliefs about such matters as the pact and the sabbat and the elaborated cumulative accounts given by the

[26] For the Azande, see the classic work by E. E. Evans-Pritchard, *Witchcraft, Oracles and Magic Among the Azande* (Oxford, 1937).

[27] For the sceptical attitudes in France, see R. Mandrou, *Magistrats et sorciers en France au XVIIe siècle* (Paris, 1968), and A. Soman, 'Les Procès de sorcellerie au parlement de Paris (1565–1640)', in *Annales: économies, sociétés, civilisations*, 32 (1977), pp. 790–814.

demonologists. My own belief is that the confessions were based very largely on an indigenous popular tradition, with relatively little contamination from élite demonology. The occasional vivid description of the sabbat is in the characteristic style of the village story-teller, manipulating elements common to folk belief in many parts of the world. Such stories must certainly have been told at the *veillées*, the winter evening gatherings often known as *poisles* in Lorraine, from the local word for the kitchen in which they customarily took place. The *poisle* appears quite often in the trials, as the scene or cause of disputes, since invitations and friendly behaviour were important signs of neighbourly feelings. These meetings were an important agency for the maintenance and development of folklore; they were also one of the occasions (alongside visits to mill, forge, and well) for gatherings at which communal action might be discussed or initiated. European folklore generally mixes only small doses of fantasy with primarily realistic elements, so it is not surprising to find that in the accounts of the sabbat given by Lorraine witches there are only a few veiled references to sexual licence or to any of the more vivid rituals found in other sources. The exiguous feasting and dancing described are little more than the transposition of the features of a village festival into a different context. Most of the active witchcraft took the form of beating water to arouse hailstorms; these were often said to have been turned aside by the timely ringing of church bells. The distribution of diabolical powder, often referred to in the trials, was generally a personal transaction between devil and witch and was rarely mentioned in connection with the sabbat. Like so much else in the theory, there was no obvious necessity for the powder at all; witches were often represented as having injured their victims without any physical agency being involved.

Such inconsistency is perhaps the most consistent characteristic of Lorraine witchcraft beliefs, which repeatedly demonstrate the adaptability of these popular traditions. They allowed villagers to articulate their hostility towards members of their society who broke communal norms too often, to isolate them amid a web of suspicion, and to drive them into dangerous threats against the potential accusers who surrounded them. Such a mechanism may well have had considerable effects on the social behaviour of individuals; when it was taken up by the legal system, it resulted in a grim toll of victims. In this as in so much else, the witches of Lorraine shared their experiences with those of many other regions of Europe. There are many reasons to study them today, and one would certainly be to demonstrate how a rather commonplace, and indeed common-sense, belief in occult power could exist through every level of an early modern society.

3

Ill will and magical power in Lorraine witchcraft

IN November 1619 Bastienne Brihey, a widow from the Lorraine village of Aydoille, was burned as a witch, one of perhaps three thousand people to suffer this fate in the duchy during the period from 1580 to 1630. Just before her execution she was asked about her accomplices in witchcraft; this was a normal procedure, which followed up earlier interrogations, and often resulted in the withdrawal of allegations against other people made earlier under torture. At least one of those she had already named, Demenge Demengeon, an elderly miller from the nearby village of Dompierre, had been ordered to be present, an invitation he chose to ignore. By February 1620 he was on trial himself, although he was to resist the torture and be released. One of the witnesses, Nicolas Jean Ferry, had often quarrelled with Demenge; he told how after Bastienne's execution the miller had come to see him, asking if she had persisted in her accusation. When told that she had 'il repartit que s'il advenoit qu'aprehendé pour cas de sortilege, et luy qui parle estoit adjourné pour deposer contre luy, scavoir ce qu'il diroit' ['he replied that he would like to know what the witness would say if he [the miller] were arrested on a charge of witchcraft, and he was called on to depose against him']. Nicolas responded 'que s'il eut esté sorcier et eu pouvoir sur luy il ne seroit aujordhuy vivant, veu l'Inimitié qu'ils se sont entreportés' ['that if he had been a witch and had power over him he would not be alive today, knowing the enmity they bore one another'].[1]

This strikingly fair-minded comment also expresses the central themes in the popular conception of witchcraft; hatred of others and the power to harm them by occult means. If the crime itself was imaginary, punishable only where the accused

[1] AD M.-et-M., B 3804, No. 2.

was coerced into a fraudulent confession, this does not mean that it can be relegated to some area of nightmarish fantasy. Apart from the brutal realities of trials and executions, the trial records plunge us deep into the everyday world of human relationships among the peasantry. Witnesses spoke of real events, although they might edit and interpret them in ways of which we must be wary. Even when the accused succumbed to pressure from their judges, telling stories of imaginary happenings, these are often recognizable as indirect representations of social realities.

The judicial arrangements in Lorraine were distinctly unfavourable to the suspects; there was no superior court to hear appeals in the manner of the French *parlements*, or to exercise effective control over the amateur judges who sat in the local courts. The review jurisdiction of the court of the *échevins* of Nancy appears to have had only a very marginal influence; when one of its members, Claude Bourgeois, wrote a treatise on criminal procedure in 1614 he recognized that there were still numerous abuses current in the duchy.[2] Bourgeois was particularly worried about the use of torture, and it was this which was crucial in securing such a high proportion of confessions. Four different types of judicial torture were in use; virtually all courts employed two of them, thumbscrews and the rack. The *tortillons*, cords tightened around the upper arm, were a much less common device, probably roughly equivalent in unpleasantness to the first two. The final torture, the strappado, was only found in a limited number of cases, and was certainly the most severe. The victims hauled into the air on this apparatus were likely to suffer dislocations of the arms or shoulders, so it is no surprise that virtually all of them broke down and confessed. There was a further important consequence of these practices; a proportion of those accused were so terrified by the prospect of torture that they admitted their guilt before it was applied. Such apparently unforced statements were bound to reinforce general belief in the reality of witchcraft, while those who made them were often tortured none the less, in order to secure answers on particular charges. Confessions made

[2] C. Bourgeois, *Pratique civile et criminelle pour les justices inférieures du duché de Lorraine* (Nancy, 1614).

under such circumstances must be treated with great caution, but they do follow remarkably consistent patterns; everyone in Lorraine seems to have had a clear notion of how a witch ought to act. As Étienne Delcambre's study of the trial records made plain, the suspects confessed to three principal offences against God and the community.[3] They had succumbed to the devil, under blandishments or threats, and had made a pact with him renouncing God. They had attended the sabbat, participating in blasphemous rituals, and performing magical acts to cause storms. Finally they had committed acts of *maléfice* against their neighbours, causing death or sickness among both humans and animals. Witnesses normally testified only to the last category; here the task of a suspect who decided to confess was relatively simple, for he or she had only to take over a selection of the allegations already heard from both judges and witnesses. With the pact and the sabbat no such help was available, so the striking uniformity of the accounts given must reflect widespread popular belief and storytelling. Yet all three aspects of the confessions share a concern with power and the desire to use it to harm others, the central themes of this essay.

Witches narrated the making of the pact in a way which generally laid little emphasis on the seductive or coercive powers of the devil, while giving a very realistic circumstantial description of their own state of mind at the time. As the judges interrogated Catherine la Rondelatte, she suddenly said:

je suis sorciere, il y a dix ans des le St Laurent dernier que retournant de Magnieres veoir ma soeur Barbon passant seule par les bois toute songearde et pensive de me veoir sy longtemps en viduité et sans enfant, mesme de ce que mes parens me destournoient de me remarier curieuse neantmoins de ce faire, parvenue que je fus à l'endroit du rond chesne au milieu des bois je fus toute estourdie et bien effraiée quant j'aperceu un grand homme noir qui saparut à moy, lequel dabordée me dit pauvre femme tu es bien pensive, soudain se recommanda à Monsieur St Nicolas, neantmoins soudain me renversa par terre jouyte de moy charnellement et au mesme instant me pinsa au front assez rudement Cela faict me dit tu es à

[3] É. Delcambre, *Le concept de la sorcellerie dans le duché de Lorraine au 16ᵉ et au 17ᵉ siècle* (3 vols., Nancy, 1948–51).

moy ne te donne point de peine Je te feray damoyselle et te donray de grands biens.

Cognut à l'heure mesme que c'estoit le malin esprit mais ne peult se retracter pour ce que à l'instant il luy feit renier dieu, cresme, et baptesme, avec promesse qu'elle seroit pour luy et luy donna une verge, disant sy elle avoit quelque hayne sur aultruy, elle pourroit se venger de laditte verge les touchant ou leur bestail deslà se disparut disant qu'elle le reveroit bien tost et qu'il s'appelloit percy.

[I am a witch. Ten years ago last St Laurence's day I was coming back from visiting my sister Barbon at Magnieres, walking alone through the woods all dreaming and thoughtful that I had been so long a childless widow, and that my relatives discouraged me from remarrying, which I would have liked to do. When I arrived at the place of the round oak in the middle of the woods I was astonished and very frightened by the sight of a great black man who appeared to me. At first he said to me 'Poor woman, you are very thoughtful', and although I quickly recommended myself to St Nicholas he then suddenly threw me down, had intercourse with me, and at the same time pinched me roughly on the forehead. After this he said 'You are mine. Have no regret; I will make you a lady and give you great wealth.'

I knew in the same hour that it was the evil spirit, but could not retract because he had instantly made me renounce God, chrism, and baptism, promising to serve him. He gave me a stick, saying that if I bore hatred to anyone, I could avenge myself by touching them or their animals with the stick, then disappeared, saying I would soon see him again, and that his name was Percy.][4]

Here the seduction is presented as a response to the personal frustrations of a lonely widow. More commonly, however, it was sheer hardship which engendered despair, as when Jaulne, wife of Claudon Curien, admitted in 1603 that she had been seduced sixteen years earlier when:

les gens de guerre frequentoient fort par deça, et de la pauvreté qui estoit par le pays, il faisoit une grande cherté, et pour lors estoit grandement molesté, à cause qu'il y avoit trois jours et nuicts qu'elle n'avoit mengé, se voyant chargée d'enfans et qu'elle n'avoit les moyens de les nourrir et entretenir, et lesquelles n'avoient mengés sinon que des racines qu'ils alloient querir.

[4] AD M.-et-M., B 4094, No. 1 (Essegney, 1608). The variation between direct and indirect speech in this passage is obviously the work of the *greffier*, trying to record what was said while producing a suitably impersonal court record.

[the soldiers were present in great numbers, and the poverty in the countryside caused a great dearth, so that she was very upset, because she had not eaten for three days and nights, and was burdened with children whom she did not have the means to keep or nourish, who had eaten nothing but roots they had been searching for.][5]

More simply still, Catherine la Malreannaix said in 1592 that she had been seduced two years previously, when married to her last husband, 'par lequel elle estoit sy mal traictée, qu'elle se desesperoit' ['who treated her so badly that she despaired'].[6] As in these instances, the devil was normally alleged to have made his approach when the prospective witch was angry or in despair on account of poverty, inability to feed her family, maltreatment by her husband, or the pressure created by informal accusations of witchcraft. The devil's promises to make witches rich were as illusory as the money he commonly proffered, which turned into dead leaves or other detritus, yet the witches did not leave the mythical encounter quite empty-handed. The diabolical powder (or analogous maleficient objects) they supposedly received gave them power not merely to kill and injure, but to assert themselves against their enemies, and to reverse the normal power relationships of village society. It symbolized a real process whereby a reputation for witchcraft conferred a certain status, however ambiguous, and could actually procure its holder greater consideration from neighbours and even enemies. The very nature of the belief is bound up with the pattern whereby witchcraft was overwhelmingly a crime of the poor and downtrodden; secret magical powers of revenge were the only way in which this group could hope to assert themselves against the harshness and brutality with which they were so often treated. Their accusers were aware of the ill will, and feared its possible effects, while the suspects themselves must often have felt guilt at their surges of hostility against their neighbours. Though a number of alternative

[5] Ibid., B 8691, No. 14 (Saint-Blaise, 1603). The soldiers were the *reiters* of the duc de Bouillon's army, on their way to support the Huguenots in France; the ravages they caused in 1587 appear frequently in the trial documents.

[6] Ibid., B 8667, No. 2 (Saint-Margarée, 1592).

scenarios can be found superimposed upon it, there is no doubt that a majority of the cases follow this pattern.

Village communities were in fact criss-crossed by lines of hostility and resentment, of which witchcraft formed a natural part. Much of the time these divisions remained latent, while ordinary social life carried on. It is no surprise, therefore, to discover from the witnesses' depositions that the suspects continued to have very close and frequent contacts with those who feared them; very few villagers apparently wished or dared to ostracize them. The possible uses of a reputation for witchcraft are suggested when witnesses claim to have made gifts or rendered services which proceeded from fear. During the trial of Jean de Socourt the *prévôt* visited nearby villages, and was told by some there that they 'le craindoient beaucoup comme faisoient plusieurs autres desdits villages ou il hantoit souvent et n'etoit refusé que bien peu de choses qu'il recherchast, tant estoit il craint et soubconné du crime de sortilege' ['feared him greatly as did many others in the said villages where he was often found, and was refused very little that he asked for, so greatly was he feared and suspected of the crime of witchcraft'].[7]

Villagers themselves sometimes believed that witchcraft had been still more advantageous to its adepts, offering it as an explanation of the economic success they had enjoyed. In a deposition which also evoked the familiar theme of family links Mongeatte Delatte reported at second hand a conversation between the accused Colas Gerard and his relative Jean Babey, 'ce tant renommé sorcier de Mandray', in which Babey had said:

que s'il le vouloit croire, et faire ce qu'il luy disoit, il ne verroit jamais son pauvre jour, depuis et delà, Iceluy Colas Gerard, estant tres pauvre, est maintenant tres riche et bien fourny de touttes sourtes de bien, à l'estonnement de beaucoup de gens, qui ne scavent d'ou ilz luy peuvent estre ainsy venues.

[that if he wished to trust him, and do what he told him, he would never know poverty. Since then this Colas Gerard, previously very poor, had become very rich and well supplied with all sorts of goods, to the astonishment of many people, who did not know how he could have acquired these riches.][8]

[7] AD M.-et-M., B 4093 (Charmes, 1607).

[8] Ibid., B 8691, No. 15 (Benifosse, 1603).

A similar theme was suggested by witnesses in the case of Janotte Raon, while she herself claimed that the accusations were made 'par envye d'autant que son marit et elle font bon mesnage (Dieu mercy) et qu'ilz marient en bons lieux leurs enffans' ['from envy, because she and her husband formed a successful household (thanks to God), and married their children well'].[9]

When they made formal confessions, however, the witches were unanimous in denying that the pact had brought them any benefit. It had been a disastrous error, which made them the helpless subjects of the devil, compelled to attend the sabbat and commit *maléfices* at his behest. As Georgette Herteman said of the devil who had seduced her, 'loin d'en avoir receu quelque courtoisie, tous malheurs et pauvrete luy sont advenues depuis qu'elle s'est donnée à luy' ['far from having obtained any benefits from him, nothing but evil and poverty had befallen her since she gave herself to him'].[10]

Judges and witches alike believed that only a formal renunciation of the devil and recital of all related crimes permitted lasting reconciliation with God and the community; even confession to a priest seems to have been thought inadequate to break Satan's hold. Intense feelings of guilt were clearly created in many of the accused, so that they accepted much of the communal view of their own behaviour. There may have been an element of self-exculpation involved, with the witch implying that she had only harmed neighbours under the impulsion of a power which was too strong for her, often claiming that she had resisted the devil as much as she could. The original act of deviance, the pact, was represented as originating from genuine hardship and despair, combined with the deceits of which the devil was the proverbial master. One might even suggest a parallel with that supposed feminine irresponsibility which Yves Castan has found used as an evasive mechanism in quarrels and violence among women in Languedoc.[11] In this way the witch, in the very act of confessing her imaginary crimes, displaced the moral responsibility for them on to her master, although the evasion was

[9] Ibid., B 2521, No. 1 (Xennenay, 1598).

[10] Ibid., B 3789, No. 2 (Bruyères, 1615).

[11] Y. Castan, *Honnêteté et relations sociales en Languedoc, 1715-1780* (Paris, 1974), 172–9.

normally partial and inconsistent. The relationship between devil and witch was presented in brutal terms of domination, with the witch attempting the same kind of evasion and non-compliance as villagers commonly employed against unwelcome outside authorities.

Attendance at the sabbat was described as an irksome obligation, offering little gratification to most witches. Only their subjugation to the devil could force them to participate in what was essentially an anti-fertility rite, with parodic acts of homage to their loathsome master followed by the corporate beating of water to arouse storms. The inversion of ordinary festivities was extended to the exclusion of any enjoyment; the food was foul, eaten by participants who were only there under compulsion, danced back to back, and discussed ways to ruin their own crops instead of protecting them. Witches naturally represented themselves as having been unwilling performers, all too conscious that they and their families would suffer if there was a dearth. As Delcambre noted, they were liable to claim that the rich had the dominant voice at the sabbat just as they did in the village, and had supported the plans to devastate the crops out of self-interest, hoping to sell their grain dearly.[12] The idea of such a gathering without a social hierarchy of its own was probably almost unthinkable, so that rich and powerful witches had to be invented by those who told stories of the sabbat, either as part of local folklore or in the deadlier context of their own imminent execution. The malevolent rich, like the enforced attendance, also formed part of the self-exculpatory structure evident in the confessions, by which the accused tried to deny responsibility for the harm they had done their neighbours. As they represented it, one slip had placed them in the devil's power, leaving them as unwilling conscripts in his forces. Women generally described their initiation in terms of a rape, committed by a stranger who was too strong to resist, in situations where they were alone and vulnerable.

It is typical of the inconsistencies in this imaginary world that it sometimes proved possible for the more numerous poor to frustrate the plans for storm-raising, at the risk of being

[12] Delcambre, *Le Concept de la sorcellerie*, i. 185-7. For additional instances, see p. 73.

beaten by the furious devil. Chenon la Triffatte claimed that she and the other poor had prevented the rich from making hail, enraging the devil, who kicked another recalcitrant witch an enormous distance.[13] Whether the poor ever tried to outvote the rich in this way in real village assemblies is unfortunately unknown, since no relevant records survive, but the closeness to everyday situations is again striking. The belief in storm-raising was certainly old-established; indeed it seems to have been the catalyst for the first major group of known witch-trials in Lorraine, in the 1480s, when the accused were blamed for a series of devastating storms.[14] No doubt it was reinforced by the general practice of ringing church bells to ward off storms, the efficacy of which was confirmed by numerous confessions, in which it was used to explain the failure of the magic used at the sabbat. In Lorraine the *maléfices* specifically associated with the sabbat were virtually always directed against the crops, emphasizing its character as a fertility cult in reverse; the devil did not use the meetings of witches to demand direct action against people or animals, as Bodin and Boguet were to suggest.[15] Although he was frequently said to have made such requirements, this was almost always during a *tête-à-tête* with an individual witch.

The whole concept of the sabbat gave the accused witches a more straightforward yet frightening opportunity to exercise power and express ill will. The judges expected them to identify and denounce the other witches they had seen there, the only supposed opportunity for them to know their companions in misfortune. In a predictable symbiotic relationship, a fair proportion of the circumstantial detail about the sabbat is concerned with the ways in which participants concealed their identity, building up a stock of routine excuses for inability to perform this task adequately. Nevertheless, the number of such denunciations was considerable, leading to horrible and pathetic scenes as those threatened sought recantations, which the convicted often refused to give even as they went to the stake. There were

[13] AD M.-et-M., B 4077, No. 3 (Charmes, 1596).

[14] L. Dumont, *La Justice criminelle des duchés de Lorraine et de Bar, du Bassigny et des Trois-Evêchés* (Nancy, 1848), ii. 26; L. Gilbert, 'La Sorcellerie au pays messin', *Pays Lorrain* (1907), 35.

[15] See p. 34.

limits to credibility, however, and there is no need to accept those rather facile suggestions that witches might have short-circuited the whole procedure by claiming to have seen their accusers or their judges at the sabbat. Accusations against enemies or hostile witnesses were very unlikely to be effective in small communities which were highly sensitive to the details of interpersonal relationships. When Zabel de Sambois accused Babey Masson of having been at the sabbat, the judges were quick to suggest that this proceeded from hatred, and pointed out that she had not reproached Babey when she appeared as a witness.[16] The obvious people to accuse, in order to satisfy the judges, were fellow villagers who also had a reputation as witches, and it was this group who were most likely to be proceeded against immediately. Knowledge of this could make them act in a highly dangerous way, as did Catherine Charpentier, according to her sister-in-law's deposition, when she believed that Jean Claude Colin would confess and identify her; she was 'toute esbahye', and said: 'chose seure, que le mechant homme la m'aura accusé, que sy une fois, il a parlé de moy, me voyla fricassée' ['that wicked man is sure to have accused me, and if he has once spoken my name, then I will be fried.'] After which she 'gagna les champs' and absented herself for a long period, in fear of being arrested.[17] A single case could in this way send shock waves through a community, and this may have been a restraining factor on persecution. Many families or groups probably had at least one member with a doubtful reputation, whom it would be difficult to protect once a formal identification was made. Claudette Saulcy may have been trying to play on such fears when she allegedly said three weeks before her arrest 'que le bruit courroit que l'on la vouloit apprehender pour une sorciere, mais que sy on la prenoit pour telle l'on en prendroit encor bien des autres' ['that word was going round that she was to be arrested as a witch, but if they took her for such they would take many others later'].[18]

Not all accusations followed this pattern, of course, and a

[16] AD M.-et-M., B 8677, No. 6 (Mouriviller, 1592).
[17] Ibid., B 8691, No. 13 (Rémemont, 1603).
[18] Ibid., B 8677, No. 7 (Entre-deux-Eaux, 1596).

particularly troubling group comprises those directed against kin, close friends, and neighbours. Where no previous antipathy emerges, these charges may have shown the witch's anger against those whom she blamed for failing to defend her adequately, preferring to desert her in the hope of preserving their own reputation. There is little sign that names were prompted by the judges, other than one very sinister case in which Marie Canot alleged that a family who hated her had procured an accusation through a relative participating in the case.[19]

Pact and sabbat were far outweighed, even in the confessions, by individual acts of *maléfice*. The judges' insistence on a full recital of such crimes appears, on one level, as part of an assumption that only a full confession could restore the lost sheep to God. On a less conscious plane, however, they may have been responding to a feeling that communal harmony would only be restored when responsibility for all these misfortunes was securely attached to the scapegoat, and purged by his or her death. Interestingly the witches often refused to admit certain *maléfices,* although confessing their crimes and the justice of their condemnation; very often this was in cases where they claimed to have felt no ill will towards the supposed victims. The devil was pushed firmly into the background; having supplied the power to do evil, he was apparently content to leave its direction largely in the hands of others. His intervention was occasionally suggested when no satisfactory motive would otherwise have existed, as when witches who had afflicted a friend claimed to have done so under direct orders, after obstinate but futile attempts to resist. Georgeatte Malley claimed she had long refused her master's insistence that she make Jean Mongeat's wife ill, in revenge for a trifling slight, since she was a neighbour and friend. She finally consented, but after her master had made the victim ill she repented, and put healing powder in a dish of millet.[20] The common provision of this healing powder at the time of the original pact is a neat expression of the general and probably very ancient belief that witches (and only they) could remove the sicknesses they had caused, although it fits

[19] Ibid., B 4500, No. 3 (La Neuveville-sous-Châtenois, 1586).
[20] Ibid., B 8715, No. 1 (La Neuveville-les-Raon, 1615).

much less well with the theological view of the boundless malignity of the devil (the sophisticated way of getting round this was to suggest that those who used force or counter-magic to procure healing made themselves vulnerable to the devil).[21] Once it came down to the detail of individual acts against others, the accused rarely tried to shelter behind the claim of obedience to orders from above; they were more likely to produce some variant on the theme of 'justified anger'. Witches sought to give a rational explanation of their supposed behaviour within a context of close yet intermittently hostile personal relationships, which normally permitted them to refer to disputes that had caused temporary ill will. Relatively few of the victims seem to have been on consistently bad terms with the accused, and this was reflected in the trial proceedings, for the opportunity to denounce witnesses as personal enemies was very little used.

The power to harm, then, was not used in any very consistent or systematic way. Rather it was employed on impulse, to react against an immediate wrong, and might often be regretted later. Thus Annon Chauderet used the powder to kill Mengeon Failly, a butcher who refused her meat when she had no money; she subsequently felt pity for him, and intended to cure him, but was prevented by her arrest.[22] Many other witches described themselves as having been smitten with remorse, so that they tried to heal their victims. Sometimes the devil had refused to co-operate, giving an easy excuse for failure, but often the healing powder he so thoughtfully provided, sprinkled over the sufferer or added to some small gift of food or drink, was said to have effected a cure. The common situation in which the victim demanded healing from the witch might be thought to create an impossible choice for the latter; to refuse was to confirm one's ill will, to accept implied that one was indeed responsible for the illness. In some cases, where the witch made a practice of semi-magical healing, the call on her services would have seemed quite natural, but in general the dilemma was more apparent

[21] For this kind of argument, see J. Bodin, *De la demonomanie des sorciers* (Paris, 1580), fos. 125r–126r, 147r, 148r–153v, and N. Rémy, *Daemonolatreiae libri tres* (Lyon, 1595), Bk. III, ch. III.

[22] AD M.-et-M., B 8715, No. 6 (Raon, 1615).

than real. The suspect who hastened round to offer a diagnosis, treatment, or comfort was putting on a display of good will which had an excellent chance of cancelling out the real or supposed ill will, and thereby much lessened the chance of a prosecution beginning. Rather more surprisingly, even when the victim's accusations were maintained until death, the kin were rarely found moving immediately to make formal charges. As long as he lived there was hope for his cure, especially if the suspect was trying to help; once he was dead it was too late to help him. Similarly, very few of the many threats to have the witch burned were followed up, unless an accusation from another quarter revived them. Many suspects must therefore have escaped prosecution, but those who did come to court faced a mass of plausible-sounding evidence about their behaviour over many years, which would have tended to blot out the logical inconsistencies within the whole scheme of conduct envisaged.

Any attempt to follow through the logic of accusations rapidly runs into the sand, as any reader of the demonologists' works will know. Given the notion of a totally malevolent devil who is the lord of this world, with a following of deluded and enslaved human beings obliged to perform his will, and equipped with deadly poisons, what individual or community could feel safe? The most excitable demonologists, such as Boguet, De Lancre, and Bodin in some moods, did indeed tend towards an apocalyptic view of the possible outcome, which lent urgency to their call for harsher repression. Rémy, probably a more representative lawyer of his time, appears less alarmed, treating witchcraft as an endemic rather than an epidemic malady of peasant society, more of a social nuisance than a mortal threat to the community.[23] In this he surely reflected the general opinion of the people, who treated witchcraft as one more hazard of everyday life, rather than a peculiar supernatural threat. There is no reason to doubt that witches were articulating common opinion, even though they were speaking in their own defence, when they sought to limit the applicability of the witchcraft explanation. Claudon la Romaine said: 'tous ceux qui sont malades, s'ils vouloient dire

[23] Rémy, *Daemonolatreiae, passim* and particularly Bk. I, ch. II, Bk. II, chs. VII–IX.

qu'ils sont ensorcelés ilz ne feroient pas bien, et que c'est dieu qui envoye les maladies' ['all those who are sick, if they should try to claim that they were bewitched, would not do well, and that it is God who sends sicknesses'].[24] Similarly Mengette Estienne, asked about the deaths of her neighbour's animals, retorted 'qu'elle mesme en a bien perdu, qu'il conviendroit bien mal que tous les inconvenients qui venoient fussent par des sorciers et sorcieres' ['that she herself had lost many, and it would be very unreasonable if all the misfortunes which occurred were caused by witches'].[25]

An expression of ferocious ill will could still be associated with a shrewd rebuttal of witchcraft charges, as when Jeannon Poirot, asked whether she had fought with Blaisette Breton

dict que cela est vray, et vouleroit l'avoir tuée, et avoir ses tripailles dans son girons, et par apres elle seroit contente d'estre pendue et estranglée. Mais elle ne vouldroit estre bruslée, par ce que l'on dit qu'une personne que n'est pas sorciere ne brusle par volontiers, et que sy lors de cete dispute ou depuis, elle fut esté sorciere, il est sans doubte qu'elle l'eut fait mourir miserablement, ou les sorcieres n'ont point de puissance sur autruy.

[said this was true, and she wished she had killed her, and had her tripes on her lap, and afterwards she would be content to be hanged and strangled. But she did not want to be burned, because it was said that someone who was not a witch did not burn easily, and if at the time of the dispute or afterwards she had been a witch, there was no doubt that she would have made her [Blaisette] die miserably, or witches had no power over anyone.][26]

She admitted saying to Blaisette's daughter Marguitte, who had subsequently died, 'qu'elle les auroit de quelle facon que se doibve estre', but her intention had been to beat or kill her in a perfectly natural way. A similar awareness of problems of causality appears in the deposition of Marie Magnien, who now believed that Petroniere Vernier had given her a sickness three years earlier, but claimed that until now, although she knew

que de longtemps on l'avoit soupconné de sortilege, mais qu'elle rebrouoit tousjours ceux qui lui en parloient, en leur allegant qu'elle

[24] AD M.-et-M., B 4077, No. 2 (Charmes, 1596).
[25] Ibid., B 8667, No. 4 (Le Paire d'Avould, 1592).
[26] Ibid., B 8689, No. 3 (Mayemont, 1602).

n'avoit point veu de mal en elle, et que sy elle eut esté telle veu leurs querelles elle luy eust fait des desplaisir, et qu'à cela on luy repliquoit que les sorcieres n'avoient pouvoir sur toutes sortes de gens.

[she had long been suspected of witchcraft, nevertheless she [Marie] always put down those who talked to her about it, telling them she had never seen any harm in her, and if she had been such, in view of their quarrels she would surely have done her harm, to which people replied that witches did not have power over all kinds of people.][27]

Highly educated and self-aware judges, like those of the French *parlements,* were made very uneasy by this sort of evidence. They were always very conscious of the dangers that the judicial system might be used as a means of settling accounts, with the bearing of false witness an ever-present threat; in witchcraft cases there could be very little control over depositions which represented interpretations rather than facts. Such worries were quickly translated into action when large-scale judicial abuse by the lower courts was detected, while a number of lawyers and clerics came to feel that all such accusations were fundamentally unsound, and that there was a basic incompatibility between judicial principles and witchcraft trials.[28] In arriving at this conclusion they were applying a more abstract and sophisticated logic to popular *mentalités,* which were bound to fail the test. Belief in the power to harm by diabolical means was part of a loose and open-ended world view, which readily employed multiple explanatory structures. The real facts of everyday life—disputes, illnesses, good and bad luck with animals—were easily shaped into the mythical facts of witchcraft, a pseudo-explanation all too frequently deployed. Such animistic ways of thought, into which we all regress at times, allow us to structure reality in line with our feelings; the very flexibility which often makes them plausible robs them of genuine explanatory power, but this is not at all easy to perceive. There was little chance that the amateur judges of Lorraine, who included a fair proportion of illiterates, would break away from a thought world they shared. It would have needed a cool head indeed to dismiss as irrelevant such a threat as that

[27] Ibid., B 3789, No. 3 (Bruyères, 1615).

[28] For the attitudes of the *parlementaires,* see pp. 45–53, and the references given there.

made by Claudon Martin to Claudon Simon: 'veux-tu avoir aussy mauvaise langue que ta soeur qui nous appelle sorciere parmy la ville, par les mant dey, si j'eusse sceu elle n'en eut eschapée à sy bon marché' ['would you have as evil a tongue as your sister who goes round the town calling me a witch; by all the gods, had I known of this she would not have escaped so lightly'].[29]

In a peasant society where wealth and status were more finely graded than modern historians always recognize, all those who had anything to lose would have shared Jean des Rechey's alarm at the threats made by Claudatte and Lambert Ferry

qu'il y en avoit de ceulx qui avoient heu de grande commodité, qui n'en avoient plus, et qu'il y en avoit encore d'autres au village qui en seroient de mesme, de sorte qu'il est arrivé plusieurs accidens et fortunes, tant à luy deposant qu'à autres dudit Vic.

[that there were some of those who had once owned much property who had lost it, and that there were others in the village who would suffer the same fate, in such a manner that various accidents and misfortunes had come upon the witness and others of Vic.][30]

When examined closely, it is evident that such threats carried no necessary implication of witchcraft at all, but in the context in which they were uttered their effect was very different. Those who made them sought to assert themselves in a society where such vigorous self-defence was quite normal, but often came fatally close to the boundaries of illicit and occult power.

Once such behaviour was assimilated to a model of maleficent witchcraft, accepted by most members of the ruling élites, those who believed that a neighbour bore them ill will and that they had suffered positive harm in consequence could invoke the machinery of criminal justice. This was the crucial link which made widespread persecution virtually certain. The consequences were tragic, and there can be no possible justification for the judicial murder of thousands of innocent persons charged with an imaginary crime.[31] One cannot

[29] AD M.-et-M., B 4126, No. 3 (Charmes, 1625).

[30] Ibid., B 8715, No. 3 (Vic [Ban-le-Duc], 1615).

[31] One might also call witchcraft a crime of the imagination, in order to make the important point that a proportion of those accused had threatened

automatically conclude, however, that the final effect on society was an entirely negative one; witchcraft beliefs and prosecutions had both functional and dysfunctional features. Their functional aspect was primarily normative, for the fear of being identified as a witch might inhibit the open display of resentment and use of threats, while the possibility of magical revenge could induce the more powerful villagers to treat dependent ones with greater consideration. In such a schema both witch and potential victim have latent power against the other, so that shared awareness may enforce respect for community norms. On the dysfunctional side one can stress the corrosive effects of suspicion and 'scapegoating', liable to build up fears on both sides which could colour everyday transactions and break down trust. This would not require legal persecution: social ostracism and minor personal violence might be more destructive of communal peace than the relatively rapid and definitive solution of a trial and execution. As in the debate over the role of hereditary and environmental influences in determining intelligence and character, no general assertions about the balance between positive and negative factors are at all safe. It may be more helpful to see trials as moments of heightened tension, releasing strong affective forces, and rich in possibilities of both kinds. The extreme cases are those panics where suspicions and denunciations snowballed, or travelling witch-finders operated, but these are absent from the Lorraine material. For all the pity and terror that the trial records inspire, one must recognize that for most of the people of the duchy witchcraft beliefs were functional. Many communities experienced periodic moments of catharsis in which they believed themselves to eradicate sources of internal pollution, and in so doing expected to gain relief from the associated misfortunes. On the everyday plane the awareness of such possibilities for action was probably comforting to all but the small minority of those who knew themselves to be prime suspects. This view is encouraged by the remarkable stability in both the numbers and nature of trials from the 1580s onwards. Beliefs about the nature of witchcraft seem to have been as difficult to change as they

their neighbours, and that in a society which believes in witchcraft such threats can prove lethal.

were to destroy, so that they probably stretch far to either side of the relatively brief period of persecution.

It would be far too easy and misleading to regard these beliefs as the natural expressions of a superstitious and backward society. According to our logic, witchcraft may look like a terrifying, virtually random form of social scapegoating, which might run amok at any moment. Given the looseness of the casual connections involved, with the everyday nature of the typical disputes and misfortunes adduced as evidence, it appears capable of virtually infinite extension. In Lorraine at least this did not occur; witchcraft scares remained small-scale and local, despite efforts by Rémy to instigate some kind of general witch-hunt throughout the duchy.[32] Accusations were normally generated through a mechanism of rumour and semi-public discussion, which it was difficult to accelerate from outside the community. Attention seems to have been concentrated on a small number of suspects; villagers may have needed at least one local witch, but apparently had difficulty with any number beyond four or five. Even those convicted witches who made a show of their willingness to accuse their accomplices rarely managed to name more than half a dozen, and those usually from more than one village. Their stories about the sabbat were often vague about numbers, but when these were specified they were commonly in the range from ten to thirty, most of whom were unknown to the accused because their villages were too distant. In any group of simultaneous trials covering several villages there was a marked self-limiting tendency for statements about accomplices to recycle the same names, until all those accused were either on trial or dead.

Given the remarkably restrained and concrete nature of Lorraine witchcraft, we can expect to interpret most of its features as transferred representations of the social matrix from which it sprang. The pact itself provides one example. As witches presented it, this was a strictly contractual affair, in which the devil obtained their services by making them an offer they could not refuse. They exchanged one master for another, a transfer which involved specific renunciation of the earlier contract, sometimes symbolically reinforced by the

[32] See p. 49 n.

devil's action in removing the 'chrism' of baptism from their foreheads, marking them, or taking possession of them sexually. To ensure their fidelity the devil often forbade them to take communion or use holy water, although they must pretend to comply with the normal ceremonies of the church. Some witches, however, found it possible to avoid the irksome labour duty of attendance at the sabbat through commutation; the devil obligingly accepted an annual offering—normally a dead chicken thrown into the garden—in lieu. Those who upheld the orthodox view that the social system was established and maintained by God might logically fear that some of its least favoured members would make an appropriate deduction, that the devil might hold out better hopes. The widespread belief that misfortune and sickness were sent by God, often invoked by suspects in their own defence, might also generate resentment, as expressed in a blasphemy attributed to Catherine la Rondelatte of Essegney. She said 'Je dois bien mauldir dieu' ['I have good reason to curse God'], and when challenged declared 'qu'il n'avoit rien faict pour elle d'avoir prins ses enfants' ['that he had done nothing for her in taking her children'].[33] The numerous acts of *maléfice* supposedly committed against children may well have had their basis in feelings of resentment or envy against the families which were luckier in avoiding both infertility and child mortality. Significantly, apart from the usual group of bewitched animals, the main charges against Catherine, some of which she finally accepted, concerned the deaths and illnesses of children. In giving his servants the power to harm others the devil was doing little more than transfer the power held by the rich in the real world; if his promises of wealth were always a cheat, with his coins proverbially turning into dead leaves or ashes, the diabolical pact was envisaged as all too efficient.

While the great majority of witches claimed to have carried out their *maléfices* with the powder supplied by the devil, witnesses rarely laid much stress on the possibilities of direct physical action. For them it was the ill will which counted, particularly if it was made explicit in a threat. The relationship between the individual will and the surrounding world of

[33] AD M.-et-M., B 4094, No. 1.

reality is a crucial subject for human psychology, in respect of both child development and later abnormality. Although today we are less inclined to think that children's wills need to be broken by the kinds of harsh discipline once used, the socializing process still has to achieve some sort of match between innate selfishness and recognition of other people's autonomous existence. The dichotomy does not go away, however, and the match is never perfect, so that awareness of malevolence, whether towards others or from them, is intensely disturbing; part of us always believes that the will and the act are one. The degree and kind of socialization must depend on relationships between the individual, the family, and the community which are plainly not constant over time, so that it would be peculiarly helpful to develop a better idea of their nature in this society. At the individual end of this spectrum the evidence is so sparse and ambiguous that little can be said, but generalizations about the community can be ventured with rather more confidence.

These peasant communities did exhibit a very striking degree of individualism, within certain limits; 'agrarian capitalism' existed long before anything that could be called a capitalist system, or that rather different phenomenon, capitalist farming. Intense conflicts of interest were therefore very common, in a community which was far from egalitarian, and whose institutions were increasingly dominated by the rich. Small communities with shared work and largely autarchic economies must of course have been highly constraining, and have made it difficult to evolve alternative ways of thought; they also probably tended to generate constant friction around any weak points in their structure, while being slow to evolve new ways of coping with these. In rural society, at least, the community appears as a repressive, conservative, but largely passive force in relation to its own members. To mobilize it for a specific purpose, whether to resolve internal disputes or to prosecute a witch, required positive initiative from one or more individuals. This was very important in restraining witchcraft prosecutions, since a would-be accuser needed wide community support in mounting a case, yet could seldom rely on this being forthcoming until after making his intentions public. Once the trial was under way, however, witnessess would normally come forward with accounts of disputes which reveal not only intrapersonal hostility, but a long process of progressive discord between the suspect and his or her neighbours. The tension between the need to preserve formal relations with neighbours and the petty frictions of everyday life can be seen as a parallel to the latent conflict between will and reality. In this sense witchcraft beliefs were psychologically linked, at some fundamental level, with the central tensions of both individual and community. Of course all communities know such strains, but the moment when they are most acute is probably when a transition is in progress from a communally based society to a more individualistic one. Demographic and economic pressures on Lorraine villagers were producing just such an effect in the very decades when persecution was at its height; while this should not be seen as the 'real' cause of the phenomenon, it is likely to have heightened the tensions on which the beliefs fed.[34]

Such deeply rooted and resonant ideas must be expected to influence people's behaviour substantially, and the trials provide plentiful evidence for this. Even substantial *laboureurs* who crossed a suspected witch were liable to be reproached by their own kin for running unnecessary risks; if misfortune followed, they resorted either to direct threats of burning or to marginally less hostile messages via third parties. Fear was the witch's chief weapon in the attempt to keep a hostile community at bay, but this was a form of social credit which it was reckless to overspend, and which invited retaliation in kind. An extreme attitude was evinced by Jean Mansuin of Clingotte, who deposed against George de Hault in 1596, claiming that he and his father had called him witch at least a hundred times; they had never suffered because it was the witch who feared them, not the other way round.[35] No doubt such brutal self-assurance was the best antidote to fears of witchcraft, although it was hardly a complete guard against the hazards of late sixteenth-century life. Indeed, Mansuin's wife had presumably been less resolute, for she was thought to

[34] For these processes see G. Cabourdin, *Terre et hommes en Lorraine, 1550–1635* (2 vols. Nancy, 1977).

[35] AD M.-et-M., B 8677, No. 2.

have been bewitched and was only healed after threats from her kin.

The study of witchcraft brings the sobering realization that the topic generates new questions faster than one can resolve the initial ones. This happens because the beliefs were so deeply integrated into the society of early modern Europe, a society whose nature and dynamics historians still find so elusive. An interpretative strategy for witchcraft, if it is to do justice to the complexity of its subject, must constantly seek to explore that integration, at both practical and symbolic levels. This requires, among many other things, that we should consider the thought structures or *mentalités* involved. When reading many trials, one is forcibly struck by the formulaic patterns they display, constantly redeploying a limited range of elements and phrases. These are the characteristics of an oral culture, in which such repetitions structure discourse, endlessly different yet the same. These cultures can be remarkably complex, and far from static, yet they tend to assimilate everything to one fundamental model of reality.[36] In this case the parallels I have been trying to draw out were the product of a basically homogeneous view of the world, which could not accommodate such modern distinctions as natural and supernatural, real and imaginary, or concrete and abstract. Broadly speaking it could be said that much of the language is bound up with concepts of power and obedience, to both superiors and communal norms. What we may see as a confusion between wish and deed was far harder to identify, both for those who saw themselves as victims of witchcraft, and for those who were persuaded that they were witches. The rather narrow 'operational' conception of witchcraft found in the trials can be related to the religious beliefs current in this society; the fundamentals of religion were quite well understood, but as a means to secure protection in this life and salvation in the next, not as an invitation to a more spiritual or selfless existence. Such aspirations could hardly flourish within the everyday pattern of village life, in which competition for scarce resources naturally tended to dominate social behaviour. One does not have to take a very pessimistic view

[36] There are some valuable observations on oral cultures in W. J. Ong, *Orality and Literacy* (London, 1982).

of peasant society in the early modern period to conclude that it was full of tensions, which must have generated hatreds and fears among its members. Witchcraft beliefs were the most extreme formalized expression of these tensions, so it should not surprise us to find that they revolved around the twin poles of hatred and power. The deep psychological forces involved are of course much less closely linked to a particular culture, and help to explain why the figure of the witch is still such a feature of the Western imaginative world, constantly reappearing in literature and iconography. If there has often been a tendency to treat the great persecution as some kind of aberration, this perhaps reflects the extent to which these ideas and images continue to threaten us as both individuals and members of society.

4

Popular revolt in its social context

THE 1590S: RELIGIOUS WAR AND SOCIAL CONFLICT

THERE can be little doubt that the middle decades of the seventeenth century, from the 1630s to the 1670s, saw the most intensive and widespread sequence of revolts in French history. The phenomenon, although recognized by French historians, received relatively little attention until the work of the Soviet historian Boris Porchnev gave rise to a lively controversy in the 1960s and early 1970s. The terms of this debate were largely set by Porchnev's ingenious but ultimately unconvincing attempt to incorporate the revolts within a Marxist–Leninist scheme of historical interpretation. For him the revolts were essentially spontaneous proletarian movements, which failed because large elements of the bourgeoisie were incorporated into the 'feudo-absolutist' ruling class through the mechanism of the sale of offices; this process was temporarily successful in diverting the bourgeoisie from fulfilling its historical role of allying itself with the proletariat against the aristocracy, which would only finally happen in 1789.[1] Porchnev's leading French critic Roland Mousnier, who was also the major authority on the sale of offices, set out to establish almost precisely the opposite case. Mousnier emphasized the participation of local élites in revolt, which he explained as a reaction against the expansion of centralized royal power. Royal absolutism and escalating tax demands had threatened the interests of both office-holders and seigneurs, who therefore encouraged attacks on *commissaires* and tax collectors. Mousnier went on to challenge the whole idea of a society divided on class lines, propounding instead the theory of a 'society of orders', whose primary

[1] Porchnev's book appeared in Russian in 1948, then in German translation in 1954. The French edition, *Les Soulèvements populaires en France de 1623 à 1648*, was published in 1963.

solidarities were determined by profession or function, not by relation to the means of production.[2]

Both historians thus incorporated the revolts into broad theses about the changing relationship between state and society, starting from such different viewpoints that the same evidence could have dramatically opposed meanings for each of them. It is hardly surprising that the original opponents should have left the matter there, but less easy to explain why the detailed regional studies by Bercé, Pillorget, and Foisil did not lead to a renewed discussion of the general issues when they appeared in the 1970s.[3] These works have advanced our knowledge enormously, yet they have not really been assimilated in general historical awareness; to take one example, in his *Historical Sociology* Philip Abrams devoted several pages to the question which make no reference to anything published after 1973.[4] The recent general survey of revolt and protest by Charles Tilly does use this more recent scholarship, from a different angle again, as part of an inquiry into revolt as a style of behaviour, which seeks to establish long-term continuities. This is a valuable and thought-provoking study, but its main focus does not lie in the seventeenth century, and Tilly passes over some important issues which are not central to his own

[2] Mousnier's views were first expressed in 'Recherches sur les soulèvements populaires en France avant la Fronde', *Revue d'histoire moderne et contemporaine*, 4 (1958), 81–113, reprinted in *La Plume, la faucille et le marteau* (Paris, 1970), 355–68. They were further developed in *Fureurs paysannes: les paysans dans les révoltes du 17e siècle (France, Russie, Chine)* (Paris, 1967), trans. as *Peasant Uprisings in Seventeenth-Century France, Russia, and China* (London, 1971). The most complete statement of Mousnier's more theoretical position on the 'society of orders' is in *Les Hiérarchies sociales de 1450 à nos jours* (Paris, 1969), translated as *Social Hierarchies, 1450 to the Present* (London, 1973). For a particularly forceful critique of the presumptions behind this work, see A. Arriaza, 'Mousnier, Barber and the "Society of Orders"', *Past and Present*, 89 (1980), 39–57. The early stages of the controversy were competently reviewed by J. H. M. Salmon, 'Venal office and popular sedition in France', ibid. 37 (1967), 21–43.

[3] Y.-M. Bercé, *Histore des croquants* (2 vols., Geneva 1974), henceforward Bercé, *Croquants*; R. Pillorget, *Les Mouvements insurrectionels de Provence entre 1596 et 1715* (Paris, 1975); M. Foisil, *La Révolte des Nu-pieds et les révoltes normandes de 1639* (Paris, 1970). There are also two more general books by Bercé, *Croquants et Nu-pieds* (Paris, 1974), and *Fête et révolte* (Paris, 1977).

[4] P. Abrams, *Historical Sociology* (Shepton Mallet, 1982), 214–18.

concerns.[5] Popular revolts therefore deserve general reconsideration, primarily in the light of the major research achievement of the French scholars who worked under Mousnier's direction. It is unlikely that we will ever know a great deal more about the facts than we do now, since the remaining gaps represent not so much the disappearance of documents as the limitations of what was actually recorded. Although some individual revolts still await full investigation, a very large sample has been subjected to exhaustive study, and there is every reason to suppose that it is representative.

The original controversy remains relevant partly because it has affected the questions asked and the way the answers have been presented. Porchnev's view of popular action as potentially revolutionary, and his emphasis on the class struggle, have prompted a series of rejoinders which quite rightly claim that these interpretations are unacceptable as they stand. His critics agree that the revolts were primarily defensive local reactions to fiscal pressure from the state, that their ideology was highly conservative, and that vertical solidarities uniting all social groups were more important than horizontal ones based on class divisions. It is possible, however, to accept most of these points as necessary corrections to Porchnev without regarding them as a wholly satisfactory interpretation in their own right. One may agree, for example, that revolts were overwhelmingly directed against royal fiscality, particularly during its period of sharpest intensification under Richelieu and Mazarin. Resistance to a tax need have no direct connection with ability to pay it; nevertheless, there must be an ultimate correlation between the wealth of the country, its distribution, and the frequency and severity of revolt against tax increases. The royal ministers did have some general awareness of this.[6] What was less apparent to

[5] C. Tilly, *The Contentious French* (Cambridge, Mass., 1986). Tilly does posit an important change around 1661 (p. 388), but the main contrast he draws is between the 'parochial and patronized' popular action of 1650–1850, and the 'national and autonomous' style of 1850–1980 (pp. 390–8). The general overview of the seventeenth century on pp. 159–61 makes a number of interesting points, nearly all of them highly contestable.

[6] See for example Bullion's letter of Oct. 1639 to Richelieu on the tax burden in Normandy, cited by Foisil, *Nu-pieds,* 62, and Colbert's extensive

them was that since the first decades of the sixteenth century the standard of living of most of the French population had been in serious decline, with wages lagging well behind prices. This had been accompanied by a polarization between the prosperous minority and the increasingly pauperized masses; the gap tended to widen steadily, with profound effects on community structures, which probably reached their peak around the early decades of the seventeenth century. This implacable if slow long-term evolution did not cause revolts directly, but it operated in combination with the more immediate effects of fiscality, and can be detected in the form of disputes about who should actually pay the taxes. The very evident traditionalism of the rebels is more ambiguous than it may at first appear, for this was a period when innovation was commonly justified by selective reference to the past and change was very rarely advocated as such. The government itself, which Mousnier very fairly portrays as the prime agency of change, was always invoking traditions which it interpreted in the most dubious fashion. The intendants and other royal agents understood very clearly the dangers which might come from a skilful manipulation of local traditions and rights; French invocations of a mythical past may not quite have rivalled such English examples as the Norman Yoke theory, but they could function in a comparable fashion.[7] Vertical and horizontal solidarities also call for some subtle distinctions which will be attempted later.

The sheer mass of information, and the number of revolts (notably in the period 1630–60, as shown by the map), makes selective illustration inevitable, but its dangers need very careful watching. There has, for instance, been a degree of over-concentration on the 1630s and 1640s, which has tended to obscure the important differences by period. The revolts of

correspondence with the intendants, to be found in P. Clément (ed.), *Lettres, instructions et mémoires de Colbert* (7 vols., Paris 1861–82), and G. B. Depping, *Correspondance administrative sous le règne de Louis XIV* (4 vols., Paris 1850–5).

[7] Examples include the *charte aux Normands* (see p. 153 below), the noble claims about the Franks and the *champ de Mars*, discussed by A. Devyver in *Le Sang épuré: les préjugés de race chez les gentilshommes français, 1560–1720* (Brussels, 1973), and the various anti-fiscal mythologies described by Bercé, *Croquants*, 607–36 and 658–61 (the latter case being that of liberties enjoyed under English rule in Acquitaine).

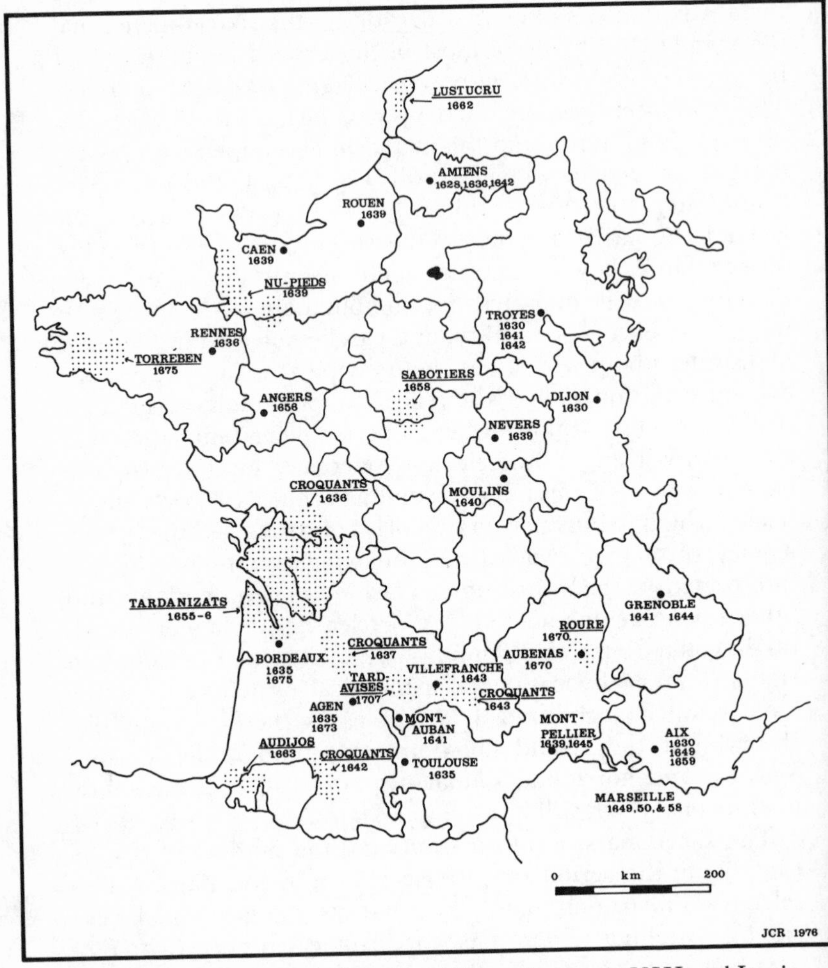

Certain urban riots and peasant revolts under Louis XIII and Louis XIV (1630–1715): the areas of major rural uprisings are identified by hatching.

the end of the sixteenth century are particularly interesting, for at that moment the burden of taxation, although bitterly felt, had yet to obliterate all other grievances in the popular consciousness. The best-known revolt of the period, the 1580 Carnival of Romans, revealed deep social divisions within a modest town in Dauphiné, which admittedly arose partly from the exceptionally complex and controversial tax structure in the province.[8] There were far more widespread movements in the early 1590s, which must be associated with the very particular circumstances of those years. The wars of religion, for so long an intermittent series of local conflicts, had flared up after 1588 into a general conflagration affecting much of the kingdom, with numerous small armies living off the countryside. In many areas the peasants assembled and took arms in self-defence, action which could lead to more significant protests and demands. The most widespread troubles were those in Acquitaine and Brittany. In the former case the assemblies of peasants who called themselves the *Tard-Avisés,* but were generally known as the *croquants,* had started by opposing the exactions of the little garrisons which lived by brigandage and ransom.[9] At least one local royalist leader, the *sénéchal* of Périgord Bourdeille, at first saw them as useful allies against the local *ligueurs*. By May 1594, however, he reckoned that there were 100,000 of them, some of whom had 'parlé ouvertement de ruiner et exterminer la noblesse'.[10] The peasant manifestos are less alarming, and indeed demanded the help of 'tous les seigneurs et gentilshommes sans reproche',[11] but the assembly at the forest of Abzac in

[8] E. Le Roy Ladurie, *Le Carnival de Romans* (Paris, 1979), Eng. trans. as *Carnival: a People's Uprising at Romans, 1579–80* (London, 1980); L. S. Van Doren, 'Revolt and reaction in the city of Romans, Dauphiné, 1579–80', *Sixteenth Century Journal,* 15 (1974), 71–100; D. Hickey, *The Coming of French Absolutism: the Struggle for Tax Reform in the Province of Dauphiné 1540–1640* (Toronto, 1986), gives a very full analysis of the background, with a specific section on the 1579 crisis on pp. 33–69.

[9] There is an account of these movements in Bercé, *Croquants*, 257–93. Some important modifications are made in the unpublished thesis by R. G. Tait, 'The King's lieutenants in Guyenne, 1581–1610' (D. Phil. thesis, Oxford, 1977), and there is a good summary of these in M. Greengrass, *France in the Age of Henry IV* (London, 1984), 125–8.

[10] Bourdeille to Henri IV, 26 May 1594, cited by Bercé, *Croquants,* 284.

[11] Circular letter of 2 June 1594, printed ibid, 707–8.

April 1594 announced the intention 'de razer plusieurs maisons de gentilshommes qui ne faisoient autre chose que courrir sur le boeuf et la vache de leurs voisins' ['of razing the houses of several gentlemen who do nothing but pillage the oxen and cattle of their neighbours'], while the *croquants* of Atur told the *maire* and *consuls* of Périgueux that 'ils estoient eslevés pour empescher les exactions et subsides, que les voleurs et gens de guerre leur faisoient paier et qu'ils estoient resolus ne le souffrir plus; ne vouloient souffrir les exactions des gentilshommes' ['they had risen to prevent the exactions and taxes which the thieves and soldiers were making them pay, which they were resolved to suffer no longer; and did not wish to endure the exactions of the gentlemen'].[12]

The most open reproaches in the manifestos were directed against the principal towns; the letter calling another assembly that April, at Chateau Missier, claimed that the peasants had not benefited from the truce 'car les villes, au lieu de la faire entretenir et tenir la main à la justice, ne se soucient de la ruine du paovre peuple, parce que notre ruine est leur richesse' 'for the towns, instead of securing it [the truce] and ensuring that justice is done, are unconcerned about the ruin of the poor people, since our ruin is their wealth'].[13] The *croquants* showed their political sophistication by demanding a permanent representation of their interests, through a 'syndic des habitants du plat pays' or 'syndic particulier du Tiers État du pays de Périgord'. While there was already a *syndic* for Périgord, he was always chosen from the officials of the three main towns, Périgueux, Sarlat, and Bergerac. At the local Estates of 1595 the demand was repeated by a party from the small towns and the rural parishes, which had already made proposals on similar lines in 1583. The local lawyers and lesser officials who drew up the representations protested that the nobles and bourgeois who lived in the three towns had taken advantage of the hardships of the wars to buy land from the peasants at low prices, while ruining them with illegal *daces* and *corvées,* alongside an unjust assessment for the *taille.* The *procureur du roi,* with the support of the Estates, responded indignantly:

Il y a des particuliers d'entre la population qui se veulent ingérer

[12] Tait, 'King's lieutenants', 197. [13] Bercé, *Croquants,* 701.

sous le nom de Tiers Estat, faire un quatrième Estat et pretendent faire des nominations de syndic, se separant du vray Tiers Estat qui sont les trois villes principales.

[There are some individuals among the population who wish to interfere under the name of the third estate, to form a fourth estate, and pretend they should nominate a syndic, separating themselves from the true third estate which is composed of the three principal towns.]

The united self-interest of the privileged towns and the crown was involved, so the initiative had no real chance of success. The very strength of this resistance, however, emphasizes how shrewdly the demand had been formulated.[14]

The social dimension of the movement was emphasized by the speed and strength of the noble reaction. Already in 1594 a league of gentlemen had been formed in southern Périgord, for the purpose of aiding Bourdeille against the *croquants,* to whom they imputed the intention to 'établir une démocratie à l'exemple des Suisses'.[15] It was only the intervention of the governor of Guyenne, the maréchal de Matignon, which deterred them from taking the field, for they were eager to 'monter à cheval à les rompre'. When the harvest of 1594 proved inadequate the assemblies recommenced. This time Bourdeille did call out the nobles, who responded with alacrity, dispersing the peasants after a few minor skirmishes.[16] The nobles considered themselves directly threatened by the assemblies, and one minor local squire, the sieur de la Brugerie, wrote to his father:

Je n'eusse jamais pensé que ses personnes la fussent tant Emansipé que d'entreprendre ce qu'ils ont faict. J'espère qu'estant par dela s'ils y font encore que je leur fere mettre de l'eau en leur vin et m'asseure que nous avons tousiours l'honneur de leur faire la loy.

[I would never have thought that those persons would have been so emancipated as to attempt what they have. I hope that if I arrive there and find them still active I will put water in their wine, and assure myself that we still have the honour of giving them the law.][17]

[14] For this episode, and the citations, see ibid., 276–7, and Tait, 'King's lieutenants', 199–202.

[15] On the 'myth of the Swiss', see Bercé, *Croquants,* 287.

[16] Tait, 'King's lieutenants', 205, 209–10. Bercé gives very little space to these activities by the nobles.

[17] Tait, 'King's lieutenants', 212.

No doubt such noble reactions were exaggerated, but it does appear that the main thrust of the revolts was against the behaviour of the leading groups in local society. An anonymous contemporary in Limousin wrote:

Ils menacent la noblesse, la dédaignent et tiennent des langages hauts même contre les villes . . . jusqu'à se faire accroire que le roi ne seroit plus leur maitre et qu'ils feroient des lois toutes nouvelles. Bref, ils donnaient terreur et épouvantement à plusieurs et *sembloit que ce fut le monde renversé.*

[They threaten the nobility, disdain them, and talk strongly even against the towns . . . to the point of giving cause to believe that the king would be their master no longer, and that they would make completely new laws. In brief, they gave terror and fright to several, and it seemed that it was the world upside down.][18]

Here the exaggeration is still more evident, for in fact the rebels appealed to the crown, and indeed Matignon protected them, even after their defeat by Bourdeille, preventing reprisals and obtaining an amnesty. Important remissions were granted on arrears of the *taille*, and the commissioners who came to seek an exceptional contribution to the royal coffers in 1597 concentrated their efforts on the towns.[19] These revolts, closely linked to the problems exacerbated or created by the religious wars, demonstrated that the village communities could distinguish their interests from those of the notables, whether these were nobles or town dwellers. While Bercé gives a generally excellent account of the episode, he is less convincing when he cites some very marginal evidence to suggest that nobles were found supporting as well as attacking the peasants, the essential conflict being between town and country.[20] The fears of the nobles are far too widely reported in numerous sources, which also emphasize the readiness of most nobles to take the field against the rebels, whose leadership clearly came from within their own ranks. Their grievances were expressed through members of the village élites, above all the petty lawyers and notaries, and the *curés*, who sought to keep the armed bands under control, avoiding any gratuitous

[18] Bercé, *Croquants,* 286.
[19] Tait, 'King's lieutenants', 216–17.
[20] Ibid., 202, makes very telling criticisms of Bercé's chosen examples.

violence. This was something of a contrast to contemporary events in Brittany, where the killings of nobles persuaded canon Moreau of Quimper that the peasants intended to slaughter the entire nobility. A series of bloody encounters saw châteaux and towns sacked and their defenders massacred, but on the other side heavy defeats for the peasant armies.[21] The same period saw peasants assembling in other regions, notably Normandy, Burgundy, and Comminges, to resist the soldiery of both sides; again these seem to have been genuine peasant initiatives, which in these cases ended up associating themselves with either the royalists or the ligueurs, depending on local conditions.[22]

RESISTANCE FROM THE ÉLITES: THE CRISIS OVER THE *ÉLUS*

With the restoration of peace and order around the turn of the century the immediate causes of conflict were much reduced, and although Sully's tax policy was unpopular, there was little active resistance. His ministry did foreshadow the main flashpoint of the next thirty years, with the protracted struggle to impose the tax officials known as the *élus* on Guyenne. The introduction of these venal offices meant the effective destruction of the provincial Estates, which would lose their main function once the assessment and collection of the taxes was removed from them. Between 1605 and 1609 the local élites fought a determined (although far from unanimous) battle to preserve their privileges, which involved lobbying at court, legal obstruction, and some passive resistance, but no actual violence. Sully persisted, so that the *élus* were in operation by 1609, only to be withdrawn by the Regency government in 1611.[23] This was a brief reprieve, for in 1622 Louis XIII

[21] For Moreau's views, and for events in Brittany generally, see H. Waquet (ed.), *Mémoires de chanoine Jean Moreau sur les guerres de la Ligue en Bretagne* (Quimper, 1960).

[22] For some of these other revolts see J. H. M. Salmon, *Society in Crisis: France in the Sixteenth Century* (London, 1975), 277–9. For Burgundy, H. Drouot, *Mayenne et la Bourgogne*, 2 vols. (Paris, 1937), ii. 283–93.

[23] The fullest description of this conflict is by J. Russell Major, *Representative Government in Early Modern France* (New Haven, Conn., 1980), 259–94. Russell Major exaggerates the general significance of Sully's actions; no doubt he would have liked to suppress provincial Estates everywhere, but

imposed the *élus* afresh, and by 1624 they were taking over the tax collection.[24] This time a revolt did break out in Quercy; according to the official account in the *Mercure françois* the instigators had passed

secrèment des avis de paroisse en paroisse que le Clergé, la noblesse et le Tiers-Estat tiendroient la main en faveur du peuple si on vouloit prendre les armes et s'assembler pour abolir ces nouvelles Eslections qui devoient estre la ruine du pays.

[secret messages from parish to parish that the clergy, the nobility, and the third estate would take part on behalf of the people if they wished to take arms and assemble to abolish these new *Elections* which would be the ruin of the country.][25]

This is very plausible, both as a picture of how the revolt began and as a description of how the élites felt—yet the actual events show that very different loyalties could come into play once actual violence began. The further explanations for the revolt as paraphrased by Porchnev emphasize some of the possible divisions; the *croquants* had risen on the grounds

that the country was going to be burdened afresh with taxes, to pay the agents charged with drawing up the lists and the other levies for the new fiscal sector, and because the richest people in the province, who paid the highest *taille* (up to 300 or 400 livres) would buy the offices to be freed of the taxes, so that it would no longer be possible to collect the total of the *taille*, and in the end the ordinary people would have to be crushed by all the old taxes and the new additions.[26]

Peasant bands assembled at the end of May 1624, with old soldiers allegedly taking the lead, to form an army said to be 16,000 strong. The leaders were named as Douat, a 'tireur d'horoscopes, physionomiste, chiromancien', and Barrau, a ruined minor noble from the town of Gramat. While the rebels demanded a reduction in taxes, their main action was against

there is no evidence of any concerted plan to this end. Guyenne was vulnerable because it was not a *pays d'états* like the others, but instead of regular meetings at provincial level, had some ten local assemblies acting independently. While this weakened their capacity to resist, it also made them a major administrative nuisance.

[24] Major, *Representative Government*, 459–70.

[25] Bercé, *Croquants*, 89.

[26] Porchnev, *Les Soulèvements*, 50.

the new *élus*; their rural properties were devastated, trees were cut down, buildings burned, crops destroyed. The insurgents surrounded the principal towns, Cahors and Figeac, demanding that the *élus* be handed over, or they would lay the surrounding area waste. Douat allegedly hoped for help from the poor inside the towns, with the ultimate objective of finding some cannon. There is no sign that this plan succeeded, and the *sénéchal* Thémines summoned the local nobility to his aid; they seem to have responded well, and a sudden attack led to the defeat and dispersal of the peasants, whose leaders were executed. Perhaps expectations of support from their superiors might explain why in the face of the noble attack they allowed themselves to be killed 'comme des bêtes, sans aucune résistance'.[27]

In 1628–9 the government launched a major attack on several *pays d'états,* establishing *élections* and other fiscal officers in Dauphiné, Provence, Burgundy, and Languedoc. In Dauphiné, where there were long-standing disputes between the clergy and nobles on the one hand, and the third estate on the other, resistance was very slight. Over the next decade the people would obtain some compensation for the general increase in taxes with the extension of the *taille réelle* or land tax to the whole province, which obliged the élites to pay a fairer share. The popular view of the estates as an institution associated with the unequal incidence of taxation would explain this quiescence. In the other three provinces, however, revolts were sparked off by the royal actions.[28]

The first, although the least serious, was in Burgundy. Here it is important to understand something of the background, as it affected the provincial capital Dijon, and its political structure. In the sixteenth century Dijon had possessed a remarkably democratic constitution, with all householders participating in the annual mayoral election—it was even compulsory to vote. Henri IV suppressed this election, then allowed the town to put forward three candidates between

[27] Ibid., 49–52.

[28] For this policy and its repercussions, Russell Major, *Representative Government,* 524–99, again provides the most extensive account, although his interpretations are often very debatable. On Dauphiné, see Hickey, *French Absolutism,* 161–78.

whom he chose. Here again there was a royal concession in 1611; the mayoral election was restored, but with a restriction on the electorate. In future this was to be limited to those who paid 4 livres and above at the two annual levies of the *taille*, the stated aim being to

esviter les abus et corruptions qui ont esté commis trop frequemment du passé à l'occasion de ce que les habitants de la plus abjecte et moindre condition du peuple ont esté ceux qui ont presque toujours esleu le dict vicomte mayeur

[avoid the abuses and corruption which have been committed too often in the past because the inhabitants of the lowest and humblest sort of the people have almost always been those who elected the said *vicomte mayeur*].

One may doubt whether this was the literal truth; nevertheless, the majority of the population, and notably such groups as the *vignerons*, were now excluded from the system, so that revolt was the only form of political expression left to them. It is worth noting that there would be further developments towards oligarchy later; in 1668 the crown reasserted its right to nominate the mayor, provoking a final series of disturbances. When municipal offices were made venal in 1692 Dijon bought its back—but only with the help of a massive loan from the titularies, who thus became effective proprietors.[29]

Reverting to the earlier decades of the century, in 1629 the crown issued an edict establishing ten *élections* in Burgundy. The estates, remembering that they had bought off an earlier attempt by Henri II in 1554, mounted the usual campaign to influence and bribe ministers, but without much success. The emissaries in Paris reported back that there was much talk of standardizing the tax system throughout France, and that the 'ministers of state' were taking this as a general maxim. Both the governor Bellegarde and the mayor warned that an insurrection would be counterproductive, but the *échevins* do not appear to have done much to prevent it.[30] Indeed, someone spread rumours of further new taxes, including one on wine,

[29] For these changes, and the citation, see G. Roupnel, *La Ville et la campagne au 17ᵉ siècle* (Paris, 1955), 164–6.

[30] Russell Major, *Representative Government*, 537.

and on 27 February 1630 a report circulated that the *élus* would be installed the next day. A crowd of *vignerons* and artisans gathered, led by the *vigneron* Champgenêt, dressed in a multicoloured cape, with laurels on his head, and being saluted as 'le roi Machas'. Here was another carnival king taking the lead in an insurrection, like Jean Serve at Romans just half a century earlier. The crowd marched to the town hall, singing a popular song with the refrain 'Lanturelu, Lanterelu', from which the revolt would take its name. While some demanded the keys of the town gates, which they were refused, others attacked the house of the financier Ganne, who was believed to be the instigator of the edict, stoning it and threatening to kill him. Probably emboldened by the failure of the authorities to intervene, and by the arrival of some reinforcements from the suburbs, who scaled the walls on ladders, the next day the rebels escalated the violence. Ganne's house was destroyed, and severe damage was caused to those of the mayor, the *premier président*, and another *président* in the *parlement*, as well as those of several financial officials. A group of *échevins* and *parlementaires* who tried to calm the riot was greeted with stones. Rumours circulated that revolt was sweeping the province, and that the peasants would be trying to get into the town. By the evening of 28 February the bourgeoisie were sufficiently alarmed for the guard to muster and attack the rioters, who were driven back into the poor *quartiers* where the revolt had originated. Spasmodic violence on the third day turned into an uneasy truce, with the rebels threatening to set light to the town if a further attack was made on their *quartiers*. Within a week news arrived that the king was furious with the feeble reaction of the authorities; the town risked losing its privileges, even its walls. This goaded the élites into action again; they found that the revolt had now lost its dynamism, Champgenêt and other leaders fled, and order was finally restored, with a couple of symbolic executions.[31] The government refused a final offer of 1.8 million livres to remove the *élections*; this was actually some 70,000 less than the price already agreed with a financier. In late April the king visited Dijon on his way to Italy, with an absence of

[31] The revolt is described by Porchnev, *Les Soulèvements*, 135–43, and by Tilly, *The Contentious French*, 14 and 402–3.

all ceremony to signify his displeasure. The municipality and other leading citizens had to make a humiliating submission, then listen to a hectoring speech on obedience from Marillac. The creation of the ten *élections* brought with it no fewer than 333 new offices, whose holders would have drawn the best part of 100,000 livres a year in salaries, and approaching 150,000 in fees. The prospect of this burden encouraged the estates to keep up their efforts to buy off the government, while they had a golden opportunity to display their loyalty in March 1631, when Orléans and Bellegarde tried to raise a revolt in the province. Dijon shut its gates on them, and within the same month the king expressed his gratitude, with the news that he was prepared to negotiate for the withdrawal of the *élections*. In May Condé negotiated a settlement, against the surprisingly low figure of 1.6 million livres, and the menace of the *élections* was finally removed.[32]

Turning to the second revolt of 1630, that of the *Cascaveoux* at Aix-en-Provence, we find both similarities and important differences. Here some of the most important background elements were environmental. Plague had struck the region in 1629, a serious oubreak which lasted ten months, and reached Aix in July 1629. Most of the rich left Aix, where tensions ran high, in an atmosphere of fear with beggars and others confined in shacks outside the town. The autumn of 1630 saw very high grain prices, the highest since the Wars of Religion. There had also been considerable fiscal pressure, notably since May 1628, with increases in the *don-gratuit,* the *taillon,* and the price of salt, alongside new creations of office. The *surintendant* d'Effiat and other representatives of the government explained that these impositions were to meet the cost of suppressing the protestant revolt, but the Estates were very recalcitrant, only willing to negotiate for a total withdrawal of the offending levies. In anticipation of such resistance, an edict had been prepared establishing ten *élections*, and the governor Guise, accompanied by the royal representatives Bullion and Dreux d'Aubray, set out for Aix in July 1629 to force its registration by the *parlement.* Their arrival coincided with that of the plague, so they hastily decamped with their

[32] For these later developments, Russell Major, *Representative Government,* 538–40 and 583–5.

work undone. Here again the government seems to have been convinced that the local élites were manipulating the system for their own benefit, and expected them to prove highly resistant to any attempt at change. The plague had a further effect, for there was no unaffected town big enough to accommodate the *parlement,* which was therefore split into two halves, one at Salon and the other at Pertuis. The *premier président* d'Oppède had come to be regarded as a client of Richelieu; now his unsuccessful rival for the job Laurent de Coriolis used his temporary position as head of the Pertuis court to build up a faction, while preparing to fan discontent and seek popular support in defence of local privileges. Meanwhile the Estates again proved uncooperative, and were suspended by the intendant d'Aubray in July 1630.[33]

When the *parlement* returned to Aix in September 1630 it was bitterly divided, while it was known that d'Aubray would soon be arriving to enforce the royal will on the *élections* and other matters. No doubt acting from a mixture of genuine conviction, self-interest, and ambition, the Coriolis faction had been stirring up local fears. Manuscript copies of a speech which a representative of the province had supposedly made to the king were circulating at Aix. According to this document, the *élus* would

> enrôler dans les tailles les personnes de nos filles en l'âge de quinze ans, et de nos mâles de vingt, les valets et les servantes à raison des gages et des salaires qu'ils reçoivent de leurs maîtres
>
> [place on the rolls of the *taille* our daughters aged fifteen and our sons aged twenty, with valets and servant-girls on account of the wages and salaries they receive from their masters].

There could hardly be any 'vie si misérable que l'on puisse être exempté'.[34]

With such ideas current it was not surprising that the actual arrival of d'Aubray on 19 September sparked off a riot; the tocsin was rung, and he had to flee the city over the rooftops, while the mob burned his carriage. The *premier consul* La

[33] There are two extensive modern accounts of the revolt, by Pillorget in *Mouvements,* 313–54, and by S. Kettering in *Judicial Politics and Urban Revolt in Seventeenth-Century France* (Princeton, NJ, 1978) 150–81; between them they constitute the basis of the present discussion.

[34] Pillorget, *Mouvements,* 325.

Barben had already been driven out by threats of violence; he was soon followed by d'Oppède and by Séguiran, *président* of the *Chambre des Comptes,* so that no serious 'royal' party was left in Aix. The Coriolis faction were less successful, however, in the consular elections at the end of September, and it was evident that they did not enjoy the support of 'les plus apparens'.

Meanwhile there was activity from the Estates; spontaneous but separate meetings of the three orders reaffirmed their opposition to the *élus,* with the nobles and the third estate going further, arranging to raise forces to defend their privileges. The governor Guise returned to the province in October, but he was now deep in the intrigues against Richelieu; he stayed in Marseille and did nothing. In Aix the Coriolis group kept the crowds active, attacking the houses of some of their opponents. An insignia was adopted, a little bell on a ribbon, called a *cascaveu,* from which the revolt was to take its name. Those who were thought willing to accept the edict were denounced as *élus,* while eventually, according to Peiresc, 'il n'y avait personne qui eut des moyens qui ne passât incontinent pour élu'.[35] As the crisis dragged on there were increasing signs of social tension, with growing hostility from the poor towards the rich in general. Although the leaders seem to have been unworried, they were more concerned about the prestige and attitude of the moderate *premier consul,* the baron de Bras. The climax of the revolt was reached in two dramatic days, beginning on Sunday 3 November when groups of peasants gathered in the villages and *bourgs* near the château of the ex-*premier consul* La Barben, notorious for his strict defence of his seigneurial forest and hunting rights. After the mass large groups marched towards Aix, where they arrived during the afternoon; their appearance provoked a large number of Aixois to join them, and a huge crowd assembled. Neither the consuls nor the *parlement* could disperse it, and the élites passed a night of terror, particularly on account of the day-labourers, whom Peiresc thought 'gens à mettre la ville au pillage en cas de grand malheur'.[36] That night the houses of two officials of the *cour des Comptes* were sacked, and the next day an enormous body of

[35] Pillorget, *Mouvements,* 331. [36] Ibid., 333.

peasants and townspeople marched off towards La Barben's property. They did not attack the château itself, but pillaged some *métairies,* and above all cut wood in the forests, to equip themselves with a commodity which was in very short supply. During their absence the *parlement* had roused itself, with the *avocat général* calling for action to end 'le progrès de cette audace effrénée, dans la suite de laquelle ne pourrait s'ensuivre que la perte et subversion de la province, et le commencement d'une vraie anarchie' ['the progress of this unbridled audacity, as a consequence of which there could only follow the subversion and loss of the province, and the beginning of a true anarchy'].[37] The foraging expedition returned late at night, the peasants having regained their own villages, to find the gates shut against them; the authorities did not dare to lock them out entirely, in view of their support in the town, but only allowed them back in as small groups.

These two days destroyed the unity of the town, always partial in any case, since the frightened élites were no longer prepared to follow Coriolis and his associates, and the majority in the *parlement* swung against them. The *garde bourgeoise* was mobilized, vagabonds were expelled from the city. An uneasy, tense situation persisted throughout November, punctuated by a brief false alarm when Guise, misled by the first news of the 'Day of Dupes' far away in Paris, conveyed the 'bonne nouvelle' that Richelieu had fallen from power. The *premier consul* de Bras was trying to organize a new party, recruited among the élites and the substantial artisans, which would maintain opposition to the *élus*, yet affirm loyalty to the king. At the beginning of December he tried a coup, driving Coriolis and his leading supporters from the city, but it turned out that he had overestimated his strength, not least because the *parlement* could not approve such an outrage against one of its leading members. Further street-fighting on 6 and 7 December forced de Bras to leave the town, while the *Cascaveoux* leaders returned in triumph. The initiative had really passed from them, however, and they took no further action; soon news came that Condé was on his way with an army, and an uneasy calm settled on the city. There was no

[37] Ibid., 335.

support from elsewhere in Provence, for the two other major centres of Toulon and Marseille had been frightened by the social aspects of the Aix revolt, and were also concerned to preserve their special position as 'terres adjacentes', not subject to the ordinary taxes, which meant that the *élus* had no great importance for them. When Condé arrived at the head of 5,000 troops at the beginning of March, he met no opposition; the rebel leaders had fled, and the city submitted meekly. The *parlement* and the other courts were exiled to various smaller towns, Aix spent two months under military rule, and Coriolis and three of his relatives were condemned in absentia. On the main point, however, the province obtained satisfaction, for the *élections* were withdrawn against a payment of 1.5 million livres, over a four-year period.

The attack on the provincial estates was also extended to Languedoc, where the crown had found them very difficult during the final phases of the war against the protestants, even insisting that their *don gratuit* should be treated as a loan. Immediately after the Edict of Grace two further edicts were issued in July 1629. One united the *Cour des Aides* and the *Chambre des Comptes* at Montpellier; a relatively anodyne move in itself, the intention was clearly to create a more powerful financial court as a counter-weight to the Estates and the *parlement.* The second established twenty-two *élections,* corresponding to the dioceses, to be staffed by 490 officials; it was promptly registered by the new court at Montpellier, and the Estates themselves were brusquely dissolved at the beginning of August, once it became clear that they would not make any special grant. The whole operation was to be carried out by the *traitant* Vanel, who had paid 4 million livres for the right to sell the offices. Although the *parlement* of Toulouse refused to verify the edicts, the crown pressed on, and the system came into operation in 1630, a year when the Estates were simply not summoned. There were predictable difficulties in selling the offices, still more in getting the holders obeyed, but the amount of taxation collected in 1630 still seems to have been greater than in any previous year. It would have needed to be; in that the *élus* were reckoned to cost some 400,000 livres a year in salaries and fees, compared with a (no doubt) optimistic 40,000 for the previous system. It should be mentioned that again there were precedents, with previous abortive creations of *élections* in 1519 and 1554.

Resistance from Languedoc took a much more coherent form than that from Provence, largely because it was led by the *syndic* of the Estates, the Toulousain lawyer Lamamye. In total contrast to Coriolis, his motives seem to have been free of any personal ambition, and he worked for unity within the province as a whole. Lamamye's technique was to call assemblies in each of the towns which sent deputies to the Estates, often attending in person, in order to register formal protests and concert action aimed at changing the royal decision. They naturally hoped for the support of the governor Montmorency, a far more influential figure than either Bellegarde or Guise, and someone who had enjoyed a generally productive relationship with the Estates over the years. There was no doubt that Montmorency's own authority would be affected if the Estates were reduced to a marginal status, so it is no surprise to find him negotiating on their behalf. By 1631 prospects had clearly improved, with the crown withdrawing the comparable measures elsewhere, so that Montmorency became very optimistic. In September 1631 the government offered a compromise; the province could buy off the *élus* by refunding Vanel his 4 millions, but a smaller group of new officials was to be created to take charge of the *assiette* in each diocese. The cost of these officials would be relatively small, and the profits of the sale were to go to the Estates, set off against the payment to Vanel. Numerous other provisions were designed to reduce the independent power of the Estates, but in principle the way was open for negotiations. The governor was permitted to reconvene the Estates in December 1631, and a protracted period of haggling began. By May 1632 a settlement had still not been reached, so the crown tried to use the *élus* to collect the taxes for that year, provoking a general movement of resistance organized by the Estates, in the form of simple refusal to accept the commissions of the *élus*.

At this point the whole situation took a dramatic turn with the appearance of Gaston d'Orléans, who had fled to Lorraine in the aftermath of the Day of Dupes, and was now trying to raise a rebellion against his brother. Montmorency, who had become increasingly exasperated by the role he had been obliged to play as intermediary between crown and province, had also been under great pressure from his wife and other members of his entourage to ally himself with Gaston. As the latter entered Languedoc with a small force of cavalry, Montmorency took the plunge; on 22 July he entered the Estates and secured a declaration associating their interests with his, giving him the right to call meetings of the Estates, and to collect the *octroi* tax in his own name. Supported by a group of bishops, and able to capitalize on resentment at the intransigence of the royal agents, he secured a unanimous vote from the third estate. Potentially this was an exceptionally dangerous situation for the crown, with a coalition between dissident grandees and provincial separatism, yet the intendant d'Emery was to be proved right in calling it a mere 'feu de paille'. Unanimous votes are often suspect, and that of the Estates was no exception; the deputies had acted from a mixture of fear and excitement, without giving any thought to the consequences. Since negotiations were now at an end, they dispersed to their home towns, where they found their actions disavowed. The province as a whole did not rebel; the *parlement* declared against Montmorency, while loyalists acted quickly to ensure control of vital towns. The protestants, utterly discouraged by their recent defeats, resisted half-hearted efforts to rouse them. Many local nobles also refused to join Montmorency, his relative and *lieutenant-général* the duc de Ventadour among them. The defeat and capture of the duke at Castelnaudary on 1 September was of relatively minor significance in comparison with his failure to raise support in a province his family had ruled for more than a hundred years. The king and Richelieu rubbed the point home by having him tried and condemned by the *parlement* of Toulouse.[38]

Both at the time and later Richelieu laid great stress on the peculiar nature of Montmorency's treason:

la faute de M. de Montmorency n'étoit pas un simple crime de rébellion, comme celui d'un autre grand qui auroit simplement

[38] For the details of this revolt and its background, see Major, *Representative Government,* 549–57 and 588–96; P. Gachon, *Les États de Languedoc et l'Édit de Béziers* (Paris, 1887); and W. H. Beik, *Absolutism and Society in Seventeenth-century France* (Cambridge, 1985), 130–1 and 199–201.

porté les armes contre le Roi en faveur de Monsieur, mais que c'étoit une rébellion accompagnée de toutes les circonstances aggravantes qu'on pouvoit s'imaginer . . . Qu'il avoit fait révolter une province par resolution du corps des États, ce qui ne fut jamais fait . . .

[the offence of M. de Montmorency was not a simple crime of rebellion, like that of another grandee who had merely taken up arms against the king in favour of Monsieur, but that it was a rebellion accompanied by every aggravating circumstance imaginable . . . That he had taken a province into revolt by a resolution of the whole Estates, which had never been done . . .][39]

One may be surprised that the Estates themselves survived, but the crown evidently decided that it was better to tame them than to provoke further resistance, particularly if more money would flow into the royal coffers as a result. In one of those displays of sovereignty which had such an important role under the *ancien régime,* the Estates, with representatives of the *parlement* and the *Comptes,* were assembled in the church of the Augustins at Béziers on 11 November 1632. Louis XIII then faced them in person, surrounded by his court, and imposed the Edict of Béziers on the province. In some respects this was a conciliatory settlement; the *élus* did not return, nor did even the smaller groups of officials envisaged in 1631. The Estates and the smaller diocesan assemblies had their meetings and rights far more strictly controlled than in the past, but they continued to levy their own taxes. The annual levy now included the customary taxes, together with a *don gratuit* of 1,050,000 livres. The total effect was to raise the tax bill from some 1.2 million in 1628 to nearly twice that sum, plus additional levies to pay off the *élus.* As things would turn out, the crown had not calculated very cleverly in all this. Languedoc was certainly paying something more like her share—though still proportionately less than such provinces as Normandy—in relation to the national situation as it stood in 1632. When the enormous costs of war forced an unprecedented fiscal offensive after 1635, however, the fixed payments imposed by the Edict of Béziers ironically turned into a defence for the province, since the Estates were still required to grant any additional sums, which they were naturally very reluctant to do. The Edict was to be revoked during

[39] Richelieu, *Mémoires,* ed. Petitot (10 vols., Paris, 1821–3), vii. 210.

the troubles of the Fronde, in 1649, and the crown never bothered to reimpose it; Colbert and his successors found direct management of the Estates a more profitable technique.[40]

Examining the linked revolts of the period 1630–2 is an instructive exercise. They form a very special group, which would have no real parallels until the Fronde, because of the way in which they combined resistance from the level of governors down to that of the poor. There was a sense in which they marked certain limits in the crown's power to coerce the peripheral provinces without unacceptable levels of disorder, so that the subsequent continuation of the provincial estates can be regarded as the fruit of this crisis. Considering them in more detail, it is also striking to see how resort to violence, with the use of mobs or noble conspiracy, was ultimately divisive. The sustained resistance of the Estates of Languedoc, which involved scarcely any violence until the unfortunate arrival of Gaston, demonstrated the benefits of a controlled agitation kept under the hand of the local *notables*. Provincial privileges were a splendid rallying-cry, yet even this could not totally obscure the very real social divisions which tended to become more prominent once a revolt was under way. The government might have been forced to compromise, but the provinces had also received some severe lessons in the risks of open disobedience. Popular agitation could easily turn against the local oligarchies, notably their richer members. Estates, *parlements,* and other bodies owed their existence to the crown, and could have their privileges savagely reduced if they stepped over the limits, particularly if open revolt brought an army or the king himself into the province. Governors who failed to control their provinces must expect deprivation and disgrace, while for Guise conspiracy brought permanent exile in Italy, and Montmorency's rebellion cost him his head in a sensational assertion of Louis XIII's determination to be obeyed. Those Languedocian nobles who had joined the governor were stripped of their titles and their seats in the Estates. Since they had generally fled or gone into hiding, they were executed in effigy or condemned to the galleys; their property was confiscated and

[40] For the Edict of Béziers and its effects, see Beik, *Absolutism and Society,* 130–46.

distributed among those who had been conspicuously loyal, and their châteaux were razed. When the crown laid hands on the elderly and half-blind *président* Coriolis of Aix he was shut up in a tower until he died, his office having already been transferred to another loyalist.[41]

There are also obvious links between these troubles and the central political crisis of the Day of Dupes. Bellegarde and Guise were involved in the intrigues against Richelieu, and the latter in particular failed to use his influence to damp down the revolt in Provence. Gaston's flight from court, which was a reaction to his mother's disgrace and the total ascendancy of Richelieu, later drew the indecisive Montmorency into his fatal rebellion. Russell Major has tried to argue more than this; for him the policy of introducing the *élections* was the work of Marillac, part of a scheme for internal reform which was to be aborted by Richelieu's preference for foreign adventure.[42] There is a curious internal inconsistency here, for Major seems to be an advocate for the Estates against the *élus*, yet praises Marillac for his commitment to internal change. I think it is evident that there is a great deal of anachronism and improper use of hindsight in such distributions of praise and blame; contemporaries simply did not see the issues in such stark terms although some of Richelieu's writings may tempt us to think they did. In any case the factual basis for Major's reconstruction is highly dubious. There is little suggestion in the contemporary evidence that there was any discord among the ministers on the subject, as we know there had been over Sully's policy in Guyenne. If Marillac was seen as the prime enemy in Burgundy (which is none too sure), it is plain that the Provençaux blamed the *surintendant* d'Effiat, while for Languedoc the blame seems to have been shared between d'Effiat and Richelieu himself. The argument that the policy was abruptly abandoned after the Day of Dupes is also hard to square with the lengthy attempt to browbeat Languedoc into compliance. Desperate for any hard evidence that Marillac

[41] These punishments were serious, but generally the more established families could hope for eventual reinstatement, particularly if they merited favour through military prowess, or if a powerful patron intervened. The long-term reappearance of some of Montmorency's allies is discussed by Beik, *Absolutism and Society*, 324–42.

[42] Russell Major, *Representative Government*, 487–621.

was the prime mover, Major has to fall back on statements which refer to other matters, such as his attempt to enforce regulations on the *parlements,* a traditional role for the *garde des sceaux*. He totally disregards the awkward fact that changes in the tax system and the attributions of the financial officials were the primary responsibility of the *surintendant*; it would have been inconceivable for the *élections* to be proposed without the full support of d'Effiat, who was a model *créature* of Richelieu.

It is broadly true that foreign policy ruled out great domestic initiatives after around 1634, but in fact between 1631 and 1633 Richelieu's prophecies about the results of keeping Pignerol had yet to come true; the crown retained considerable room for manœuvre, as the royal intervention in Languedoc demonstrated. If we make the natural assumption that the government's motivation was primarily fiscal, then it had gained a good deal from the crisis; some 7 million livres in compensation for not introducing the *élus,* a revision of the tax system in Dauphiné, and a vastly increased annual contribution from Languedoc. During the 1630s it does seem that the government became more cynical in its attitude, as when it threatened the small Pyrenean provinces with *élections* in order to force additional payments out of them. We still need to explain, however, why the original offers to buy out the *élus* were resisted in 1629–30, when they would have given a substantial immediate gain. The economics of the whole business look very dubious, since the cost of introducing the officials was so high in annual terms. Here I think there is little doubt that the king and his ministers were acting from a mixture of common convictions and ignorance. The same justifications were repeatedly used when *élections* were established; the local tax-collecting bodies were levying large sums for purposes of their own, while making very little contribution to the treasury, and burdening the people unfairly by rigging the system. Richelieu wrote that the Estates of Languedoc were raising 4 million a year, virtually none of which reached the king.[43] This was plainly untrue, yet

[43] Richelieu, *Mémoires,* iv. 474–5. The intendant Machault wrote to the Cardinal in August 1631 informing him of various abuses in the finances of Languedoc; P. Grillon (ed.), *Les Papiers de Richelieu* (Paris, 1975–), vi. 513.

Richelieu probably believed it. All the things the crown alleged had some basis in fact; the error was in supposing that vast sums were involved, and that there were great reserves of untapped wealth. One is reminded of Olivares's policy towards the Catalans, based on very similar convictions. We should also remember that few people in the early seventeenth century had much capacity for the kind of economic reasoning needed, even had the facts been more readily available.

THE GREAT INSURRECTION IN THE SOUTH-WEST, 1635–7

The declaration of open war against Spain in 1635 was accompanied by a renewed fiscal offensive. The *taille* was raised to unprecedented levels, while a wide range of new indirect taxes were created and sold to financiers. If the country already felt it was being overtaxed, these new burdens showed that the people of France had seen nothing yet, compared with the demands the king would now make on them. Whereas the revolts of the previous period had been primarily directed against the *élus,* with the aim of preserving local rights, and therefore possessed a certain coherence, the exceptional number of rebellious acts which characterized the war years seem extraordinarily diverse. We can hardly doubt that they were all reactions to royal fiscality, with the multifarious abuses to which it gave rise, or that they genuinely expressed the despair of people tried beyond endurance. They can be very roughly divided into three categories. The first is the most spectacular, the major revolts which involved numerous towns or villages, and required repression by armed force. They were less numerous than one might think; a group of revolts in the south-west between 1635 and 1637, the *Nu-pieds* in Normandy in 1639, the Rouergue *croquants* of 1643. Then come some serious individual urban rebellions, such as those at Moulins in 1640 and Montpellier in 1645. Finally one must consider the extensive rural violence directed against the tax officials, yet not constituting identifiable movements of revolt in the same sense.

There can be no doubt that all the revolts in the south-west were linked, although this is not to advance any kind of conspiracy theory, nor to suggest that they followed a single pattern. From the riots in Bordeaux in May 1635 to the great rising of the *croquants* in the summer of 1637, however, the region was in a state of almost continual effervescence, which proved highly contagious, and which encouraged a general acceptance of violence as a political weapon. The target was of course the wave of new or increased taxes. The *taille* had already been increased sharply in 1634; by 1635 it was double the level of 1628.[44] In 1634 the intendant Verthamon was threatening Bordeaux with the loss of its exemption from the *taille,* and obtained an *arrêt du conseil* restricting the existing privilege, for land outside the town, to properties managed directly by the owners. There were other fiscal pressures on the city, but the initial spark came from an apparently minor tax, that on the *cabaretiers,* which was only expected to raise 10,000 livres a year from the whole *généralité* of Bordeaux. Since 1577 anyone wishing to open a *cabaret* had been obliged to purchase letters of permission from the crown; in 1627 this was turned into a tax whose payment rendered the letters hereditary. This generated little money, so in December 1632 the council created a new annual tax, at the rates of 6 livres for those in the towns, 5 for the *bourgs* and those on major roads, and 4 for villages. Lists of those who sold wine without actually running a *cabaret* were also to be drawn up. Of course this was not the first tax on wine, and the standard duties payable as *aides* would have been much more substantial charges on most *cabaretiers,* but it was a new levy which might be intensified in the future. The local influence of such men, at the centre of networks of sociability, also posed an obvious danger to the enforcement of the tax. There had been plenty of local resistance to earlier increases in duties on wine, so it is surprising that the council did not foresee the results of an innovation whose total projected yield was only a few hundred livres. Already in 1629 there were murders and attacks on the houses of those who were involved in the new wine tax of 5 *sols* per *muid,* which affected Angoulême, Saint-Jean d'Angély, Saintes, and the *bourgs* of the region, and had to be repressed with troops by the intendant de la Thuillerie. The nobility

[44] Bercé, *Croquants,* 68. The account of the revolts of the towns of Guyenne in 1635 is based on Bercé's work, pp. 294–363.

assembled and sent syndics to plead for the withdrawal of the tax, but without success.[45]

Troubles soon broke out in Poitou and Normandy as a result of the 1632 edict, while the various *cours des aides* had to be forced into registering it by *lettres de jussion.* This process took a good deal of time, so it was not until August 1634 that the *traitant* made a first attempt to levy the tax in Bordeaux. Within eight days a mob, in which inn-keepers were naturally prominent, attacked the bureau, and soon the *traitant* had left the city, abandoning his operations. He complained to the council, which responded by an *arrêt* of December 1634 making the royal officials and the *jurats* responsible for enforcing the tax, on pain of being liable to pay the assessment personally. This *arrêt* did not become known in Bordeaux until the beginning of May 1635, then on 10 May a royal *archer* arrived in the city to enforce it. On 14 May, as he tried to carry out his commission, a crowd of several hundred people gathered; the *archer* fled to the town hall which was then besieged by a rapidly growing number of rebels. By the evening the building had been captured, the *archer* and six other people massacred as they tried to flee, and the authority of *jurats* and *parlement* completely undermined. Ironically the revolt had come at the moment when the captains of the bourgeois guard were meeting in the town hall to consider how best to maintain order in the city; now the *notables* were barricaded in their houses as the mob rampaged through the city, which it continued to do the next day. On 16 May the *parlement* managed to send a message to the provincial governor, the octogenarian duc d'Épernon, begging his help, and he arrived the next morning from his château at Cadillac with a small body of guards and loyal gentlemen. He stripped the *jurats,* personal enemies installed during his spectacular quarrel with Archbishop Sourdis the previous year, of their functions. According to Épernon's secretary and biographer Girard the rebels had drawn up a list of 400 leading citizens

[45] R. Mousnier (ed.), *Lettres et mémoires adressés au Chancelier Séguier (1633–49)* (2 vols., Paris, 1964), 1094–5 (Mémoire touchans les émotions arrivés depuis 1629 jusques à 1643 ez provinces de Xaintonge, Engoulmois et Poictou . . .).

whom they regarded as *gabeleurs,* and whose houses they intended to attack, but the governor's arrival temporarily pacified the city. He had only some fifty men with him, while Girard was almost certainly correct in arguing that many of the bourgeois sympathized with the rebels. On 15 June a new revolt broke out, and numerous barricades were erected in the streets. Épernon immediately attacked these with his tiny force, and succeeded in clearing them; four of his men were killed, and at least thirty rebels. It does seem plain that this was less than a full-scale insurrection, or that the people were reluctant to oppose the governor, since his men could not possibly have overcome large numbers of determined opponents. Whereas the troubles in May had been directed against a clear target, and were probably largely spontaneous, those of June responded to no such provocation, and must have been organized. Further rioting broke out on 30 June and 1 July, again repressed by Épernon, this time with some modest assistance from the bourgeoisie; on this occasion a number of houses were sacked. This was the last serious violence, and in the autumn the king issued letters granting an abolition for all the rebels, against the assurance that the taxes would now be paid, excluding that on the *cabaretiers* which was now withdrawn. The repression had been minimal; five of those captured during the riots were hanged, although in 1636 one of the known leaders, Lureau, was also executed for involvement in a smaller subsequent riot. The misbehaviour of the bourgeois, who had done so little to suppress the revolt, beyond keeping control of the city gates, was not punished at all.[46]

This revolt might seem rather unimportant in itself; its significance lay in its effect over large areas of the south-west. The news of the rising in Bordeaux in May spread quickly, but it was the second revolt in June which lit the powder train. There were riots in a dozen of the smaller towns along the Garonne and nearby, with attacks on officials and *commis* identified as *gabeleurs;* this wave reached as far as Toulouse, where the *parlement* and the *capitouls* reacted quickly to stifle the troubles. In Périgueux the mayor, suspected of complicity in the tax farms, had to flee the city, while the intendant

[46] For the revolts of 1635, see Bercé, *Croquants,* 294–363.

Verthamont found himself virtually a prisoner of the mob, forced to march at their head, promising the removal of the the tax on the *cabaretiers,* and unable to prevent the murder of the mayor's secretary. Serious though these repercussions were, they were overshadowed by the ferocious revolt at Agen, the most important city on the Garonne between Bordeaux and Toulouse. Here the revolt began on Sunday 17 June, when the *vice-sénéchal* tried to respond to a request from Épernon that he send his *archers* by river to reinforce the authorities in Bordeaux. Threats made to the unwilling boatmen encouraged popular fears that the whole province was about to be punished; the tocsin was rung, the mobs attacked the houses of leading members of the *élection* and the *cours des aides,* both highly unpopular as nests of *gabeleurs.* No fewer than fifteen people were killed, some of the bodies being afterwards mutilated. Three of these victims were killed by the peasants of the surrounding villages, who had reacted to the sound of the tocsin, while a number of others had their rural property devastated, or their town houses pillaged. The exceptional virulence of this attack united the bourgeoisie; the guard seized control of the gates, preventing any union between the rebels and the peasants, and barricades were thrown up at strategic points. By the evening of the 18th the rebels, conscious that they could not resist their better armed opponents, and having obtained a declaration from the *chambre de l'édit* that the new tax would be withdrawn, had dispersed. It is interesting to note that the *garde bourgeoise* was theoretically able to muster 810 men in this middling-sized town. Once again the repression was very limited, although numerous rebels must have been killed during the street-fighting; only five executions followed, some of them not until January 1636.

Resistance to the tax on the *cabaretiers* had been intensified because the delay in levying it meant that three years' worth became due at a single payment in the spring of 1635. The attempt to collect the tax in the *généralités* of Poitiers and Limoges had already caused the farmer to report nineteen separate attacks on his *commis* to the council between January and April. In July the troubles spread to the rural parishes and the *bourgs* of Saintonge, where the *sergents* and *huissiers des tailles* were trying to deliver the normal assessments; they had to flee for their lives amid cries of *gabeleur.* The anger of the tax-payers was increased because the additional sums payable since 1628 were identified separately as the *rentes des droits aliénés,* so that the additional burden was plain to all. The *vice-sénéchal* of Saintonge claimed he could do nothing since his small forces were quite inadequate for the task. Elsewhere there was not even a pretence of following the normal procedure; the *élus* of Périgueux, Condom, and Agen had fled, and did not resume their posts until October, with the arrears of 1634 still unpaid and no tax demands even issued for 1635. Guyenne was exempt from the *aides,* but in Poitou and the western provinces the *commis* had all fled, so that the farmers received nothing from them; in Saintonge and Angoumois receipts did not begin again until the autumn of 1637. In short, this was a fiscal débâcle on a disastrous scale, and the rebels had achieved a great deal at relatively little cost to themselves. The weakness of the authorities had been starkly revealed, together with the unwillingness of the local élites, nobles and bourgeois alike, to protect the tax officials and agents, widely seen as traitors to their region.

The position remained very difficult for the crown in 1636, although there was far less open violence. The *aides* were still not being collected, while the *droits aliénés* were widely resisted, and could only be collected by force, on a very patchy basis. In Saintonge and Angoumois the resistance of the previous year continued, with parishes assembling to drive out any tax collectors who dared show their faces, or in some cases kill them.[47] In May there was an alarm when it was reported that the peasants intended to come to Angoulême in large numbers for the three-day fair which began on the 22nd, and the authorities promptly cancelled it. The Comte de Jarnac, who commanded the château of Angoulême, went to meet representatives of the peasants, who agreed to send a deputation to the king. In June and July anti-fiscal assemblies continued to meet, while the council decided to use force, dispatching three regiments from Normandy. They were only in the area for a week, unable to identify any enemy in any case, before the great crisis of the Corbie campaign led to their departure for

[47] For the events of this year, Bercé, *Croquants,* 364–402.

the Picardy front. On 20 August the intendant Villemontée and the comte de Brassac, governor of the Angoumois, met the syndics of the local *châtellenies* at Saintes, promising concessions, and succeeded in restoring order. This was a very orderly movement, therefore, which can only be termed a revolt in the sense that the assemblies were unauthorized, and that they were supported by an effective tax strike. The grievances and other documents it produced are of the greatest interest, since they go beyond the immediate rejection of new taxes to discuss the deeper problems of the tax system. The leaders were drawn from the local seigneurial judges and the *curés*. Two of them stand out above all; a lawyer from Barbézieux, the sieur Gendron, and a seigneurial judge from the *bourg* of Berneuil, Simon Estancheau. These two men seem to have collaborated in drawing up grievances, Estancheau being the more prominent. Later he would write a series of memoirs for Richelieu, the most important of which is a discussion of the abuses of the *tailles* sent in January 1637. This document can be associated with the *ordonnance des paysans de Poitou* (which is in fact misnamed, and originated from the same region) drawn up the previous spring. In essentials, Estancheau blamed the assessment process for much of the hardship and protest associated with the *taille*. At the first stage the distribution between parishes was grossly unfair, because the *élus* protected those parishes where they owned property, or from which they took bribes. Worse still, when the levies were made within the parish the rich perverted the whole system, protecting their tenants and sharecroppers, or of course their own property, so that the whole weight of the tax fell on the poor. It was impossible to use the legal system against such abuses, because of complicity, the lack of means among the oppressed, and the fact that the sufferers were commonly in the debt of their richer neighbours. Estancheau also denounced the use of violence by the lesser nobility to oppress the peasantry, abuses over weights and measures, and the exploitation of monetary difficulties in connection with the payment of rents and dues. In effect, the inequalities which disfigured the system had become insupportable now that the rate of taxation had become so much higher. His suggestions for reform were twofold. Firstly, the tithe should be used as a basis for the charge demanded for each parish—an interesting predecessor of Vauban's famous plan for a *dîme royale*. Secondly, the *taille réelle* should be introduced, since any attempt to right the abuses of the *taille personelle* would not last more than a few years at best.[48]

These remarkable documents give us an exceptional insight into the grievances of the peasantry, and reveal far greater social tension in the countryside than we might have expected from the apparent complicity of the nobles and others in revolts. It is notable that Bercé, although he cites the documents frequently, rather underplays this aspect, preferring to stress other characteristics of the peasant attitudes. His depiction of the mythology of what he calls 'un fond antifiscal commun à tous' is very skilful, with its themes of the age of gold, the king misled, the evil ministers, and the king robbed.[49] This is very convincing in its own terms, and helps to explain how groups with rather different interests could unite, but it tends to gloss over the extent to which these rebels had marked themselves off from the local rulers. It is not just a question of Estancheau's personal views, for the themes are clearly apparent in some of the earlier documents drawn up by the assemblies in the spring of 1636. As the 1636 manifesto puts it:

Et pour les Riches de chacune paroisse qui ont achevé de ruyner le peuple, au temps advenir Il ne fault permettre qu'ilz se meslent des affaires des tailles, et fault leur en donner leur bonne part, et aussy aux mestayers de Monsieur cestuy cy, Monsieur cestuy là, qui possèderont la meilleure partie du bien de leur paroisse.

[And for the rich of each parish, who have managed to ruin the people, in the future they must not be permitted to meddle in the affairs of the *taille*, but must pay their fair share, as must the sharecroppers of Monsieur this and Monsieur that, who possess the best part of the property of their parish.][50]

Bercé quite rightly cites the *cahier* of the nobles drawn up in 1649 as evidence for the common interests of nobles and

[48] Estancheau's *mémoire* is printed by Bercé, *Croquants*, 746–9, with other documents concerning the movement on pp. 736–45.

[49] There is an extended discussion of these themes in Bercé, *Croquants*, 604–73.

[50] Ibid., 739.

peasants, but does not mention the presence in the later document of an article complaining about the imposition of *taxes d'office* on the *métayers* and tenants of the nobility; this was the standard way in which the intendants and their agents tried to compensate for the manipulation of the local assessment.[51] There is no shortage of other evidence, from other revolts and from the criminal records of the time, notably those of the *grands jours* of 1634 and 1665, to show that the relations between nobles and communities were often far from harmonious.[52] I will return to this theme later, and to the whole question of vertical or horizontal solidarities.

The outcome of 1636 was another success for the rebels. Villemontée and Brassac had returned with instructions which had been drawn up in terms of the anti-fiscal ideology, and which proved successful in damping down the revolt. The arrears of the *taille* from 1632–4 were effectively written off, and ambiguous promises made about future rates of tax, with statements that those who had profited from the taxes would be prosecuted. This was a cynical operation, since the crown and its agents knew that the peasants would expect far more than was actually to be given, but it served its turn in getting the *taille* functioning again from the beginning of September. Once again the government had been humiliated and its lack of coercive power exposed, so that no real solution to the tax problem in the region seemed in sight. By the end of September Villemontée had been forced to grant similar remissions to Poitou, while pressure was mounting from Anjou and Touraine in October. Far from respecting its earlier promises, the council was preparing a new fiscal offensive for 1637, to which there were to be added some special local levies. The duc d'Épernon and his son the duc de la Valette had to provide for the army of Bayonne, with virtually no financial assistance from the treasury. Winter quarter for the troops had already exacerbated local feeling, but now the dukes, without clear authority from the king, tried to impose two special taxes. The first was a levy of grain, called 'les

[51] Ibid., pp. 490–2; R. Mousnier, J.-P. Labatut, and Y. Durand, *Deux cahiers de la noblesse pour les états généraux de 1649–51* (Paris, 1965), 88 (art. 18).

[52] A. Lebigre, *Les Grands Jours d'Auvergne* (Paris, 1976) is a good starting-point on this question.

rations de l'armée de Bayonne', ordered by Épernon in December, and begun in February. Communities were to borrow the necessary grain from merchants and religious houses, then use local taxes to repay the debt. The second was a special levy, on the same basis as the *taille,* of 150,000 livres from the *généralité* of Bordeaux, under the title of the 'crue pour l'armement de Guyenne'. Meanwhile the crown had reorganized the *taille* for 1637, with the assessment for each *généralité* divided into two equal parts. The first was to be levied on the countryside, the second on the towns and *bourgs*, even those which enjoyed an exemption—this was covered up by calling it the 'subvention sous forme de prêt et emprunt'. The preamble to the edict pretended that the council had taken pity on the *plat pays* and reduced its liability, but both Villemontée and Verthamon protested that in fact the burden had increased—the former claimed it had done so by a third.[53] As a final blow, an *arrêt* of 10 March 1637 ordered the reimposition of the arrears of the previous two years and revoked all the concessions, including that of the *droits aliénés,* which the peasants regarded as a total abuse benefiting only the officials and tax farmers. The use of *sergents* to levy the 'rations de l'armée de Bayonne' in Périgord in April was the catalyst for the new revolt. Two *sergents* were killed on 22 April, and others were killed or driven out over the next few days. The speed with which the revolt developed startled contemporaries, but it probably resulted from the tight-knit character of the area in which it started, the parishes of the Paréage, which were located in the large *forêt de Vergt* just outside Périgueux. A local gentleman Antoine de Ribeyreix stirred up the revolt, which saw 4 to 5,000 men assemble, and arrive outside Périgueux on the first of May. They demanded that the gates be opened, cannon handed over to them, and the *gabeleurs* delivered up for punishment, but naturally the gates were shut and manned against them. They named a local nobleman of good family, La Mothe la Forêt, as their chief, and called a rendezvous for all the communities of the region in the forest on 7 or 8 May, at which 30,000 men are said to have been present. La Mothe la Forêt selected 8,000 of

[53] Villemontée to Séguier, 26 June 1637, in Mousnier, *Lettres et mémoires,* pp. 390–2; extracts from Verthamon's letters, ibid., 405–6.

the strongest and best equipped to form his army of the Communes of Périgord, with which he then marched on the important town of Bergerac. This was a protestant stronghold, whose walls had been slighted some years earlier, so it was unable to resist the peasants, who occupied it on 11 May. Fortunately for the town it was not the centre of any financial activity, and the peasants behaved in a disciplined, orderly fashion.[54]

La Mothe la Forêt plainly felt the momentum had to be kept up, and he planned to move towards Bordeaux, raising the countryside as he went. In military terms his great need was for cannon, without which he could not hope to assault any of the larger walled towns, and it was partly in search of artillery that his forces attacked Saint-Foy on 15 May. This was another protestant town, an old *place de sûreté,* still well fortified, and a handful of Épernon's guards had already arrived there to stiffen resistance. When the rebels summoned it to surrender the inhabitants made a sudden sortie, captured Ribeyreix, and defied their attackers, who had to retreat. This threw the general's plans into disarray, and it was only after several days that the *croquants* moved southwards from Bergerac, probably hoping to attack Agen or Cahors. At the end of May they were in the Agenais, where they captured the bourgs of Eymet, La Sauvetat, Miramont, and Lauzun. By now, however, Épernon had summoned his son the duc de la Valette from Bayonne; he had brought some troops with him, and collected up the elements of some other regiments which had been recruiting in the province to give him an army of 3,000 foot and 400 horse. He made straight for the nearest detachment of the rebels, which was holding La Sauvetat, and assaulted them on Whit Sunday, 1 June, without even waiting for his cannon to come up. It was a bloody encounter, with 200 dead on the royal side, and 1,000 to 1,500 of the rebels. La Valette then marched towards Bergerac, and sent one of the great protestant nobles of the region, the marquis de Duras, to negotiate with the Communes. On 6 June La Mothe la Forêt agreed to disperse his army, on the terms that there would be no pursuit, and that la Valette would personally seek their pardon from the king. One of the popular leaders, a Périgueux

[54] This section is based on Bercé, *Croquants,* 403–62.

doctor named Magot, tried to resist what he saw as a sell-out, but La Mothe la Forêt had him killed. The peasants might have fought again, but their general was an experienced soldier, very conscious of their military weakness, especially now that the royal artillery had arrived. In any case he had always seen the revolt as a prelude to negotiation, and would have found it very difficult to resist the mediation of men of his own class.

This was not only one of the greatest revolts of the *ancien régime,* but also one of the most complex in terms of its leadership. La Mothe la Forêt is frequently cited as though he were an archetypal noble leader of peasant revolt, but in fact he stands out as an exceptional figure. Among the other nobles involved only Ribeyreix resembles him, as an authentic local gentleman acting at least partly out of sympathy for the peasants. The other principal noble leader, Madaillan, was an exotic figure who had served in the Thirty Years' War, conspired with Gaston, and would end on the scaffold for an impressive list of crimes including incest. The rebels hoped they would be joined by a more prestigious noble in the marquis d'Aubeterre, son of a marshal of France, but he failed to arrive in time. He was in any case a very dubious figure, whose violent and irresponsible behaviour had already led to his disinheritance in favour of his younger brother, so for him the revolt was an opportunity for personal reassertion.[55] Among the other leaders we know there were at least four *curés,* and the doctor Magot already mentioned. One other person seems to have had considerable influence; this was de Jay d'Ataux, the *lieutenant-général* of the *sénéchaussée,* and the man who had actually made the assessment for the offending tax. Captured by the rebels before he could reach Périgueux, he had been saved from death by La Mothe la Forêt, and then appears to have played a very skilful game, advising the general on the terms of his manifestos, and perhaps finally

[55] For Aubeterre's position, see his mother's views as reported to Séguier in 1636, Mousnier, *Lettres et mémoires,* 298. She was already complaining to Richelieu of his behaviour in 1629; Grillon, *Les papiers de Richelieu,* iv. 578. Further correspondence in 1631 reveals that he had escaped from prison, and the reluctant duc d'Épernon was being ordered to pursue him; ibid., vi, pp. 317, 632, 639.

persuading him to abandon the struggle. What is very remarkable is that once the revolt had broken out no more nobles joined it, despite pleas from La Mothe la Forêt. If they would not respond when invited by an authentic nobleman of his stamp, they were clearly not prepared to join a peasant revolt on almost any terms. An anonymous correspondent, probably from Bordeaux, did comment:

Avant que ledit sieur de la Valette eut réduit La Sauvetat à soy, il n'avoit pas six gentilshommes du pays de Périgord à luy, mais après cette défaite, il eut en moins de trois jours plus de deux cents volontaires à sa suite.

[Before the said sieur de la Valette captured La Sauvetat, he did not have six gentlemen from Périgord with him, but after that defeat, in less than three days he had over 200 volunteers with him.][56]

The oddity of this piece of evidence does not seem to have struck Bercé. La Sauvetat is in the Agenais, just beyond the southern border of Périgord, and la Valette approached it at high speed from the south. How could the nobles of Périgord have known anything of his movements, and how in any case could they have left their houses and crossed a hostile countryside? After the battle he was moving through Périgord towards Bergerac, while it would have become much easier to leave home now that the peasant bands were dispersing.

The defeat of the army of the Communes did not mean an end to the troubles. Madaillan had escaped from La Sauvetat, and made his way towards Quercy, where troubles had already begun in Cahors when the passage of officers from the regiment of Biscarrat aroused fears of troop lodgings. The disturbances turned into an attack on the *élus*, which was repressed by the bourgeois guard, led by the consuls and the *sénéchal*, the comte de Cabrerets. Madaillan then appeared, seized a castle near Cahors, and tried to raise a new peasant revolt. His plans were frustrated when la Valette sent a detachment of troops after him and drove him out, but meanwhile a more serious movement had begun in Haut-Quercy, led by the same parishes of the *causse* which had been prominent in 1624. Several thousand peasants seem to have taken up arms, and attacked the *bourg* of Gramat. They

[56] Bercé, *Croquants*, 430.

occupied it after the flight of the *sénéchal,* the comte de Favières, and his brother the comte de Gramat, killed two local lawyers, and after a certain amount of pillaging carried off the legal documents certifying the privileges of the town. The other initial activity of the rebels took the familiar form of attacks on the property of *gabeleurs,* the *élus* above all. They then marched to the important local town of Figeac, where the bourgeoisie, led by two local noblemen, made a vigorous sortie in which some eighty rebels were apparently killed, while the others soon dispersed. Bercé points out that the local nobility here, far from supporting the revolt, put it down. He attributes this to their awareness that the general situation was now unfavourable, but it seems more plausible to look at the nature of the revolt itself, which put forward no such programme as we find in Périgord, and appeared to seek primarily the typical forms of popular justice on the *gabeleurs*.[57] Resort to arms and attacks on persons and property quickly alarmed the local rulers into action.

THE *NU-PIEDS* IN NORMANDY, 1639

The next major insurrection was that of the *Nu-pieds,* which broke out in Normandy in the summer of 1639. It was well-known that the province was heavily taxed, and the 1630s had seen the burden increase to unbearable levels. The *généralité* of Caen was assessed for approximately 1.1 million livres of *taille* at the beginning of the 1630s; in the four years 1636–9 the demand, including extraordinary additional taxes, ran at 2.25, 2.4, 3.5, and 2.8 million respectively. The inevitable result was massive arrears, so that in June 1640 the *élection* of Coutances showed arrears of 10 per cent for 1637, 48 per cent for 1638, and 88 per cent for 1639. It is evident that the exorbitant demands of 1638 had caused a complete breakdown of the system, with arrears rising to such levels that they could never be met.[58] Those towns exempt from the *taille*

[57] Bercé, *Croquants,* 438.

[58] M. Caillard, 'Recherches sur les soulèvements populaires en Basse-Normandie (1620–40) et spécialement sur la révolte des Nu-pieds', *Cahier des Annales de Normandie,* No. 3 (Caen, 1963), pp. 23–152, gives a very detailed account of the financial pressures of the 1630s. There is additional information in Foisil, *Nu-pieds,* 62–101, and in the long account of the revolt by Porchnev, *Les Soulèvements,* 303–502.

suffered too; in 1637 Rouen was required to pay a forced loan of 300,000 *livres,* itself a royal concession after an initial demand for 400,000. This was part of the general attempt that year to pass some of the burden over to the towns, making them pay half the sum expected from the *taille* in the form of the forced loan. The royal officials had suffered from increases in their *gages* which actually amounted to another kind of forced loan, since they had to be purchased; the *gages* were in practice never paid in full or on time, while the payment for the *paulette* was increased. The crown milked the system further by the creation of new administrative divisions, of which the most important was the *généralité* of Alençon, in order to sell batches of new offices. These practices led to bitter struggles between old and new officials, and to great confusion within the administration. It was perhaps not surprising that nothing effective was done to equalize the burden on different communities. In the *élection* of Caen over the years 1632–8 the demand per household in three villages increased from approximately 2 livres to 8, from 12 to 22, and from 4 to 15. Such divergences led to migration between villages, and depopulation was made more serious by the effects of the plague, which reached an intense level in 1636–9, with mortality rates probably around 20–25 per cent. Industry was in recession, with the cloth industry suffering from Dutch and English competition, and the number of paper mills falling from 100 to fifteen over the decade. The nobility had been subjected to a very unpopular *recherche de noblesse* in 1634; in the attempt to uncover tax evasion by fraudulent claims to nobility many genuine nobles were harassed because they could not provide suitable evidence on paper. The *recherche* was also associated with the edict on the *taille* of 1634, which limited noble exemption to one estate. Nobles found their tenants deserting their holdings under fiscal pressure, leading to one exceptional case in 1636;

Estant veritable que les écclésiastiques et gentilhommes de lad' parroisse de Saint Ouen le Brizou se sont volontairement et par charité cottizez à la somme de 300 livres payables aux recettes en diminution des imposts et pour obliger les contribuables à retourner habitter lad'paroisse qu'ils avoient entièrement abandonnées, les heritages de laquelle sont demeurés incultes et laissés en nonvalleur depuis deux ans encor qu'ils aient esté diminués à la foulle des autres parroisses.

[It is true that the clergy and gentlemen of the said parish of Saint-Ouen-le-Brizou have voluntarily and charitably assessed themselves for the sum of 300 livres payable to the receivers to diminish the taxes and to oblige the taxpayers to return and live in the said parish which they had entirely abandoned, so that the land has remained entirely uncultivated and been left producing nothing for two years, although their taxes have been reduced at the expense of other parishes.][59]

In numerous other cases the reaction to the tax burden was the more typical one of violence against collectors, or simply evasion of payment.

The situation was therefore already explosive in the summer of 1639 when the government decided to impose further burdens on the province. A new tax on leather affected local industries, while the cloth industry in Rouen and the surrounding areas was threatened by an edict establishing inspectors to enforce new rules about dyeing. Most dangerous of all, the *gabelle* was to be extended to the previously exempt coastal area around Avranches, where some 10,000 salt workers were concentrated, and which was a traditional centre of salt smuggling. Although it is not clear whether the edict was ever officially issued, news of it had reached the province, and in June a check on all strangers had been organized through the innkeepers of Avranches. The revolt broke out on 16 July in a bizarre and accidental fashion, when the *lieutenant* of the *présidial* of Coutances, Poupinel, arrived in Avranches on a quite unconnected judicial errand. He was identified as the carrier of the hated edict—one contemporary source implies that this may have been the doing of a personal enemy—and murdered by the salt workers. An *armée de souffrance* was organized, and mounted a series of operations against *gabeleurs* and other people associated with the tax system, damaging their houses and property. The revolt was confined to two relatively restricted areas at the base of the Cotentin peninsula, those around Avranches and Domfront; other areas likely to be affected by the new taxes did not participate openly. The two centres of revolt, although only some thirty

[59] Caillard, 'Recherches', 119.

miles apart, do not seem to have had much contact with one another, while the governor of Avranches, Canisy, kept control of the town gates, so that the rebels made their headquarters in the suburbs. The specialized basis of the Avranches movement among those connected with the salt industry (from whom it took its name, because they worked on the tidal flats in bare feet) gave it unusual cohesion, yet limited its capacity to expand. The *armée de souffrance* had no plans beyond its punitive expeditions, and making sure that no *agents du fisc* came anywhere near the Cotentin. Its leadership remains rather mysterious; one minor local noble, Jean Quetil sieur de Ponthébert, seems to have played a considerable role in the first agitation, but soon withdrew, declaring 'qu'il ne croyoit pas que ses discours eussent faict une si grande playe'.[60] He evidently took refuge within the town itself:

> Bien que les mutins mesmes se voyant abandonnez de luy criassent tous les jours aupres dez murailles qu'on leur rendit leur général Ponthébert qui s'estoit renfermé parmy les hiboux. Ainsy appeloient-ils les bourgeois d'Avranches, qui n'osoient pas sortir en plein jour.

> [Although the rebels, seeing he had abandoned them, cried every day around the walls to be given back their General Ponthébert who had shut himself up among the owls. This was what they called the bourgeois of Avranches, who did not dare to come out during the day.][61]

Five other nobles can be identified as having participated to some degree, out of nearly 500 known to have been living in the *vicomtés* of Coutances and Avranches at this period. A small group of priests also emerge as leaders, probably in a more committed fashion than most of the nobles, alongside some minor officials and artisans. As Foisil says, where anything can be discovered about the leaders they came from 'the lower level of each order; modest *vicaires*, poor lesser nobles, needy magistrates, envious barristers. Small men who probably lived in close proximity, sharing the same feelings, uniting their efforts against the *horzain* (outsider), the tax collector, those who threatened their privileges'.[62]

[60] Foisil, *Nu-pieds*, 208. [61] Ibid., 223. [62] Ibid., 228.

The extreme localism of the original revolt did not prevent it from setting off complementary outbreaks across the province, notably in Rouen and Caen. At Rouen a *commis* who had arrived to collect the new duties on dyed cloth was killed on 4 August, almost certainly as a result of a prearranged plan among the clothiers. The authorities took no effective action to punish the murderers, and soon a more extensive agitation was under way, directed against a wider range of indirect taxes. According to the memoirs of the *parlementaire* Bigot de Monville:

l'advis de ces désordres arrivez en la Basse-Normandie redoubla le courage du menu peuple de Rouen . . . Ces émotions estoient le subject ordinaire des discours du peuple qui les publiait comme des actions héroïques.

[the news of these disorders happening in Lower Normandy redoubled the courage of the common people of Rouen . . . these disturbances were the normal subject of talk among the people, who spoke of them as heroic actions.][63]

Between 20 and 23 August various bureaux for the collection of taxes were attacked, as were the houses of a number of highly unpopular *partisans*. One of these, the *commis des gabelles* Le Tellier de Tourneville, and his household defended themselves by firing on the crowd, but were eventually driven out when the house was set alight on 23 August. Although the bourgeois guard was present by now, they actually fired on the fleeing members of Le Tellier's entourage, ten of whom were killed. Despite this extraordinary episode, there had been a rapid evolution in the attitude of the bourgeoisie. They had originally refused to muster, on the grounds 'qu'ils ne prendroient point la querelle des monopolliers'; by 22 August, however, the rebels had become threatening to the élites in general:

les séditieux tournoient au butin, sans espargner ceux qui n'estoient suspects d'aucun party et imposition nouvelle . . . Le peuple mutiné s'adressant non plus aux partisans, mais aux plus riches marchands, ils menacèrent le logis du sieur Raye et d'autres.

[the rebels turned to plunder, not sparing those who were not suspected of involvement with any new tax or contract . . . no longer

[63] Foisil, *Nu-pieds*, 231.

attacking the financiers, but the richest merchants, the mutinous people threatened the houses of the sieur Roye and others.][64]

The *échevins,* to whose appeal the bourgeois had remained deaf the previous day, revisited their houses and found them 'en assez bonne disposition de prendre les armes pour la conservation de la ville'.[65] Barricades were erected to control the streets, with the notable exception of that where Le Tellier's house was. The attitude towards him was very different, for he was seen as a traitor to his community:

Les plus factieux des bourgeois disoient que il leur estoit fascheux d'exposer leur vie pour l'interest de ceux qui ne vivoient que pour ruiner leur pays par les advis qu'ils donnoient au Conseil et ordres qu'ils exécutoient dans la province.

[The most factious of the bourgeois were saying that it made them angry to risk their lives in the interest of those who only lived in order to ruin their country by the advice they gave the Council and the orders which they executed in the province.][66]

The sack of Le Tellier's house on 23 August marked the culmination of the revolt, and the crowds then seem to have melted away, although the town remained in a highly excitable state. The *parlement,* which the crown expected to punish the ringleaders and re-establish the normal collection of taxes, dragged its feet on both counts. The *premier président* Faucon de Ris appears to have been weak and vacillating, despite the advice of his bitter rival and enemy, *président à mortier* Bretel de Grémonville:

qu'il faloit que le roy restablit son autorité ou que le Parlement la restablit. Que ce dernier point ne pouvoit estre que par la punition des coulpables. Qu'il seroit aisé de les punir, veu que les bourgeois estoient désabuséz et ne favorisoient plus les séditieux, desquels ils avoient veu l'insolence.

[that it was necessary for the king to re-establish his authority or for the *parlement* to do so. That this last point could only be secured by the punishment of the guilty. That it would be easy to punish them, seeing that the bourgeois were disabused and no longer favoured the rebels, whose insolence they had seen.][67]

64 Ibid., 232, 238, 251.
65 Ibid., 251.
66 Ibid., 252.
67 Ibid., 260.

In fact the *bureaux des droits* were not reinstalled until December, after intense pressure from Paris, while no real progress was made with the prosecution of a group of rioters being held in prison.

August also saw troubles in other Norman towns, most notably in Caen, where again there were two phases, with a first disorder going unpunished, then succeeded by a more serious one. Four days after the first outbreak in Rouen, the attempt to collect a tax on skins led to an attack on the *commis* by the tanners. News of the second riot in Rouen seems to have been important in triggering the parallel action in Caen, this time six days later; it followed a remarkably similar course, with attacks on the houses of prominent tax collectors. The bourgeoisie of Caen reacted in the same lukewarm fashion as those of Rouen, but a group of gentlemen and officials took arms from the start, and were aided by the arrival of the comte de Matignon, *lieutenant-général* for Basse-Normandie, who ultimately succeeded in pacifying the town. The government had every reason to be dissatisfied with the local officials and municipalities, weak in their responses to the crisis, then reluctant to punish the offenders or enforce normal tax collection. The sanctions might have been relatively modest, however, had not the continuing presence of the *armée de souffrance* around Avranches made military intervention essential. Whereas urban revolts tended to be short if violent, and to force the élites into some kind of limited repression, this more broadly based movement showed exceptional staying power. Whatever their responsibility for instigating it, the local nobility certainly showed no interest in trying to put down the revolt, nor did Matignon make any attempt to organize them for the purpose. As the armies prepared to go into winter quarters Richelieu ordered Gassion into Normandy with 5,000 troops from the army of Picardy; he received his orders on 10 November, arrived in Caen on 23 November, and met the rebel forces outside Avranches on 30 November, defeating them completely. His speed of movement may have helped him to achieve this with only 1,200 foot and 500 horse, for one of the contemporary accounts suggested:

Si on eust encore différé quatre jours de les attaquer ils grossissoyent leur corps jusqu'à dix mille hommes et s'ils eussent résisté au premier effort, tout le pays se feust hautement déclaré; car un chacun en attendoit le succès.

[If there had been a further four days' delay in attacking them they would have swollen their army to ten thousand men, and if they had beaten off the first attack the whole country would have declared openly for them; for everyone was hoping for their success.][68]

Such estimations may have represented fears or hopes rather than realities, but Richelieu was now determined, once having been obliged to use regular troops, that an example should be made. The chancellor Séguier himself was sent to Normandy with full delegated powers, and undertook a savage repression. There were numerous summary executions, with many more condemnations in absentia, while troops were billeted with no regard for exemptions, and the towns were forced to pay indemnities to the victims of the revolt. The officials were severely blamed:

Ces rebellions ne seroient pas venues au point où on les a vus dans la dite Province, sans la connivence ou faiblesse de ceux qui ont l'autorité et le pouvoir de les empêcher qui ne s'y sont pas opposez avec la vigueur et le courage que recueilleroit nostre service et qu'ils estoient obligé de faire ayant nostre auctorité.

[These rebellions would not have come to the point that has been seen in the said province, without the connivance or weakness of those with the authority and power to prevent them, who did not oppose them with the vigour and courage required in our service, and which they were obliged to show as holders of our authority.][69]

Many of the courts were suspended from their functions for more than a year, their work being undertaken by temporary *commissaires*. The *parlement* of Rouen, blamed less for complicity in the revolt itself than for its failure to re-establish royal authority afterwards, was suspended for nearly two years, and then re-established as a *parlement-semestre*, with two groups of judges sitting for alternate six-month periods, and with many of the old councillors excluded. Only in the very different conditions of the regency period after 1643 did the magistrates succeed in whittling away the humiliating conditions imposed on them.

[68] Foisil, *Nu-pieds*, 292. [69] Ibid., 316.

In his account of the revolt Porchnev suggested that it might have spread much further and faster without the strong action taken by the government, even reaching other provinces, but this is very unconvincing. One of the most striking features of the major revolt around Avranches is precisely its failure to benefit from the period during the summer and autumn when no action was taken against it. Once the *armée de souffrance* was in existence its leaders had no idea what to do with it, beyond local punitive expeditions against tax collectors and their property, while their mentality was so local that they made no known attempt to link up with the other Norman rebellions. Numerous minor revolts elsewhere in France in 1639–40 responded to the same fiscal pressures as operated in Normandy, but with no suggestion of the slightest cross-influence. Nor is it possible to argue that the rebels displayed any strong social motivation; they were primarily concerned to defend their communities and their privileges, and their main targets were men concerned with the most hated aspects of the tax system. There was a sense in which the rebels were almost too successful in achieving these limited aims, so that there was no obvious way in which they could take further action unless they became overtly revolutionary. Close analysis of the revolt therefore gives little support to Porchnev's position, yet it is hardly more favourable to the alternative thesis of Mousnier. The meticulous study by Foisil produces very little evidence that there was significant participation by the élites, although they had been seriously alienated by royal policies over the previous years, and had little sympathy for the *agents du fisc.* Their initial attitude was highly equivocal, but in the longer run revolt forced them into action to preserve public order, at least in the urban context. The bourgeois of Rouen, Caen, and Avranches were eventually prepared to take up arms against the rebels, for fear of more general pillage, and perhaps also because of an awareness that failure to act might lead to royal action against them and their privileges. The revolt of the *Nu-pieds* was essentially a popular movement, most of whose leaders were humble men; it fell very far short of uniting Normandy in its cause, whatever degrees of complicity or sympathy it may have enjoyed outside its own ranks. It is fascinating to see how deep and general resentment of the government and all its works could not break through the political, social, and mental structures of the seventeenth century into really effective action. The greatest blockage of all was probably the ideological one; the surviving manifestos are vigorous calls to action 'pour la défense de la patrie oppressée par les partisans et gabeleurs', but never confront any of the real issues about the relationship between provincial rights and central power. They invoke a semi-mythical past, that of 'les ducs Guillaume' and the 'charte aux Normands' of 1315, to suggest that kings *ought* to behave differently, while never mentioning Louis XIII, Richelieu, or their war policy.[70] To have done so would of course have been futile, since it would have emphasized the underlying conflict of loyalties, and forced the leading nobles and officials into ostentatious support for the crown. In this situation the tax collectors served as a kind of lightning-conductor for popular fury, the expendable group whom everyone could agree to denounce.

THE 1640S AND THE FRONDE

This was particularly true with a levy as unpopular as the *taxe des aisés*, a forced purchase of *rentes* which the intendants were trying to operate in 1639–40, and which directly threatened the élites. In the small town of Moulins, capital of Bourbonnais in central France, the collector of the tax Jacques Puesche and some of his associates were murdered at an inn in the *faubourgs* on 23 June 1640, and the substantial sum of money they were carrying was stolen. There is no clear evidence of complicity by the leading inhabitants, beyond general accusations by the intendant Chaponay, yet there must naturally be considerable suspicion that the action was prompted in some degree by those who had most to lose. It was followed by a lengthy crisis, recorded in a large group of letters to Séguier; the bourgeoisie were patently unwilling to act against those responsible, who for their part kept the *faubourgs* in a state of aggressive self-defence against any attempt to punish the crime, and forced the abandonment of other forms of tax collection. The provincial governor and the mayor of the town blamed one another for the inaction, while the government

[70] Foisil, *Nu-pieds*, 188–94.

vainly tried to compel them to restore order by force. Deputations visited the prince de Condé at his château of Montrond early in August, to be told that they must recover the stolen money, punish the murderers, and secure full obedience before they could hope for an amnesty. It seems that several thousand livres were repaid, while the élites must have become increasingly aware that severe reprisals might result if they failed to comply. Only towards the end of August did the governor finally succeed in capturing and executing the principal leader of the revolt, with a somewhat reluctant bourgeois guard defending the gates against attacks from the *faubourgs*. This passage of arms proved sufficient to discourage the rebels, who dispersed instead of carrying out their threat to blockade the town. The main punishment suffered by Moulins was the quartering of two or three regiments for the winter, while the intendant made largely vain attempts to identify other persons involved in the revolt, in the face of persistent local opposition. Like the events in the Norman towns the previous year, this case brings out many of the social and political ambiguities characteristic of urban revolts. So long as the rebels posed no serious threat to the rich and their property in general, it was very hard to mobilize the *garde bourgeoise* for positive action beyond securing the gates; if royal troops were unavailable, the local authorities could only play for time until they found a good opportunity to set up a conflict in which they could polarize social groups.[71]

The 1640s did not see any more peasant rebellions on the same scale as those of the late 1630s, but a high level of general unrest and resistance persisted. The most interesting individual episode was the revolt of the *croquants* of Rouergue in 1643.[72] The province was something of a special case, since

[71] The revolt in Moulins is discussed by Porchnev, *Les Soulèvements*, 190–213, with some of the relevant documents printed on pp. 604–13. Large numbers of further documents are in Mousnier, *Lettres et mémoires*, 437–52, 454–62, 465–6, 468–9, 475–82, and in A. Lublinskaya (ed.), *Lettres et mémoires adressés au chancelier P. Séguier (1633–49): provinces du nord et du centre* (Leningrad, 1980), 157–88. There is a briefer discussion of the revolt by A. Leguai, 'Les émotions et séditions populaires dans la généralité de Moulins aux 17e et 18e siècles', *Revue d'histoire économique et sociale*, 43 (1965), 45–65.

[72] This revolt is described by M. Degarne, 'La révolte du Rouergue en 1643', *17e siècle*, 56 (1956), 3–18. Her account is rather incomplete, and

the *taille* was *réelle,* imposed on land rather than individuals, although the *élus* had been introduced in the 1620s. This was a notably poor agricultural region, which had been hit by a bad harvest in 1642, so that grain prices in 1643 were two to three times the level of 1640, and there was an agitation for special tax reductions; according to the intendant Charreton this began in February 'à la suscitation de quelques officiers et habitans de Villefranche.'[73] The return of an unsuccessful deputation to court led by the *juge-mage* of the principal town, Villefranche-de-Rouergue, coinciding with the arrival of the prominent *partisan* Duperret, seems to have been the catalyst for a revolt on 2 June, when some 1,200 armed *croquants* entered Villefranche. The recent deaths of Richelieu and Louis XIII may well have encouraged hopes of concessions from the government, while causing some disarray among the authorities. The intendant, caught in the convent of the Cordeliers, was compelled to authorize a temporary reduction of the *taille* to the level of 1618, and in a curious piece of legalism the rebels required that: 'on leur bailla et veriffia comme ils n'avoient rien fait contre le service du Roy et avoient vécu dans l'obéissance' ['it should be notified and attested that they had done nothing against the service of the king, and had lived in obedience'].[74]

The *sénéchal,* the comte de Noailles, tried and failed to pacify the province by negotiation, so that troops had to be sent in. Before their arrival Noailles seized two of the leaders, provoking a siege of Villefranche by the *croquants*; by early October they had been dispersed, and the troops began a more general repression. The reports by the intendant laid great stress on the misbehaviour of the officials in Villefranche, alongside that of the *parlementaires* of Toulouse, who had tried to argue that the commissions of the intendants had expired with the king's death. The *juge-mage* had allowed the whole process to start by authorizing an illegal assembly, at which it had been decided to offer only half the *taille* demanded for 1643, and had later maintained relations with

should be supplemented from the documents printed by Mousnier, *Lettres et mémoires,* 493–4 (this letter clearly belongs to 1643, not 1642), 504–5, 576–8, 1112–33.

[73] Mousnier, *Lettres et mémoires,* 1114.

[74] Degarne, 'La révolte du Rouergue', 4.

the rebels. As usual it is hard to detect what degrees of weakness, constraint, or deliberate trouble-making were involved; the *taille réelle* certainly gave the privileged groups a common interest with the people, and the officials had grievances of their own over recent manipulations and exactions by the crown. On the other hand the named leaders were relatively humble inhabitants of Villefranche, the mason and *cabaretier* Lapaille, the master saddler Lafourque, and the surgeon Petit; others known to have been implicated were artisans, valets, and *laboureurs,* although the intendant alleged that the rural *croquants* had been led by several 'petits cadets de gentilshommes ou se disants dels'.[75] It is also interesting to note other accusations made by the intendant against the local élites, to the effect that they had been manipulating the tax system in two important respects. Firstly, they had been getting the *cour des aides* to declare their own property noble land, which was exempt, on the basis of fraudulent titles. Secondly, Villefranche itself claimed to be outside the jurisdiction of the *élus,* and the town authorities had deliberately assessed the *taille* at a higher level than was required, in order that they and those they protected should be able to avoid paying at all. This latter charge, very plausible in itself, is completely at variance with the events of February 1643, when the *juge-mage* was said to have collaborated in the halving of the tax levy. There was also the suggestion that personal rivalries had been important, with leading members of the *présidial* court who had interests in the financial system, and were family enemies of the *juge-mage* and others, as targets of the original disorders.

The total effect of revolts before the Fronde was very serious indeed, not because they threatened the government as such, but through their disruption of the tax system, combined with the constraints they placed on any new taxes. Although in a sense the big revolts could be defeated militarily, they left a legacy of resistance which was much harder to deal with. One of the clearest examples is that of Pierre Greletty, a member of a substantial peasant family who had been involved in the Périgord revolt of 1637. The following year he emerged as the leader of a group of bandits in the *forêt de Vergt* outside

[75] Degarne, 'La révolte du Rouergue', 11.

Périgueux, original centre of the 1637 revolt, evading various attempts to capture him, and retaining popular support because his prime victims were the *gabeleurs*. He was eventually removed from the region only at the end of 1641, by a deal under which he became a captain in the royal army and took a hundred of his men with him.[76] There were numerous other regions which became centres of resistance, while on an even wider scale simple non-payment of taxes built up arrears virtually everywhere. The situation was in many ways even worse with the indirect taxes, since the farmers were expected to enforce these themselves, with whatever help they could obtain from the local authorities and the courts—this was why it was necessary to register the edicts establishing new taxes before anyone would bid for them. The immediate damage to the government was therefore slight when a particular excise duty proved unenforceable, since the loss fell on the financiers, but the long-term result was to render such contracts saleable only at drastically reduced prices. Major attempts at innovation, notably the *sol pour livre* sales tax and the *taxe des aisés,* made in 1639–40, only to be abandoned in the face of widespread resistance.[77] The ultimate response to these problems was to jack up still higher the *taille* and associated levies for troops, hand over the assessment to the direct control of the intendants, and equip them with special brigades of *fusiliers* to collect the money by force. Expensive and only partially effective in straight fiscal terms, this strategy also carried a very heavy political cost, which became fully apparent in the crisis of the Fronde.

We may note for a start that the crisis of 1648 has one striking resemblance to that of 1630, in that the royal officials themselves were alienated by a further attempt to blackmail them over the renewal of the *paulette.* To this must of course be added such factors as the dispossession of the financial officials, the *trésoriers* and *élus,* in favour of the intendants, and the creation of new offices, notably such examples of misplaced ingenuity as the *parlement semestre* at Aix. This

[76] Greletty's career is described by Bercé, *Croquants,* 456–8.

[77] For the failure of these taxes, see R. J. Bonney, *Political Change in France under Richelieu and Mazarin* (Oxford, 1978), 327–30, and *The King's Debts* (Oxford, 1981), 167, 183–5.

time, however, there were prominent scapegoats available in the form of the intendants, unmistakably identified with the suppression of local rights and the use of fiscal terrorism in the service of the financiers. Their removal soon became inevitable, with disastrous consequences for the government's revenues; all the old hopes of an end to extraordinary taxation were revived, so that the summer of 1648 brought a virtual tax strike over much of the country. Now that the collection of the *taille* was back with the ordinary officials, with the *fusiliers* disbanded and often in hiding, all the techniques of evasion and resistance developed over the previous years could be deployed with impunity. The reduction of 20 per cent granted by the crown in October fell far short of popular expectations, above all because it failed to mention the arrears of 1647, although those for 1646 and earlier had already been written off by the declaration of 17 July. In that month the intendant of Guyenne, Lauson, had warned the council that

les bruitz qui viennent de Paris et qui continüent portent tel prejudice au service du Roy que je ne sçay si on osera demander au peuple la taille ny de 47 ny de 48 tant il est prevenu d'une descharge generale comme si un nouveau Perou s'estoit respandu dans l'Espargne au moment que les Assemblées des Compaignies de Paris ont faict esclater leur zèle pour le soulagement du peuple.

[the rumours which come from Paris and are continuing cause such damage to the king's service that I do not know if we will dare demand the *taille* for either '47 or '48 from the people, since they are so persuaded of a general discharge, as if a new Peru had flowed into the Exchequer at the moment the assemblies of the companies of Paris burst out with their zeal for the relief of the people.][78]

Meanwhile the *bureaux* of the *commis* who levied the indirect taxes had been attacked in many places, leading to a general flight of their collectors, so that salt, wine, and other commodities were traded free of tax across much of the country.[79] While some sort of tax collection seems to have resumed in 1649, it was plainly on a patchy and inadequate scale, and receipts did not get back to their former levels until the mid-1650s. It should be added that fiscal terrorism did not

[78] Mousnier, *Lettres et mémoires*, 841.

[79] Bercé, *Croquants*, 463–516, gives a vivid account of the period in the south-west.

cease entirely; there was a notable example at the beginning of 1650 when Foullé, who had already distinguished himself as a 'terrorist' intendant, was sent with a commission from the council to deal with extensive resistance in Bas-Limousin. A particular centre of violence against tax collectors was the *bourg* of Saint-Bonnet-Elvert, on the upper Dordogne, perched on top of a steep hill. Foullé, who was using the troops quartered in the province for the winter, took the *bourg* by assault in March, massacred all the inhabitants who had not fled, and declared these others guilty of *lèse-majesté* and to be hanged by the roadside if they were found.[80] This campaign in Limousin seems to have worked in its fashion, but it was hardly conceivable that these methods could ever have been employed on a wide enough scale to be effective nationally.

THE POLITICAL AND SOCIAL DETERMINANTS OF REVOLT

The end of the Fronde saw the re-establishment of the system of coercion by intendants and *fusiliers,* to the accompaniment of widespread local resistance. There were few major revolts in this period, however, the most significant being the *guerre des Sabotiers* in the Sologne in 1658. This was a typical enough peasant movement, with the usual suggestions of marginal involvement by the local nobility; perhaps its chief interest lies in the treatment of the intendants at Orléans and Bourges, Sevin and Gallard, who were recalled to Paris in disgrace for having failed to control it.[81] Their fate emphasizes the delicacy of the intendants' position in the face of popular revolt, something which should always be kept in mind when reading their own accounts. Our prime source for the history of the revolts lies in the reports sent to the government by its local representatives; governors, royal officials, bishops, and intendants. None of the authors could be truly impartial, for they had too many motives for fitting the evidence into various preconceived schema. As leading opponents of the revolts, they commonly had only a very limited knowledge of their motives and organization. In general the members of the local

[80] Ibid., 474–5.

[81] L. Jarry, *La Guerre des sabotiers de Sologne et les assemblées de la noblesse, 1653–1660* (Orléans, 1880). For the disgrace of the intendants see Bonney, *Political Change*, p. 233.

élites showed a tendency to minimize the importance of specific revolts, advocating concessions to popular demands, and trying to exculpate their colleagues. On the other hand, the intendants blamed the officials, the municipalities, even the governors themselves, as being guilty of anything from feeble indulgence to direct instigation. Above all they accused the rural nobility of habitual complicity in those disorders which prevented the collection of taxation. There is no doubt that many such accusations were well founded, yet it remains probable that the intendants were a little quick to take refuge in stereotyped explanations. There are occasional striking conflicts of evidence. In the 1640s the two intendants in Languedoc, Bosquet and Baltazar, were at loggerheads over their respective precedence and the right way to treat the province; their reports on the serious revolt in Montpellier in 1645 contradict one another directly over the blame or credit due to the *cour des aides* and the *juge-mage*.[82] Baltazar again, in long dispatches to Paris, placed great emphasis on alleged Huguenot conspiracies as the hidden cause of the troubles in Languedoc; their general implausibility apart, the ostentatious loyalism of the Huguenots during the Fronde gives the lie to such claims.[83] It is easy to imagine the antagonisms which would arise between such Parisian interlopers and the local officials or *hobereaux*, but there were other practical considerations which might influence their presentation of the facts. The intendants were under constant pressure from a government which never wished to recognize its own responsibility for the dissension provoked by its policies. Richelieu himself gave an extraordinary example of such double thinking when he waxed indignant at the policies of his *surintendants des finances*, who should, he thought, have taken more account of the likely results of their decisions. The government in Paris was always looking for a scapegoat, so the wise intendant provided a candidate other than himself. In line with his natural suspicions of an alien provincial world, he tended to

[82] This episode is discussed by W. H. Beik, 'Two intendants face a popular revolt: social unrest and the structures of absolutism in 1645', *Canadian Journal of History*, 9 (1974), 243–62.

[83] For some examples, see A. Lublinskaya (ed.), *Lettres et mémoires adressés au chancelier P. Séguier* (Moscow–Leningrad, 1966), 95–103, 184–6, 189–90.

exaggerate the complicities of the *personnes de condition,* with the added bonus that these would conveniently explain his inability to master a popular revolt. Just as there was a popular mythology of the king misled and cheated, so there was an official mythology of an obedient people led astray by malicious and self-seeking members of the élites.

Beyond the possible biases in the evidence, there remains the classic problem of the relationship between individual examples and general evaluations. In citing instances taken over a period of twenty or thirty years, or of a few individuals in a large province, one can easily give a solid appearance to highly fragile intellectual constructs. This has clearly tended to happen in the debates over the relative importance of the vertical and horizontal articulations of society, where there has also been a tendency to set up a facile opposition which underestimates the great complexity of human societies. At the cost of stating a necessary truism it must be emphasized that all more complex societies function through numerous fine balances between what are usually classified as horizontal and vertical solidarities. In addition these forces are variable according to the occasion, with alliances and oppositions being temporary and highly unstable. This can well be seen as a social mechanism which prevents the formation of deep and regular divisions which could threaten society's capacity for common action, and might lead to civil war or revolution. Such consequences are liable to follow if the society is subjected to intense pressures which are perceived as originating from outside its traditional categories. A brutal change in governmental style and demands, of the kind realized by the French monarchy under the Cardinal-Ministers, would naturally provoke a temporary coalition among groups which all regarded themselves as threatened. The logic of this situation would be to make vertical solidarities seem stronger, so that they might mask the horizontal divisions which would nevertheless persist, and eventually tend to reassert themselves. A more detailed examination of the different types of revolt does seem to reveal such a pattern.

Urban seditions were highly conditioned by the military organization of towns. Protected by walls, hundreds of larger and middle-sized communities also possessed a militia. These bourgeois guards, despite some dilution by the participation of the servants from bourgeois households, represented a very effective means of keeping order. Together with the moral authority of the elected municipal governors, the *échevins* or consuls, and of the companies of judicial officials, the firearms of the guards were in principle capable of crushing any plebeian revolt. One must therefore conclude that most urban revolts were at least tolerated by the élites, and may well have been encouraged by them. The arrival of an emissary charged with imposing some new tax or charge would readily provoke a riot, which commonly took the carnivalesque form of *la conduite du gabeleur*. The magistrates would respond by calling out the guard, with nicely calculated slowness, to see that the mob did not get out of control; then they would seek a deal with the tax farmers, which might reduce the level of the imposition, and be accompanied by *lettres d'abolition* from the crown which guaranteed them against any repression or action for damages. This manœuvre was widely and successfully executed, calling for nothing more than a few words in the right place, followed by shrewd timing of the appropriate delay in mobilizing the guard. Its efficacy ultimately depended on two conditions; a willingness to negotiate on the part of a government which lacked any real repressive power, and confidence that the militia could control a riot if it threatened to turn against the rich. One has the impression of watching a series of rituals, well understood by all contemporaries, but allowing face to be saved all round. Sometimes things went wrong, either because excessively violent treatment of the victim brought strong reaction from the government (as at Moulins), or because the mob refused to disperse and pursued popular vengeance against a group of citizens too deeply implicated in the tax system (as at Agen and Rouen). In such cases the disturbances became prolonged, and once they lasted more than a few hours the danger of real fighting between the bourgeois guards and the rebels increased sharply. The unity of the towns in the face of the government was very relative; a protracted revolt tended to become progressively more divisive, more hostile to the rich and powerful. The urban élites found themselves simultaneously harshly criticized by the government, which threatened to remove their privileges, and denounced them for their cowardice in not keeping the people under control. The ministers knew very well that a town was capable of suppressing a revolt, and that it must be compelled to do so. The continued existence of the medival walls was important here, not least because many of the poor lived in the *faubourgs*, and could therefore be excluded from the town, as happened at Avranches and Moulins. Although one must recognize the equivocal behaviour of the élites, it was very rare for their members to be directly implicated in violence. In the cases of persistent revolt it is usually possible to identify genuine popular leaders, either artisans or marginal members of the professions, who admitted allegiance to no superiors; the authorities showed a keen awareness of their power in seeking either to negotiate with them or to eliminate them. These men (and occasionally women) seem to have owed their position to a strong, often aggressive, personality, rather than to any special social position; they commonly functioned in and from *cabarets* and other centres of popular urban sociability.

The urban revolts seriously reduced the fiscal possibilities open to the government. It was therefore necessary to sell yet more offices, suppress the payment of *gages* to existing officials, and so on; it was hardly surprising that the officials sulked, sympathized with the rebels, and even supported or instigated their actions. The ministers also succumbed to the temptation to raise the demands for the *taille* beyond any level at which there was a realistic hope of collecting it, ensuring continued resistance from the countryside. Here too it was less the spectacular major revolts which threatened the government—they could always be suppressed by military power—than the kind of guerrilla warfare which became endemic over vast regions. The intendants and their *fusiliers de la taille* were often swamped by the mounting tide of arrears, drawn into innumerable petty conflicts which also provide interesting evidence on communal solidarities. Violence was normally sparked off by the arrival of an outsider, seeking to enforce the payment of the overdue taxes. He might be carrying an *arrêt* empowering him to seize goods, to impose the *contrainte solidaire*, or to imprison the village collector. Greeted by a volley of stones or bullets, he would try to flee, but if unlucky might be wounded or killed. His mission was enough to unite the community against him; the various constraints threatened the village élites, often tenants of the seigneur, and by extension the seigneur himself. The seizure of flocks and herds, the most mobile and accessible form of wealth, clashed with a complex network of interests. One of the provinces most subject to these problems was Normandy, where even the exemplary punishment meted out to the *Nu-pieds* seems to have had no lasting effect. A long series of letters from the intendant Du Boulay-Fabvier complained of his difficulties in this area; in July 1643 he wrote from Conches of 'ceste misérable eslection pour laquelle réduire je crois qu'il faudrait une armée'.[84] In February 1645 he reported:

J'ay apris qu'il s'estoit faict une assemblée de soixante ou quatre vingt gentilhommes dans St Pierre sur Dive, eslection de falaise . . . sur des plaintes qu'ilz pretendoient avoir a faire de ce qu'on executoit leurs bestiaulx, de ce qu'on contraignoit solidairement leurs fermiers . . . et de ce qu'on vouloit obliger les fermiers de paier la taille auparavant qu'ils eussent acquité tous leurs fermages . . . Ils est vray qu'il ni a pas une personne de grande condition mais Ilz y en ont voulu enbarquer qui les ont rebutez et ce qui me faut encore voir leur procedé plus extraordinaire c'est qu'en ma conscience Ilz n'ont aucune subiect de plainte au contraire Ilz agissent comme s'ilz estoient maistres de tout et disposent de la taille, de l'imposition et de la collecte comme Il Leur plaist, bastent, frapent, excèdent, outragent et violentent tous les sergens et les collecteurs s'ilz ne font ce qu'ils veulent, et au fonds pas un d'iceux ne va a la guerre et ne font aucun mestier que d'user de violence envers tout le monde que si je veux proceder extraordinairement contre eux je ne puix avoir un tesmoing . . .

[I have learned that there has been an assembly of sixty or eighty gentlemen in Saint-Pierre-sur-Dive, election de Falaise . . . on account of the complaints they pretended to be due over the seizure of their animals and the imposition of the *contrainte solidaire* on their tenants . . . and that it was intended to make their tenants pay the *taille* before they had paid all their rent . . . It is true that there is no person of high rank among them, but they have tried to recruit some who have refused them, and what makes me regard their proceedings as still more extraordinary is that on my conscience they have no cause for complaint; on the contrary, they behave as if they were

[84] Mousnier, *Lettres et mémoires*, 533.

masters of all and could dispose of the *taille,* the assessment, and the collection at their pleasure. They beat, strike, outrage, and attack all the *sergents* and collectors if they do not do as they wish, and at bottom not one of them serves in the armies, nor has any trade but to employ violence against everyone; if I want to take extraordinary legal action against them I cannot find a single witness . . .][85]

These petty nobles were taking the characteristic line that they had justified grievances, represented chiefly as infractions of their privileges. The same tone is found in the 1649 *cahier* of the Angoumois nobility, which does indeed harp on the abuses of the tax system, but almost exclusively in terms of the particular problems of the nobles. There was a good deal of activity of this sort, with nobles assembling to discuss the defence of their common interests, and in 1658–9 such assemblies took on a decidedly seditious character in Normandy and parts of western France.[86] It is a striking fact, however, that these movements remained entirely self-contained; their leaders apparently had no thought of seeking assistance from any other social group, and there is no record of any real connection with popular revolt.

The royal agents considered that the local nobles, like the town élites, were under an obligation to maintain order. This is very apparent during the repression of the Nu-pieds; in the diary kept for Séguier it is noted that:

La noblesse de la campagne a souffert qu'on aye battu tous les jours le tocsain dans leurs paroisses pour l'assemblée des rebelles, et que leurs tenanciers se soient souslevez pour cest effect, sans y avoir apporté remède quelconque.

[The rural nobility have allowed the tocsin to be rung every day in their parishes to summon the assembly of the rebels, and their tenants to rise in that cause, without doing anything to stop this.]

In the general ordinance of 8 March 1640 emphasis was laid on the need:

d'obliger les gentilshommes chacun en l'estendue de leurs terres de contenir nos subjects dans l'obéissance et les empescher de faire aucune assemblée contre Notre service, ce qui leur ait aisé veu le

85 Ibid., 719.

86 This agitation is very well described by J.-D. Lassaigne, *Les Assemblées de la noblesse de France aux 17ᵉ et 18ᵉ siècles* (Paris, 1965), 85–127.

pouvoir qu'ils prennent ordinairement sur leurs tenanciers auxquels ils font bien exécuter leurs volontés lorsqu'il s'agit de leur intérêt particulier.

[to oblige the gentlemen, each on his own estate, to maintain our subjects in obedience and prevent them holding any assembly contrary to our service, which should be easy for them in view of the power they ordinarily hold over their tenants, whom they very readily oblige to carry out their will when it is a matter of their private interest.][87]

These examples, and similar instances from other provinces, emphasize the different situation which arose in the countryside, where there was no repressive force comparable with the town militias. The *maréchaussée* was even less formidable in the seventeenth century than it was to be in the eighteenth; even had they been many times more efficient, a mere 3,000 men scattered over the whole country could hardly have achieved much. Such force as there was lay with the nobles; in some regions and contexts it was very effective, as one sees in the state of affairs reported from the Auvergne by d'Argenson in 1633, before the great crises. Writing of various excesses committed by the young chevalier Destaing d'Argenson complained that

il est difficile de faire chastiement de ces personnes la, qui sont en bon nombre dans le hault Auvergne, commettans force concussions, exactions et violences sur le peuple, mesme contre les ecclesiastiques des dixmes desquels ilz jouissent par force meprisans tous les juges de la province, si le roy n'y met la main par une chambre des grands jours, ou en nous donnant, comme nous avions ces annees dernieres, quelque autre force que les prevostz des mareschaux, qui tous par foiblesse, crainte ou malice ne font rien contre les puissans des provinces ou ilz sont establis.

[it is difficult to punish such persons, who are numerous in the upper Auvergne, and commit many assaults, exactions, and violences on the people, even against ecclesiastics, whose tithes they seize by force in defiance of all the judges of the province, unless the king takes them in hand through a session of the *grands jours,* or by giving us, as we had some years ago, some armed force other than the *prévôts des maréchaux,* who all by feebleness cowardice, or malice do nothing against the powerful in those provinces where they are established.][88]

[87] Foisil, *Nu-pieds,* 212.

[88] Lublinskaya (ed.), *Lettres et mémoires,* 147.

In most cases the nobles could call on a group of servants, sometimes even bands of gipsies, and then a more extensive network of relatives and friends who would back them up. Frequent acts of violence resulted; it was only during the personal rule of Louis XIV that the crown finally succeeded in taming this rural anarchy, notably in central and western France. It was therefore predictable that a village revolt should often involve the local seigneur—but one must remember that at least a third of all communities had no resident seigneur, and that many others were basically town dwellers who spent little time on their rural properties. The most banal and common situation, the arrival of a *sergent,* did in fact pose a very difficult problem for a noble who happened to be present. According to the government his duty was to support this intruder and protect him in carrying out his duties, an extravagant piece of altruism which must have been exceedingly rare. For a start, the seigneur would almost certainly damage his own interests by impoverishing his tenants or neighbours in obedience to a judgement issued by those *messieurs de la ville* whom he heartily detested and despised. Even had he wished to do so, it is not clear that he had the power; it is likely that all but his closest servants would have evaded, even refused, such an order. He would also have risked inflicting irreparable damage on those links of protection and influence which secured his local predominance. The prudent course of action was clearly to stay at home and turn a blind eye, yet this was really to give tacit consent to the resistance. The seigneur was thus placed in an almost impossible position; that his loyalties were largely on the side of the village seems very likely, and is borne out by the numerous cases where seigneurs took a leading part in the violence offered to intruders. This did not mean that, at least in his own eyes, he was disloyal to the crown, for it was always possible to claim irregularities, abuses, or corruption on the part of the tax collectors in self-exculpation. Faced with an organized revolt involving numerous parishes and the establishment of a peasant army, most nobles seem to have backed away, so that it was often possible for the provincial governor to mobilize them to help in the repression. In such circumstances they came under pressure from both sides, since there could also be threats from the rebels of the kind found in the 1636 manifesto of the Angoumois peasants: 'Enjoignons à chacune paroisse de faire marcher tous les gentilzhommes et donner les armes, à peine de brusler leurs maisons et de n'estre payez de leurs rentes et agriers' ['We call on each parish to make all the gentlemen march and provide arms, on pain of having their houses burned and their rents and dues withheld'].[89]

The need for such menaces hardly suggests that the nobles were rushing to take the lead in this movement, whose considerable element of anti-seigneurial feeling has already been mentioned. While Porchnev's usage of the 'front des classes' as an explanation for revolt is certainly unconvincing, the government did implicitly invoke an analogous concept in its attempts to force the notables, urban and rural, to rally to the defence of the established order. If vertical solidarities were crucial in starting many revolts, horizontal divisions generally played the vital role in controlling and ending them.

THE TAMING OF REVOLT: LOUIS XIV AND COLBERT

In principle the government's ultimate answer to the rural guerrilla war was a military one, built around the intendants. The special brigades of *fusiliers de la taille* were created in the late 1630s, paid from the tax revenues; they were usually mounted dragoons, and the difficulty of their task was recognized by paying them at roughly four times the rate allowed for an ordinary soldier. They operated in units of ten to twenty, although in some areas of aggravated revolt, notably in the south-west, it was necessary to assemble hundreds of *fusiliers* for what amounted to small campaigns.[90] The council's *arrêt* of August 1642 effectively transferring control of the tax system to the intendants was a further move towards the direct extraction of taxes by force.[91] These expedients were successful up to a point in that they prevented a complete collapse of the royal finances, but the correspondence of the intendants alone makes it quite clear that it was a case of damage limitation, and that the basic problems were no nearer solution. Not until the 1660s did the government achieve some kind of mastery over the situation, despite the

[89] Bercé, *Croquants*, 738. [90] For the *fusiliers*, see ibid., 108–12.

[91] This decision is discussed by Bonney, *Political Change*, 183–90.

maintenance of what was virtually a wartime level of taxation. How was this achieved? It was certainly true that the occasional rebellions of these years were punished with greater severity, as with the Lustucru of the Boulonnais in 1662, nearly 400 of whom were sent to the galleys, or the followers of Audijos in the Pyrenees, remorselessly hunted down for hanging or the galleys by the intendant Pellot.[92] The deterrent effect of such measures should not be underestimated, but it cannot really explain the great reduction in small-scale resistance to taxes. Two factors were probably crucial here. Firstly, with the end of the war the problems associated with winter quarters for the armies were enormously reduced; disorders apart, the *subsistence* tax had at one stage represented a third of the total burden on the peasantry in some areas.[93] Secondly, Colbert's everyday management of the taxes was intelligently conceived precisely to reduce the difficulties experienced earlier. Much effort went into collecting existing taxes as equitably as possible, into preventing the accumulation of arrears (those of the final years of the war being formally cancelled), and into making the most unpopular forms of coercion a last resort.[94] Although the revenue from indirect taxes was greatly increased this was achieved without creating new levies. The significance of this policy was emphasized when financial pressure was increased after the renewal of war in 1672, with the new duties of 1675, the *marque d'étain* and the *papier timbré*, setting off the last major rebellions of the *ancien régime*. Once more riots in Bordeaux were followed by related outbreaks elsewhere in the smaller towns of Guyenne; in Brittany the *Torrében*, protesting against the same taxes, again developed into the kind of anti-seigneurial movement the province had known in the 1590s.[95] The events

[92] There is a good summary of the Boulonnais revolt in Tilly, *The Contentious French*, 148–9. For Audijos, Bercé, *Croquants*, 599–600, and J. H. M. Salmon, *Renaissance and Revolt* (Cambridge, 1987), 267–92.

[93] Bercé, *Croquants*, 46–7. Parishes which paid the tax, and then had troops billeted on them regardless, might of course suffer even more.

[94] The best study of this process is still that by E. Esmonin, *La Taille en Normandie au temps de Colbert, 1661–1683* (Paris, 1913).

[95] For Guyenne, Bercé, *Croquants*, 517–23, For Brittany, Y. Garlan and C. Nières, *Les Révoltes bretonnes de 1675: papier timbré et bonnets rouges* (Paris, 1975).

of this year bear out Tilly's point that there was a traditional repertoire of rebellious gestures, with its provincial singularities, lurking just under the surface ready to respond to any serious provocation. As before, these larger revolts proved no more than a momentary embarrassment to the government, which took the chance to destroy some tiresome local privileges as part of the repression. It was the virtual disappearance of the everyday attacks on tax collectors which was crucial for the tax system as a whole.

A satisfactory explanation of this change must go further than the better management of the taxes, to analyse the ways in which this may have affected the local communities and their behaviour. Some of the main themes running through earlier protests are relevant here. Much emphasis was placed on the arbitrary and unequal assessment as between communities, which the intendants blamed on corruption and manipulation by the financial officials. When it came to the assessment of individuals, there was again general agreement that the system was heavily loaded against the poor, with the richer and more powerful men using their position to evade paying their share. In 1637 Simon Estancheau developed the criticisms made the previous year of 'les Riches . . . qui ont achevé de ruyner le peuple' at much greater length, with such comments as:

les officiers des terres et les hommes riches, puissans et d'othorité dans les paroisses, et qui tiennent la plus grand part des habitans soubz leurs pates et en leurs liens par les moyens de plusieurs sommes qu'ils leur doivent, rejetteront tousjours leurs tailles sur le pauvre peuple.

Et qui pis est, pour surcroit d'oppression, c'est que depuis quelques années les gentilshommes dans les paroisses se meslent et acistent à l'esgallement des tailles et empeschent la liberté des asseyeurs et qui ne baillent à leurs mestayers et autres gens qui se jettent sous leurs ailes, la taille que leur faculté pourroit porter, et par inthimidation et menaces empeschent que lesditz mestayers et autres gens qui leur font des présents n'entrent en charge de colletage dans les paroisses, ce qui apporte une grande foulle au peuple . . .

[the officials who manage the land and the rich, the powerful, and the holders of authority in the parishes, those who have the majority of the people under their thumb and in their dependency by reason of various sums they owe them, will always push off their own taxes on the poor people.

And what is worse is that as an additional form of oppression, for several years past the gentlemen in the parishes have meddled and been present at the assessment of the *taille,* preventing the assessors from acting freely, so that they do not impose on their sharecroppers and other people who have thrown themselves under their protection the *taille* their means would justify, while by intimidation and other means they prevent the said sharecroppers and others who make presents to them from having to serve as collectors in the parishes, which causes great damage to the people . . .][96]

The original response to this problem was a typically authoritarian and *ad hoc* one, with the intendants taking charge of the *assiette,* and imposing arbitrary *taxes d'office* on those whom they thought undertaxed in such ways. While these techniques were never entirely abandoned, from the 1660s the regular tax officials were brought back into the system, under much closer supervision; it appears that although such abuses naturally continued, their scale was greatly reduced. The effect was to transfer a significant proportion of the tax burden to the richer peasants; this may well eventually have damaged French agricultural productivity, but it meant that there was a greatly enhanced prospect of villages paying their assessment. There had been a marked trend over the previous century, particularly in northern and eastern France, for oligarchies of 'rural bourgeois' and *laboureurs* to establish a dominant position in the local community, both in its general capacity and in its specialized sense as a formal political organization.[97] This contrasted with the situation in the rest of the country, where relatively few peasants built up capital on such a scale, larger holdings were usually worked by sharecroppers, and the community retained more egalitarian and democratic features. These regional variations, which may also be associated with regions of concentrated or dispersed habitation, may well help to

[96] Bercé, *Croquants,* 747.

[97] The best general introduction to this subject is probably J.-P. Gutton, *La Sociabilité villageoise dans l'ancienne France* (Paris, 1979), which includes a helpful discussion of tax assessment and collection.

account for the very uneven incidence of revolt; where the *laboureurs* dominated, they were generally able to protect their own interests fairly satisfactorily until the 1660s, and therefore had strong motives to avoid the high-risk tactic of revolt. Where such local hierarchies were weaker, it was easier for men with less to lose to emerge as leaders, in communities which were inherently more volatile. During Colbert's ministry the position of the *laboureurs* was doubly affected, by more stringent tax assessment and by a sequence of good harvests which resulted in low grain prices. Since these trends favoured the mass of poor peasants, there was little prospect of uniting communities in revolt behind an unpopular minority of the rich, whose difficulties were in any case only relative.[98]

The crown's policies towards the upper levels of society must also be taken into account. Few at the time or since have doubted that Louis XIV was determined to reduce the higher nobility to a politically marginal role, above all by drawing them to court and depriving them of any effective power in the localities. The later seventeenth century therefore saw a sharp decline in the clientage systems which had previously offered a considerable degree of protection to those who defied the royal officials. Misbehaviour by the local nobility was also attacked directly, the most notorious area in central France being subjected to the *Grands Jours de l'Auvergne* (which extended over a considerably wider area than that province).[99] The *recherches de noblesse,* with the subsequent regulation of noble status by the intendants, provided an important mechanism by which lesser nobles could be disciplined.[100] As a more positive feature the professionalization of the army and navy offered poor nobles the chance of military careers, tying them much more effectively into the fabric of the absolutist state. A similar mixture of toughness and reassurance was shown towards the officials. Those who stepped too far out of line now risked permanent deprivation, but for the great majority the Louis-quatorzien régime meant confirmation of their status and

[98] For the situation of the *laboureurs* during this period, see P. Goubert, *Beauvais et le Beauvaisis de 1600 à 1730* (Paris, 1960), 151–89; J. Jacquart, *La Crise rurale en Île-de-France, 1550–1670* (Paris, 1974), 632–6, 707–15, 746–7.

[99] See Lebigre, *Les Grands Jours d'Auvergne.*

[100] The best study of the *recherches de noblesse* is by J. Meyer in *La Noblesse bretonne au 18ᵉ siècle* (2 vols., Paris, 1966).

privileges, albeit at a level some way below their rather unreal aspirations of earlier decades. In the case of Languedoc Professor Beik has demonstrated most convincingly how the later seventeenth century saw the creation of a much more stable partnership between élites and crown, in which the former's increased docility was rewarded with a greater share in the profits of the tax system.[101] While in institutional terms a major *pays d'état* must be regarded as untypical, such alliances between royal authority and local privilege were certainly crucial to the smooth functioning of the *ancien régime* as a whole. Another very important group to be brought under better control was the clergy; the *curés* who had been prominent leaders of earlier revolts were now being subjected to seminary training and to much tighter episcopal discipline.[102] These processes affecting nobles, officials, and clergy were all gradual, and certainly far from radical, yet they were enough to tip what has always been a very fine balance of loyalties decisively towards obedience to the crown and its agents. Among these last the intendants were crucial, as they became accepted in the role of brokers between the government and the provinces. One of their most important achievements, under Colbert's direction, was virtually to eliminate the enormous burden of communal debt which had built up before 1660. Towns and villages had borrowed recklessly to pay the taxes and forced loans, so that beneath the direct royal debt there lay this second concealed layer of liabilities. The urban debts, in particular, had commonly been set off against local excise taxes, designed wherever possible to fall on outsiders or the poor; this hidden financial structure was an important factor in exacerbating the hostility between towns and countryside which so often appears in the revolts.[103]

CONCLUSION: VERTICAL AND HORIZONTAL SOLIDARITIES

The interaction between vertical and horizontal solidarities is

[101] Beik, *Absolutism and Society*, esp. 244–339.

[102] E. Le Roy Ladurie, 'Révoltes et contestations rurales en France de 1675 à 1788', *Annales: économies, sociétés, civilisations*, 29 (1975), 6–22, stresses this point, although I do not find his link with the Revocation of the Edict of Nantes convincing.

[103] For an urban example, see P. Deyon, *Amiens, capitale provinciale* (Paris, 1967), 461–3; for rural debts, Gutton, *La sociabilité villageoise*, 111–14.

indeed a central part of any general interpretation of these revolts. To make it work properly, however, it is vital not to conceive these solidarities in a static form, expecting to demonstrate that one or the other was dominant in seventeenth-century France. Historians who follow this course end up describing real phenomena in a strangely partial way, which leaves out so much that it distorts reality severely. Much of the difficulty disappears if we employ a dynamic model of society, in which both kinds of solidarity operate with variable force according to circumstance. Even within a revolt lasting only a few days it is often possible to observe quite sharp shifts of position by the various groups involved. Furthermore, both Pillorget and Bercé have rightly emphasized a different kind of solidarity, that of the community or corps. Here again we need to grasp the double-sided quality of such human organizations. A body such as a *parlement* could feel a common interest against royal intervention; it was also the theatre for bitter personal jealousies and animosities, such as divided the *parlement* of Aix and were crucial to the revolt of the *Cascaveoux*. Similarly, the local community did cohere against outsiders, yet it was the scene of intense exploitation of the poor by their richer neighbours. In rural revolts the leaders were rarely the rich peasants or the elected heads of the community; where they can be identified they seem characteristically to possess a marginal status, men with a degree of authority and education, such as the *curés,* the local lawyers, and doctors. Leaders themselves could be very ambivalent, like La Mothe la Forêt, whose claim that he had joined the revolt under duress is perfectly compatible with his later vigour as its leader, and his final decision to disband his forces. Men like this were struggling between conflicting loyalties, just as were some of the English nobles involved in the Pilgrimage of Grace. In urban revolts the thesis of complicity between different social groups is highly plausible, but I have tried to suggest reasons why it was possible for the urban oligarchies to play with fire in this fashion, and there are enough cases where the revolt acquired serious social overtones to demonstrate that such alliances were temporary and highly fragile.

What the revolts reveal to us, then, is a complex and divided society, placed under intense pressure by the demands of royal fiscality. The effects of this pressure were as much divisive as cohesive, because the natural reaction of those who possessed political power was to divert the tax demands towards their subordinates. The grievances of the 1636 *croquants* are merely the most explicit reaction to such processes, which constantly undermined the unity of resistance to taxation. Sometimes deliberately, but more often unconsciously, the crown drove a series of additional wedges between its potential opponents, through its manipulation of the machinery of privilege and exemption. This operated as between different towns and provinces, and between social groups. The ultimate result was a society full of anomalies and injustices. These, rather than the absolute burden of taxation, were the prime causes of revolt—yet they were also the reason why it could never succeed on a large scale. For all the tremendous bitterness and hostility royal policies aroused, their opponents mistrusted one another as much as they did the government. They also disabled themselves in advance, by using the convenient fictions of the king led astray by evil ministers and robbed by dishonest officials, which left them with no defence against direct assertions of the royal will. Revolt therefore functioned best on the very local level. If the rural parishes revolted, it was rarely across a whole province; when this did happen, the peasant armies attacked the towns. Urban revolts were essentially isolated; even when they sparked off similar outbreaks elsewhere there was no attempt at concerted action. Popular revolt, then, could not stem the crown's drive to increase its revenue. It could certainly hinder and inflect it, and in this more limited sense the rebels did score many successes, which was perhaps all that most participants expected. Despite their bursts of savagery towards tax collectors, the term 'peasant furies' is a misleading description for movements which generally showed considerable awareness both of their own limited objectives and of the complexities of the world in which they took place. The contemporary usage of 'émotions populaires' is preferable, for it reflects the outbursts of 'justified anger' which normally began revolts. Those who made violent assaults on *gabeleurs* did not think they were challenging the established order of society, rather that they were maintaining it. So long as they stuck to these limited targets, they could expect considerable sympathy from local officials and judges, so that the crown regularly found it necessary to grant *évocations* which allowed prominent royal agents and financiers to have their lawsuits tried before courts in other provinces. Rebels themselves often pointed up their own view of their role by mimicking the formal procedures of justice, while the punishments they sought to inflict on those identified as traitors to the community were very like the penalties used by the authorities themselves. To drag the guilty through the streets at the head of a procession and to devastate their property was to subject them to the kind of formal humiliation which characterized much of *ancien régime* justice. Actual killings of *gabeleurs* were relatively rare, partly no doubt because they were quick to flee; the vast majority of fatalities occurred in the course of direct military action instigated by the government.

In political terms revolt functioned through the relationship between largely spontaneous popular violence and the tolerant attitude of disaffected élites. This was the essential reason why it was most successful when it took the form of numerous brief episodes, sufficient to drive away tax collectors without threatening any wider disruption. Such limited co-operation needs to be interpreted with great caution, leaving room for the great differences in viewpoint and behaviour as between the groups involved, and not aggregating evidence of élite complicity as if this were a single and consistent pattern. In general the dynamics of popular revolt were self-limiting, so that none of the larger provincial outbreaks seriously threatened to turn into a French equivalent of the revolt of the Catalans. There was no significant connection with the noble conspiracies against the government, as Montmorency's fiasco of 1632 demonstrates. The rural nobility had their own forms of activity, holding assemblies to co-ordinate the defence of their particular interests, but again these seem to have been quite independent of more popular movements. Virtually every kind of protest and disobedience occurred simultaneously during the Fronde, only to demonstrate once again that they were incapable of coalescing into a real challenge to the government; ultimately they merely led to a chaos which virtually necessitated the return of strong royal control. After 1660 a combination of better management, more effective repression, and favourable circumstances largely broke the pattern of revolt which had established itself over the previous decades. This was a self-reinforcing trend, because once violent resistance became more exceptional, measures to control and punish it (themselves developed over the years) became much easier to deploy. By the end of the century it was possible to introduce major new taxes, the *capitation* and the *dixième,* without any significant reaction. There had been a subtle but crucial change in the structures of privilege on which *ancien régime* society rested, in line with the far greater demands the crown was now making on the country as a whole. A relatively lightly stressed system, with a great deal of local autonomy, in which privileges were granted to provinces, towns, and broad social groups, had given way to a much higher pressure one. This was characterized by stronger central control and the selective use of privilege to benefit small key groups, whose co-operation was the crucial factor in stabilizing local society. The huge increase in taxation which provoked the revolts was also the central feature of this important structural change in French society; once the crown had beaten down the immediate resistance it was able to create a 'spoils system', based on these flows of wealth, which appeared to resolve the problems of linking centre and periphery. As the differences of interest which had always operated to limit the effectiveness of revolt were strengthened and formalized, so revolt ceased to be a reasonable option, and virtually disappeared from the political scene.[104]

[104] Violence against tax collectors did of course continue in the eighteenth century, notably with such brigand-heroes as Cartouche and Mandrin. For the reappearance of agrarian uprisings in one area from the 1760s onward see P. M. Jones, *Politics and Rural Society: The southern Massif Central* c. *1750–1880* (Cambridge, 1985), 170–7.

II

Agencies of Control

5

Church and state from Henri IV to Louis XIV

THE abjuration of Henri IV was one of those acts in which political and religious elements were inextricably mingled. It is understandable, in view of its immense political significance, that relatively little attention should have been paid to the equally profound consequences for the relationship between the crown and the church. When the first Bourbon king finally abandoned protestantism he reasserted his position as *le roi très chrétien* and as *le fils ainé de l'église*. In so doing he effectively recognized that church and state existed in a symbiotic relationship, which had been gravely damaged by the disastrous period of religious and civil war, and which must now be restored. In this respect, as in many others, the conflicts of the later sixteenth century opened the way for changes which greatly strengthened the authority of the crown and its control over the major institutions of French society. The abjuration also symbolized, however, the extent to which such gains were only possible if the crown worked within certain parameters, established by the interests and opinions of its more powerful subjects. Once the French protestant movement had lost its initial impetus, and with it all hope of ousting catholicism as the dominant religion of the country, an effective king simply had to be a catholic. Quite apart from the danger of resistance to a heretical monarch, as exemplified by the Catholic League, the crown needed the church as both a source of patronage and a major prop to royal power. Henri IV understood these basic truths very well; the delay between his accession and his abjuration was essentially a tactical one, which allowed him to time an inevitable move so as to gain maximum benefit. Once he had taken the decisive step, he proved remarkably adroit in manipulating the situation. Powerful forces which had apparently threatened to tear the nation apart were harnessed to support the crown in the work of restoring hierarchy and stability. It was a tragic irony that the king who had served the church so well should die at the hands of a deranged catholic zealot. Ravaillac was a throw-back to a past nearly all French catholics wished to forget, however, and whereas the assassination of the impeccably catholic Henri III in 1589 had brought widespread rejoicing among the Leaguers, the reaction in 1610 was one of revulsion and horror.[1]

French catholics certainly had reason to regard the wars of religion as a disastrous period for the church. Immense physical damage had been inflicted on churches and monasteries, ordinary religious observances had been massively disrupted, and attempts to introduce the Tridentine reforms had made little impact. Catholics as well as protestants had evaded paying their tithes, while there had been substantial alienations of church lands to help the kings pay for their largely ineffective campaigns against the heretics.[2] After forty years of debilitating conflict the Edict of Nantes was a bitter pill to swallow, however grudging its concessions to the Calvinist minority might really be. The most powerful expression of catholic feeling, the League, had turned into a profoundly embarrassing movement, whose social and political radicalism had ultimately driven away all but its most fanatical supporters. It had also had an unfortunate side-effect, in

[1] The valuable book by Roland Mousnier, *The Assassination of Henri IV* (Eng. trans. London, 1973) can mislead the unwary on this point. Mousnier explains with great skill why a catholic extremist might distrust the king, even to the point of convincing himself that to kill him would be a godly work. He does not suggest that such views were at all common; although there had been frequent attempts on the king's life earlier in the reign these had largely died out by the turn of the century.

[2] For the tithe, see E. Le Roy Ladurie, *Les Paysans de Languedoc* (Paris, 1966), esp. 375–89 (and information on land sales, 359–71); J. Goy and E. Le Roy Ladurie, *Les fluctuations du produit de la dîme* (Paris, 1972); E. Le Roy Ladurie on both tithe and land sales in *Histoire économique et sociale de la France* (Paris, 1977), vol. I, t. II, pp. 699–726. For the land sales see also I. Cloulas, 'Les aliénations du temporel ecclésiastique sous Charles IX et Henri III (1563–87): résultats généraux des ventes,' *Revue d'histoire de l'église de France,* 44 (1958), 5–56, and C. Michaud, 'Les aliénations du temporel ecclésiastique dans la seconde moitié du 16e siècle: quelques problèmes de méthode', ibid., 67 (1981), 61–82.

compromising the Jesuits, the most active members of the early catholic reform movement in France. The damaging associations with pro-Spanish and ultramontane attitudes, acquired during these decades, proved very durable; they help to explain why the Society was always a divisive element within the French church. Gallican sentiments, always strong, had been reinforced by the ineptitude of papal diplomacy, so that when the king promised to secure the reception of the Tridentine decrees, as part of the price for his absolution by the pope, he was safe in the knowledge that the *parlement* would block any such move.[3] Henri was perfectly willing to see the church follow the general lines established at Trent, but not to allow any reduction of the rights he enjoyed by virtue of a combination of tradition and the Concordat of 1516.

In some respects the Concordat had itself been an attack on those nebulous Gallican liberties, so easy to invoke and so impossible to define. François Ier had needed papal support for his Italian ambitions, and despite Marignano could not dictate his own terms; instead he struck a crafty bargain which made concessions to the pope, while strengthening royal control in the vital area of ecclesiastical appointments. It needed a major confrontation to force registration by the *parlement* of Paris, the perpetual watchdog of Gallicanism. The immediate significance of the Concordat should not be exaggerated; in practice the kings had been able to disregard the system of capitular elections prescribed by the Pragmatic Sanction of Bourges (1438) when they chose, forcing their own nominees on the chapters. Yet in the long term the crown had gained a great deal by the Concordat. The papacy, which had never accepted the Pragmatic Sanction, was firmly associated with the new arrangements, and had strong motives for upholding them. The royal power of appointment was now unambiguous, not only eliminating a good deal of minor

[3] The question of the king's sincerity over this issue cannot really be resolved; many historians have assumed that he genuinely wished to receive the Council. My own more cynical view is based on the fact that he never applied serious pressure to the *parlements* on this issue, as he did over the Edict of Nantes; he must have known the Council would never be received unless he did this.

friction, but rendered far more secure against any period of temporary weakness on the crown's part. Although the last period of the wars of religion did produce some attempts to by-pass the king in nominating bishops, these proved largely unsuccessful, not least because the papacy preferred to equivocate by delaying the dispatch of bulls. One may also doubt whether the kind of understandings which had been effective in the fifteenth century could have survived the more formal legalism of the sixteenth and seventeenth centuries. Overall it appears to have been a very prudent move to obtain a durable settlement at a moment when the monarchy was exceptionally strong both at home and abroad, and the Bourbon kings had every reason to maintain the position as they inherited it.[4]

Ultimately the best guarantee of the Gallican liberties lay in the ingrained attitudes of French administrators and churchmen, and these seem hardly to have changed down the centuries. Even those Frenchmen who can be described as ultramontanes normally hedged about their claims for papal authority with major qualifications, but academic discussions of such issues, even when they led to voluble debates or paper wars of censures and declarations, are often misleading as guides to practical reality and general opinion. The papal nuncios were rarely in much doubt over their position, feeling themselves to be in a hostile land, where the most informal discussions with French bishops needed royal permission, and where the king's ministers generally treated them with studied indifference. The need to register papal bulls and other enactments with the *parlement* might occasionally prove a nuisance to the crown, but this modest price was well worth paying for the immense countervailing benefit of a filter to keep out any unwelcome Roman initiative. Although in theory such international orders as the Jesuits might have turned the Gallican defences, it is very hard to find serious evidence in support of their innumerable critics. Effective power lay with the provincials in France, whose own self-interest was generally enough

[4] The most helpful modern discussions of the Concordat are those by R. J. Knecht in 'The Concordat of 1516: a re-assessment', *University of Birmingham Historical Journal*, 9 (1963), 16–32, and in *Francis I* (Cambridge, 1982), 51–65.

to produce a sturdy independence of central direction, reinforced by an awareness of how exposed their position was. As the seventeenth century wore on, the French Jesuits would find it not only convenient but essential to deny that they approved of everything written or done by other members of the Society.[5]

This whole complex structure of virtual caesaropapism had been thrown into temporary disarray during the wars of religion, when ideology briefly threatened to overthrow the nexus of convention and self-interest which made it so effective in quieter times. The resulting tensions could hardly be dispelled immediately, and Henri IV was in a particularly difficult position over the extension of toleration to the protestant minority. The Edict of Nantes had to be registered by the *parlements,* and to a large degree enforced by them. The king was bound to be suspect both to the protestants he had deserted and to the catholics he had joined; the necessary interventions by the royal council and other agencies of government, to ensure the pacification of local disputes, were liable to arouse fresh resentments. To minimize such difficulties Henri seems to have aimed at an even-handed policy, under which a concession made to one side would be matched by a reciprocal favour to the other. Critics could then be met with an impressive list of royal decisions given in their interest, apparently demonstrating the king's sympathy for their cause. There was a difference, of course; Henri made it plain that he favoured the peaceful conversion of the protestants, so when he intervened on their behalf it was to redress precise grievances over the application of the edict. He was active in sponsoring missionary activity for reconversion, and above all in pressurizing leading nobles to follow his own example. In at least one case, that of Sully, this seems an ill-judged zeal, for

[5] A startling expression of this, at an early date, came when the Jesuits were questioned by the *parlement* over the Santarelli affair in 1626, and frankly admitted that they adjusted their views according to whether they were in Rome or Paris. The exchange is quoted in A. G. Martimort, *Le Gallicanisme de Bossuet* (Paris, 1953), 106; this is the outstanding work on seventeenth-century Gallicanism in general. The complex relations between French and Roman Jesuits in the later decades emerge from G. Guitton, *Le Père de La Chaize* (2 vols.; Paris, 1959), and H. Hillenaar, *Fénelon et les Jésuites* (The Hague, 1967).

the minister was far more valuable as a lukewarm Huguenot *politique* than he could possibly have been as a dubious catholic convert, his presence in the government reassuring the protestants without really threatening the catholics.[6]

One incidental benefit of the chaotic period surrounding Henri's assession to the throne was that an unusually high number of bishoprics were vacant by the late 1590s. This allowed the king an exceptional opportunity to remodel the episcopate, and he took advantage of it to alter the dominant pattern of appointments. During the decades of religious warfare most sees had been filled from among the powerful noble families of the province concerned, or occasionally by their clients, so that they were integrated with the local power structure. It was virtually second nature for the crown to react against such practices whenever it had the power: indeed François Ier had installed numerous bishops whose primary loyalties were to the monarchy, drawing mainly on prominent families from the legal and administrative spheres. Henri revived this policy, which was to become a permanent feature of the French church, in the sense that after 1600 bishops were very rarely natives of the region in which their see was situated. Great court families did not dominate the recruitment, however, for such cases actually became less common with the passage of time, and by the early eighteenth century the dominant group was that of the leading provincial nobility, above all families from south of the Loire. The crown's management of appointments emphasized, beyond its instinctive hostility to local power blocs, the traditional political role of the bishops, traceable back at least to the Merovingians, but soon to be reduced dramatically.[7]

Episcopal power operated on a number of levels, which tended to reinforce one another. Within his diocese, and especially his cathedral town, the bishop could rely on a combination of wealth, patronage, traditional rights, and

[6] On the immediate problems of registration and enforcement, see J. Garrisson, *L'Édit de Nantes et sa révocation* (Paris, 1985), 13–27. On Sully, see D. J. Buisseret, *Sully* (London, 1968), 197–9.

[7] M. C. Peronnet, 'Les Évêques de l'ancienne France' (service de reproduction des thèses, Lille, 1977) is an invaluable work on the history of the episcopate, which deserves to be made more widely available. For the points discussed here, see esp. pp. 467–549, 626–37, 666–8.

religious authority. The precedence accorded him on virtually all ceremonial occasions was not empty form; no mere bourgeois could compete with him, while he treated on equal terms with such bigwigs as the provincial governor or the *premier président* of a sovereign court. He expected to organize poor relief, both in ordinary times and in emergencies, to mediate in local disputes, and to intercede with the royal government on behalf of his diocese. Where there were local estates an archbishop or bishop commonly presided over them, with his colleagues attending as of right—the most powerful of all provincial estates, those of Languedoc, were dominated by the twenty-two local bishops.[8] This situation generally suited the crown very well, for after 1600 most bishops proved loyal and reliable. Self-interest alone must have pushed them in this direction, with hopes of promotion to more prestigious sees, the award of rich abbeys *in commendam*, or royal patronage for their relatives. A good deal of the bishop's local power also rested on his privileged status with government and king, so that if he were known to be out of favour a sharp loss of influence would follow. Normally he would be an assiduous correspondent, acting as a general government informer in the style later associated with the intendants, and thus occupying a pivotal role between central and local authority. Long periods of residence in Paris would not necessarily damage this function, since most of the local duties could be effectively delegated to *vicaires généraux*. These absences usually reflected the bishop's wider responsibilities as a member of the ruling élite in both church and state who might be called on to advise the king or perform executive tasks. Some of them would be to attend the Assemblies of the Clergy, in which half the seats were reserved for the bishops, the remainder being held by *vicaires généraux* who knew their place and normally kept it. Membership of this great corporate entity was a further source of strength, for the episcopate as a whole would be quick to defend the interests of individual members in any but exceptional cases.

[8] For Languedoc see W. H. Beik, *Absolutism and Society in Seventeenth-century France* (Cambridge, 1985), 121–4 and *passim*. Peronnet, 'Évêques', 645–6, reckons that forty-nine bishops in all had regular positions in Estates or similar institutions.

Cardinal Richelieu was not only the most striking example of a cleric who rose to great political power; he had also gone out of his way to advocate such appointments.[9] Although his propagandists had to look back to Louis XII's minister the Cardinal d'Amboise for a satisfactory prototype, there had been numerous examples over the intervening century, some of which (such as the Cardinal of Lorraine) it was wiser not to invoke.[10] Richelieu himself continued the tradition, so that we find the bishop of Mende active in army provisioning, the Cardinal de la Valette employed as a general, and Archbishop Sourdis of Bordeaux as an admiral. Although his foreign policy advisers Père Joseph and Mazarin were not bishops, they must be added to the clerical element in Richelieu's governmental team. There was also some recruitment into the episcopate from among those classic royal servants the intendants; Marca, Bosquet, and Villemontée are cases in point. Both Richelieu's precepts and his conduct show that he saw church and state as mutually reinforcing structures of authority, whose membership might often be interchangeable. His early reform proposals lay great stress on the restoration of clerical discipline, through the implementation of the tridentine rules, as a means of restoring order in the polity as a whole.[11] It was during this period that the renewal of the episcopate, begun under Henri IV, was completed, with the result that nearly all the 116 French dioceses of the time experienced a serious attempt to apply the catholic reform.[12] This proved no easy task, not least because bishops were far

[9] For Richelieu's speech at the closing ceremony of the Estates General in 1615, when he advanced this view, see *Mémoires du Cardinal de Richelieu*, Société de l'histoire de France (10 vols., Paris, 1907–31), i. 340–65.

[10] For the use of the Cardinal d'Amboise as a parallel, see [J. Sirmond] *La Vie du Cardinal d'Amboise* (Paris, 1631), and E. Thuau, *Raison d'état et pensée politique à l'époque de Richelieu* (Paris, 1966), 224.

[11] This is particularly apparent in the important proposals drawn up in 1625, printed in P. Grillon (ed.), *Les Papiers de Richelieu,* (Paris, 1975–), i. 244–5, 248–57. For Richelieu's views on the relationship between church and state, see W. F. Church, *Richelieu and Reason of State* (Princeton, NJ, 1972), 81–92.

[12] Peronnet, 'Évêques', esp. 688–9, and for a selection of reforming bishops P. Broutin, *La Réforme pastorale en France au 17ᵉ, siècle* (2 vols.; Paris, 1956).

from enjoying total control over their own clergy. The exemptions enjoyed by the regular clergy were a long-standing grievance, while cathedral chapters also enjoyed virtual independence, aided by the normal pattern of recruitment among the ruling families of the region. Annoying though these privileges might be, they were less important than the dispersal of rights of patronage with regard to the basic parochial unit, the cure of souls.

Many cures had been attached to religious houses since their creation, while others had become so over the centuries, or passed under the control of other corporations, even of laymen. By the seventeenth century it was a lucky bishop who appointed a third of his *curés;* in most cases it was far less. Even where the bishop did appoint, he still had to reckon with deep-seated attitudes to benefices as a type of private property. The holder could resign in favour of a named party, commonly a relative, drawing a pension from the income of the benefice.[13] There was no established system of training for priests, so that it was virtually impossible to judge their competence in advance; the only real qualification required was a property one, but this was satisfied by possession of the benefice itself, and in any case had no relevance to pastoral ability. Although most bishops made a serious attempt to exercise their visitation rights, in the manner advocated by the Council of Trent, such intermittent and partial checks were not very efficient on their own. However immoral or ignorant the *curé,* he was only likely to be disciplined if his parishioners chose to denounce him, and even then it might need further litigation and expense to deprive him definitively. The impression from diocesan studies is that a handful of highly scandalous priests were removed in this way, but that a far larger body of mediocre incumbents had little difficulty in retaining their comfortable berths.[14] If the standard of the parish clergy

[13] The complexity of the situation over benefices was far greater than these brief remarks can convey, but the overall effect was to weaken episcopal power still further. For Richelieu's comments, and his ingenious idea of an examination system, see *Testament politique du Cardinal de Richelieu*, ed. L. André (Paris, 1947), 197–9.

[14] Regional studies include those by J. Ferté, L. Pérouas, J.-F. Soulet, R. Sauzet, and P. Hoffman. There is important information in R. Sauzet, *Les*

does seem to have improved greatly over the century, this resulted rather from changes in the attitudes of new entrants, and above all from the introduction of seminary training as the norm (essentially an achievement of the years after 1660). Better education did not automatically lead to greater docility; there were still many ways in which a bishop might clash with his *curés,* with no certainty that he would emerge the victor. Despite all these reservations, the church still possessed by far the most efficient linear power structure in France. The number of bishoprics meant that they were small enough to be practical administrative units; most important of all, the 30,000 *curés* provided a network which allowed contact with virtually the whole population. Sunday mass and parish meetings cemented French society both symbolically and politically. It was therefore possible to use the church as a channel of communication, with the *curé* making announcements (which might be entirely secular, e.g. the assessment for the *taille*) in association with the Sunday mass. This was also the occasion for the *monitoire,* issued at the request of the royal courts to force witnesses to testify on pain of excommunication. From the time of Colbert onwards the administrative monarchy would employ the system in the other direction, treating the *curés* as one of the best sources of information for its numerous inquiries into the state of the country. To this one may add the role of the church in organizing public celebrations and processions, interceding with the deity for protection against natural disasters, but also helping to sustain the charisma of the monarchy. By the reign of Louis XIV this had become a virtual propaganda campaign, whose dangers for both parties would only very slowly become apparent.

Hierarchical structures within the church were emphasized by the routinization of clerical taxation from the sixteenth century onwards. Inevitably finance was a crucial part of the relationship between church and state, and did much to determine the precise form this took. Until this point the kings had been very sparing in their use of assemblies of the clergy,

Visites pastorales dans l'ancien diocèse de Chartres pendant la première moitié du 17ᵉ siècle (Rome, 1975), and in J. Godel (ed.) *Le Cardinal des Montagnes: Étienne le Camus* (Grenoble, 1974).

called only for exceptional purposes. The most celebrated was that of Bourges, where the notorious Pragmatic Sanction was proclaimed in 1438. Another had been called at Tours in 1493 to consider a project for clerical reform, while those at Tours in 1510 and Lyon in 1511 were held in preparation for the schismatic council of Pisa and Milan, a notably damp squib in the long sequence of French anti-papal manœuvres. François Ier had taxed the church throughout his reign, with varying degrees of consent from *ad hoc* national or local assemblies; he was able to invoke precedents going back to at least the twelfth century, and originally connected with the Crusades. These taxes, known as *décimes,* were levied on the nominal value of benefices. The pope granted François the right to raise a *décime,* without the consent of the clergy, in 1516, and this was the occasion for drawing up a *département général des décimes* which was to remain in use until the 1640s, and survive into the eighteenth century as the basis for part of the levy; a notable example of the administrative inertia of the *ancien régime,* and the source of endless complaints. At first the king usually contented himself with a single *décime,* although there were naturally exceptions. In 1527, for example, the clerical representatives in an assembly of notables agreed to contribute 1.3 million *écus* towards the ransom for the king's sons. In return they obtained a promise of sterner measures against heretics, a theme which was to become familiar. It reappeared at the assembly called by Henri II in 1552, although on this occasion the king's financial needs led him to return jurisdiction in heresy cases to the church courts, reversing his earlier decision to hand them over to the *parlements.* In practice this meant a decrease in persecution, but the clergy could hardly be expected to abandon traditional rights of jurisdiction, whatever the reasons of expediency in a particular case. The crown's desperate shortage of funds had already led it to collect four *décimes* every year since 1545, some 1.6 million livres annually. The even greater problems associated with the religious wars would see this practice institutionalized on a new basis.

Before the fighting had started, it was at the Assembly of Poissy, which met in July 1561, that the decisive step was taken. On the surface it might have appeared as a partial success for the clergy, for although the four *décimes* were to be collected for another six years, after that the contribution would be reduced to 1.3 million livres a year for a further decade. Moreover, the king promised to exempt the clergy from 'toutes autres décimes ou emprunts particuliers', apparently a major and highly valuable concession. Too valuable, in fact; such general renunciations were always highly suspicious, and the government started to break the terms of the Contract of Poissy within a year of its conclusion. The 22.6 million livres were supposedly to be used to redeem the *rentes* already issued on the security of the city of Paris, but the clergy looked even more attractive as a security, and the receipts were soon alienated against new *rentes*. On top of this, a whole range of additional levies were made, so that the clergy claimed in 1577 that it had paid a total of some 62 millions in sixteen years. This was part of a general claim that, having met its side of the bargain, the clergy should now be exempt. This was hardly a realistic option in the circumstances of the time, and the assembly of 1579 agreed to a new contract, for ten years from 1580, at the rate of 1.3 millions a year. On top of the straightforward taxes, there were five separate sales of church property between 1563 and 1586, which may have involved capital to the value of 20 million livres, although it is probable that benefice holders often bought the land themselves, or repurchased it later—the crown rather illogically accorded special facilities for such repurchases, which must have discouraged buyers and kept prices down. These operations were carried through with the consent of the *parlement* and the pope, over the protests of most of the clergy, although they seem to have been initiated by a small group of cardinals and courtier bishops. Nor did the church escape the effects of the sale of offices, for from 1554 onwards the crown repeatedly established financial offices connected with the various levies on the clergy; since the latter were very anxious to keep control over their own affairs, they generally tried to buy up these abusive creations, if necessary raising additional funds of their own for the purpose. The kings had certainly sold their catholicism dear, and their treatment of the church may help to explain the widespread clerical support for the League, whose programme was an explicit criticism of their failure to deliver the other side of the bargain, the elimination of heresy. The collapse of royal power in 1588 was naturally followed by the *de facto* cessation of clerical taxation, but from 1594 the system was restored, this time on a basis which would last, in its essentials, until the Revolution.[15]

The General Assembly of 1595 revised and clarified rules drawn up in 1580. Henceforward there were to be General Assemblies every ten years, interspersed with periodic Assemblies of Accounts. A further decision in 1625 established what was to prove the definitive rhythm, a regular five-year cycle alternating General Assemblies and Assemblies of Accounts. Each ecclesiastical province (archdiocese) elected four members to the former, two to the latter; half of the places were reserved to the bishops, the other half to the lower clergy. The electoral system was indirect, with diocesan assemblies sending representatives to a provincial assembly, which made the final choice. The nominal purpose of the General Assembly was to renegotiate the contract for the *rentes,* while the Assembly of Accounts (or Little Assembly) was to verify the operation of the contract by the *receveur général du clergé* halfway through its course. The periodic assemblies were supplemented by a system of permanent representatives; the syndics of the clergy, who operated between 1561 and 1579, were thought to have become too much the creatures of the crown, and were replaced by the *agents généraux,* each pair of whom normally served for single five year periods. Bishops were not allowed to hold this office, appointment to which was made by two provinces in rotation. As with all such institutions, the crown expected to interfere when it wished, manipulating elections to the assemblies to ensure a greater degree of compliance, sometimes imposing *agents généraux* of its own choice.[16] It is evident that the assemblies would quickly

[15] For these sixteenth-century developments, see L. Serbat, *Les Assemblées du Clergé de France: origines, organisation, développement (1561–1615)* (Paris, 1906), and R. Doucet, *Les Institutions de la France au 16e siècle* (2 vols.; Paris, 1948), 831–59. References for the sale of ecclesiastical land are given in n. 2 above.

[16] On this point see A. Cans, *L'organisation financière du clergé de France à l'époque de Louis XIV* (Paris, 1910), 141–4.

have become redundant if they had been confined to the formal business associated with the contract to finance the *rentes,* which had settled into a routine by the early seventeenth century, with the interest payments covered by 1.3 million raised in *décimes* every year. There were other reasons why both the king and the clergy found the institution useful. It provided a forum for discussion of a range of issues, some of them relating to the hotly disputed frontier between clerical immunities and the royal courts, others to questions of moral theology, others again to relations with Rome. While this was obviously attractive to the participants, it was also potentially useful to the crown, which could employ the opinions of the clergy as a weapon in various causes of its own. By far the most important reason for the continuation of the assemblies was however the financial one; the government was rarely short of excuses for further raids on the wealth of the church, and the assemblies had by now established an effective principle of 'no taxation without representation'.

The Assembly of 1615 had decided that Assemblies of Accounts should meet at two-year intervals until the next General Assembly in 1625. These meetings concerned themselves primarily with reviewing the operations of the *receveur général,* but in both 1617 and 1619 they made representations to the king on the religious situation in Béarn. By the time the deputies assembled again in 1621 the 'expedition de Béarn' had taken place, giving rise to a major Huguenot rebellion. To the government the logical consequence was clear; since the civil war was in the cause of religion, the clergy should bear some of the cost. To put further pressure on the Assembly, Louis XIII sought the aid of the pope, whose known willingness to sanction further clerical taxation helped to persuade the deputies that they must comply with the royal demand for 3 million livres. Their hopes of making this grant, financed by the sale of offices, a final one were soon dashed. In 1625 they were persuaded to disgorge another 1.75 million, while a special assembly called during the siege of La Rochelle was subjected to intense pressure by Richelieu and Louis XIII, who eventually extorted 3 million livres towards the cost of the siege. On this occasion the clergy pointedly ignored a papal brief exhorting them to be generous, apparently not recognizing that in so doing they were tacitly accepting the fact that the king had a right to call on their help without papal permission. When they next met, in 1635, it was in the context of a general European war, so that under the circumstances a *don gratuit* of nearly 4 million was perhaps less than might have been feared. The following meeting, deferred to 1641, was notable for real resistance to the government's demands; its two presidents, the archbishops of Sens and Toulouse, had to be excluded by *lettre de cachet* before a shocked Assembly conceded an unprecedented 5.5 million livres. Such a level could hardly be maintained under the Regency, and the figure fell to 4 million in 1645, then a derisory 500,000 in 1650, recovering to 2.7 million in 1655.[17]

Under the personal rule of Louis XIV the *don gratuit* became an accepted and regular phenomenon, and for three decades it remained at a fairly modest level. Before 1690 only the Assembly of 1675 met during major hostilities; its offer of 4.5 million livres was much the largest, the other quinquennial grants remaining within the range of 2.2 to 3 million. The clergy's behaviour emphasized how the atmosphere had changed, for they needed little encouragement to make their contributions, showing no resentment at the perpetuation of extraordinary taxation in peacetime. The crown did offer some excuses; in 1665 it invoked the campaigns against the Turks, in 1670 naval operations against the Mediterranean corsairs. The real reason for the clergy's docility, however, lay in the implicit bargain which each Assembly constituted. In return for relatively modest payments the crown maintained clerical exemptions and took note of clerical grievances. This was particularly important in view of the constant attack by the secular courts and the tax officials on the exemptions and privileges involved. The need for royal protection was all the greater because the government was largely run by men drawn from the same groups of lawyer-administrators who led the attack; the prospect of substantial financial grants was an indispensable counterweight in this context. The dangers for the church only became fully apparent after 1690, when the

[17] The assemblies of the period 1615–65 are covered in the admirable study by P. Blet, *Le Clergé de France et la monarchie* (2 vols.; Rome, 1959), on which this section is based.

two great wars saw unprecedented demands on French resources. Whereas between 1660 and 1690 they had contributed 36.9 million, more than half of this being the ordinary *décimes*, at an annual average of 1.23 million, the last twenty-five years of the reign found them paying 160 million, of which the *décimes* comprised barely 10 per cent, at an average rate of 6.4 million a year. While the true level of the church's total income remains uncertain, a reasonable estimate would be in the range of 60 to 80 million, so that exactions on this level would have been decidedly uncomfortable if they had been met by straightforward taxation. Fortunately the clergy's credit remained good, so that they were able to borrow large sums, notably the two enormous payments of 24 million and 8 million agreed in 1710 to purchase outright exemption from the *capitation* and the *dixième*. These were to prove remarkably good bargains in the longer run, since the crown honoured the deal until the Revolution.[18]

Despite the exceptional pressure of these last decades, the clergy had been remarkably successful in defending their patrimony over the century between 1610 and 1715. A total contribution of around 280 million livres represented an average of less than 3 millions a year, with no forced sales of church land. There was a large-scale operation, beginning in 1690, to sell off surplus church silver; Louis XIV, who had already melted down his own plate, told Archbishop Harlay that he was informed that 'il y a beaucoup d'argenterie dans les églises au delà de ce qui est nécessaire pour la décence du service divin'. No accounts survive, but the *Nouvelles ecclésiastiques* estimated the value of the objects sold at 100 million livres, and the net product at 17 million.[19] The profit was to go to the churches, not the government, but in practice the latter benefited, since the resources made available facilitated the payment of clerical taxation. The assemblies had proved one of the most successful of all the monarchy's institutions, perhaps because they had evolved through a lengthy dialogue, rather than being imposed from above. They may sometimes

[18] For this period, see Blet, *Le Clergé*, and *Les Assemblées du clergé et Louis XIV de 1670 à 1693* (Rome, 1972); also Cans, *L'Organisation*, and *La Contribution du clergé de France à l'impôt pendant la seconde moitié du règne de Louis XIV (1689–1715)* (Paris, 1910).

[19] Cans, *La Contribution*, 10–13; the citation is on p. 10.

have been difficult to manage, but even during the Fronde they had never proved a threat to the government. At the same time, the discussions between their representatives and the ministers had been serious ones, marked by a certain mutual respect, which had kept the level of taxation within the clergy's capacity to pay, and provided some sort of check on the enterprises of the lawyers. As so often, the crown had been able to exploit the long-running conflicts inherent to the polity in order to play the role of arbiter, while never fully satisfying any of the parties. One must also note the importance of the protestant factor during the crucial period from 1560 to 1630, which saw the assemblies—and their grants—become routinized.

This picture of the generally smooth operation of the assemblies should not, however, be allowed to obscure the numerous ambiguities which persisted about their nature and function, features we would expect to find associated with any *ancien régime* institution. From the point of view of the papacy they must have seemed a thoroughly tiresome device, endlessly reiterating unwelcome claims about Gallican liberties, fighting off Roman jurisdiction, and threatening to take over the role of a regular national council of the French church. Where earlier nuncios had been able to welcome the assemblies for their defence of the church against heresy and their support for the tridentine decrees, their successors had to chronicle the progression towards the Four Articles of 1682, the high water mark of royal Gallicanism in this period. In the Jansenist controversy too the assemblies came increasingly to pronounce on matters of doctrine, forcing Rome into some thoroughly awkward positions. The relationship between the assemblies and the French church as a whole could also be an uneasy one. They were essentially a mouthpiece of the secular clergy, and one of the most privileged groups within that enormous body. The persistent hostility to the regular clergy, which caused numerous conflicts with Rome, was one expression of the situation, since the interests of religious orders were effectively unrepresented. The ordinary *curés* fared no better; it was unthinkable for them to be elected to the assembly, while proposals to raise the *portion congrue,* to give them a greater share of the tithes, or to decrease their liability to the *décimes,* were always confronted by the self-interest of the rich benefice-holders. The whole manner in which the assembly was constituted and operated tended to turn it into a pressure group for vested interests, a bulwark of hierarchy and privilege. To follow an account of the proceedings is often to find oneself mesmerized by the narrow-minded legalism of the participants, manipulating their arsenal of precedents from canon and civil law to the virtual exclusion of wider considerations. The bishops were generally enthusiastic supporters of the catholic reform, but it was fatally easy for them to identify this with their own authority and the maintenance of traditional structures. While the assemblies gave the church a voice, their very nature ensured that it would never give expression to radical views of any kind. Here again one recognizes the curiously negative quality characteristic of the *ancien régime,* with the government itself standing out as the only organism really capable of promoting significant change.

Although changes certainly did occur in the church during the seventeenth century, they largely took two forms. The first was the application of reform diocese by diocese, following the general precepts of Trent and the example of Borromeo; this was largely an exercise in moral discipline and clerical training. The second was the introduction of new orders, which often brought new types of activity with them, ranging from education and the care of the sick to the promotion of new forms of devotion. Both created difficulties which could involve the state in various ways. Where campaigns were launched against popular superstition and immorality, communal resistance might lead to disorder, or call for firm intervention by the authorities if anything was achieved. Bishops had a hard task establishing control over their clergy, since the principle of the benefice was so entrenched in everyone's mind; the *appel comme d'abus* allowed the secular courts to intervene in such cases, so that a kind of legal guerrilla warfare could be initiated by any attempt to interfere with a recalcitrant *curé,* let alone a religious house. The 1695 edict on clerical jurisdiction was the culmination of a long process by which the state identified its authority with that of the bishops, at least in respect of the secular clergy.[20] In the

[20] The text of the edict is conveniently published in L. Mention, *Documents relatifs aux rapports du clergé avec la royauté de 1682 à 1705* (Paris, 1893), 113–34.

great quarrel between seculars and regulars the government was more cautious, conscious that the exemptions of the latter proceeded from the papacy, so that in practice it was wise to leave them untouched whenever possible. The new religious orders were also valued as an essential part of the catholic reform, providing additional manpower as well as new pastoral techniques. The rivalries and jealousies between them, on the other hand, combined with the endless problem of their relationship to the bishops, could at times make them disruptive as well as creative. Their multiplicity rendered the church still more unwieldy and fragmented, while many of the doctrinal quarrels of the period were initiated and fuelled by their members. The effects of these processes were inevitably complex. In some respects they both increased the utility of the church as an adjunct to the state, and gave the latter many more occasions to intervene directly in clerical affairs. They also, however, generated vested interests, conflicts, and strongly rooted opinions which no amount of royal mediation could do much to reduce, and which greatly complicated the task of managing the church.

Beyond their basic spiritual and pastoral concerns, the clergy dominated educational provision. Humble village schoolteachers were under the control of the *curé* and bishop, while the colleges which revolutionized secondary education were run above all by some of the new orders, Jesuits, Oratorians, and Doctrinaires.[21] The government largely opted to devolve both financial and intellectual responsibility on to the church, making only occasional interventions itself. Many members of the élites, conscious of the urban poor as a threat to public order, were keen supporters of schemes to socialize them through basic instruction in Christian belief, literacy, and trade skills. The seventeenth century saw an enormous charitable campaign, in town after town, seeking to combine the salvation of souls with the imposition of order in place of disorder. There is no conceivable way of measuring the impact of this educational drive, but it does seem likely that it

[21] For education in general, see R. Chartier, M. M. Compère, and D. Julia, *L'Éducation en France du 16ᵉ au 18ᵉ siècle* (Paris, 1976); the Doctrinaires are the subject of an excellent study by J. de Viguerie, *Une œuvre d'éducation sous l'ancien régime* (Paris, 1976).

was considerable, with the combined efforts of the church and the municipalities doing much to buttress the social fabric of the country in a manner which was highly convenient to the crown. There was less agreement over the desirability of educating the peasantry. Reforming clerics were very conscious of the connection between illiteracy and superstition, but royal administrators often felt that schooling was irrelevant, even detrimental, for peasant children, and were hostile to the imposition of communal taxes to support schools. This attitude prefigured that of many *philosophes,* being based on the notion that peasants could perform their tasks perfectly well without formal schooling, which was liable to render them dissatisfied with their lot, seeking for positions above their natural station. After 1680 the problem of the *nouveaux convertis* brought about a temporary change of heart, expressed in the royal declarations of 1698 and 1724, whose provisions, although expressed in general terms, were mainly intended to enforce catholic education on the nominally converted protestants. By the middle of the eighteenth century this phase would be over, with 'enlightened' intendants trying to reduce the numbers of village schools.[22] This was a very interesting case where the ideals of the catholic reform, extended to the whole population, led to a potential clash with the socially restrictive ideas of the élites and the state. In practice the conflict was substantially muted by the other limitations on rural education, since many peasants were unwilling to sacrifice the ancillary labour of their children, and even by the early nineteenth century only about half of the *communes* had schools; the seventeenth century figure must have been far lower. There were enormous regional variations; in around 1710 74 per cent of the parishes in the diocese of Rouen had a school for boys, 26 per cent a school for girls. Some other northern dioceses, such as Reims, and some of the richer dioceses of the Mediterranean plain, were also very much in advance, while great areas of central and western France remained dismally underprovided.[23] The church had done a great deal, but it had perhaps never given the issue the prior-

[22] For these vagaries, see Chartier *et al., L'Éducation en France,* 11–16, 36–41.

[23] For these patterns, ibid., 16–26.

ity it truly deserved. If so, this was primarily because the *curés*, the men who were most aware of the problems, had so little chance to make their voices heard, whereas the bishops as a group were less convinced on the subject. For them it may well have seemed more important to enforce the teaching of basic Christian doctrine through the catechism, a duty which could in principle be assigned to the *curé* himself.

Education was only one of the most important among many 'public service' functions largely assumed by the church, ranging from the Capucins' role as unpaid firemen to the establishment of houses for reformed prostitutes. Most of this activity can be loosely summed up as charitable; the classic ways to support it were from bequests in wills, donations by the pious, and collections on specific days. Preachers endlessly reiterated the spiritual and practical benefits to be won by such generosity, most crucially its importance for salvation. The ultimate way of giving expression to these charitable impulses was to join one of the numerous orders (mostly feminine) such as the Filles de la Charité, which often performed prodigies in caring for the sick and destitute. On a more mundane level, virtually all clerics, from bishops to *curés*, took an active role in organizing direct local aid for the poor. Here there was substantial lay provision, in the form of the hospitals organized by municipalities under the general supervision of the state, but these institutions had multiple problems. Chronically underfunded in most cases, they quickly became primarily agencies of repression, whose main purpose was to clear the streets of itinerant beggars, orphans, and madmen. For the great mass of the poor, who struggled on within their own communities, the informal aid organized by the local clergy was vital. The haphazard nature of the evidence precludes any statistical precision, but most historians have concluded that if these arrangements often achieved a great deal, they still fell far short of meeting the need over most of the country. It is perhaps a moot point whether France could have sustained a formal system of poor relief on the English model, or whether it would have been advantageous for her to have done so; what seems certain is that the general catholic vision of charity was inimical to such routinized systems of relief. The giving of alms by individual volition was central to a religion posited on good works as the road to salvation, not least because it provided the classic opportunity for the rich to overcome the disadvantages of their position within the moral economy of salvation. The government was therefore relieved of considerable responsibilities in this area, and could concentrate on the (still substantial) problem of locking up and controlling the wandering or disorderly poor. By the middle of the next century the inadequacies of this policy were becoming plain; a major attempt at reform in 1764, with legislation introducing a communally based system of relief, was, however, foiled by the opposition of the *parlement* of Paris.[24]

The dominant theme of the new foundations was that of intervention in the world, through teaching, works of charity, and missionary activity. On occasion this was symbolized by the decision to limit vows and observances so that the orders remained technically secular rather than regular. The great religious foundations of earlier centuries had been primarily closed orders, whose members served God through prayer and devotion; these activities were also a benefit to the community, since they might be expected to obtain divine favours both for the living and for the souls in purgatory. Many of the greater houses were rich and socially exclusive, retaining much of their prestige even where their religious standards had fallen. The inevitable erosion of harsh disciplinary rules, coupled with the widespread damage inflicted during the wars of religion, does seem to have left many in a relatively poor state around the beginning of the seventeenth century. The current of reform was running quite strongly among the regulars by this time, with notable congregations such as those of Saint-Vanne and Saint-Maur gathering support for a return to earlier and stricter rules of life. The Assemblies of the clergy, and the Estates-General of 1614, naturally called for the pope and the king to sponsor and advance this movement. In 1622

[24] There is a large literature on poverty; particularly helpful recent contributions are those by J.-P. Gutton, *La Société et les pauvres: l'exemple de la généralité de Lyon 1534–1789* (Paris, 1971); O. H. Hufton, *The Poor of Eighteenth-Century France, 1750–1789* (Oxford, 1974), and C. Jones, *Charity and Bienfaisance; the Treatment of the Poor in the Montpellier Region, 1740–1815* (Cambridge, 1982). Although the last two books are concerned with the eighteenth century, they provide a great deal of relevant information for earlier periods.

a serious initiative emerged; the Cardinal de la Rochefoucauld received a papal brief granting him exceptional powers of visitation over the major monastic orders for a period of six years. Unfortunately he turned out to be a well-intentioned bungler, who was too busy to make visitations or hold proper consultations, relying instead on zealous self-appointed reformers within the various orders. His hasty decisions polarized opinion, alienating many monks who were basically in favour of more gradual reform. This was disastrous, because there were elaborate constitutions providing for elections within houses, and for general meetings of delegates for whole orders. This complex democratic system was a gift to obstructionists, still more so with the *parlements* always ready to intervene through their *arrêts*. Matters became still worse when the ageing cardinal was granted a second brief in 1632, this time for only three years; his hasty interventions provoked more trouble, swelling the flow of protests to the crown. The whole affair provided gilt-edged opportunities for a far more dangerous and predatory figure, none other than the Cardinal de Richelieu. Already in 1627 he had secured election as coadjutor to the abbot of Cluny, a retiring figure who hastened to vacate the place the next year. Richelieu then pressed for the union of the reformed congregations, Saint-Vanne and Saint-Maur, with Cluny; Saint-Maur agreed in 1634. In the same year the cardinal was recognized as general administrator by the Congregation of Chazal-Benoît. In 1635 he extended his power to the Premonstratensians and the Cistercians; in the first case troops had to occupy the abbey to secure his election. With the Cistercians, who had appealed to him against the clumsy activities of La Rochefoucauld, he removed the abbot of Cîteaux, Nivelle, by appointing him bishop of Luçon, them calmly had himself elected in his stead. Since the cardinal was also the commendatory abbot of some fifteen other rich monasteries, he now controlled the vast majority of the French monastic establishments. It was a fantastic and unparalleled coup, bringing in an enormous annual income (some 300,000 livres p.a. by 1640), and conferring such power than Richelieu appeared poised to effect a general reform of the regulars. There can be no doubt that he wished to achieve this, and that he would have regarded his massive pluralism as a quite justifiable technique for cutting through the obstacles which had frustrated his predecessor.[25]

As in numerous other areas, the cardinal's ambitions turned out to be in excess of his power, which had a decidedly paradoxical quality to it. Only a dominant first minister could ever have aggregated so many key appointments, yet such a man had little hope of using them properly. If la Rochefoucauld had been a busy man in a hurry, Richelieu represented an extreme instance of the same phenomenon. Like so many of his contemporaries, he thought monks could only justify their existence by keeping strictly to the original rules for abstinence and intensive religious observance; he therefore placed all his authority behind the groups of uncompromising reformers who had been appearing in each order. For their part, these men saw a golden opportunity to secure definitive reform by authoritarian means, and felt no sympathy for their weaker brethren, who now seemed to face a choice between compliance and expulsion. The immediate result was a polarization, which made it uncomfortably clear that the large majority of monks were hostile to the proposed changes. This was not just a matter of self-indulgence and sloth; the changes in the rules had generally been made through proper procedures and with papal consent, and it was under these conditions that individuals had decided to enter the religious life. Now they were faced with aggressive and intolerant minorities, who seemed quite prepared to subvert the traditional constitutions to secure their ends. There were many monks who had favoured reform by consent, but who were thoroughly alienated by the tactics now employed; they included some of the shrewdest and best-connected abbots and others, men who often had strong family ties with the grandees and the *robe* nobility. Those who opposed Richelieu, in what was sure to become an immensely complex legal battle, could count on many allies among the judges. Public

[25] On monastic reform in this period, see P. Denis, *Le Cardinal de Richelieu et la réforme des monastères bénédictins* (Paris, 1913), a valuable study but one which tends to assume that reform was automatically a good thing, and the much more balanced views of L. J. Lekai, *The Rise of the Cistercian Strict Observance in Seventeenth-Century France* (Washington, DC, 1968). For Richelieu's accumulation of benefices, see J. Bergin, *Cardinal Richelieu: Power and the Pursuit of Wealth* (New Haven, Conn., 1985).

opinion, originally inclined towards the reformers, was rapidly alienated by the use of force, the more so when outsiders were bound to see this as quite disproportionate in arguments over the minutiae of dietary rules. The last and greatest obstacle was beyond even Richelieu's reach; the pope, bitterly hostile to French foreign policy, was also the natural resort of those who knew the government was against them. Pope Urban VIII refused to install the cardinal as Abbot General of either the Cistercians or the Premonstratensians, despite intense diplomatic pressure. These were international orders, and the foreign houses naturally rallied behind the pope against the strict reformers. In the case of the Cistercians Abbot Claude Largentier of Clairvaux led opposition within France, finding that the government dared not act against him for fear of angering the pope still further, so that disobedience was encouraged as knowledge of the papal attitude spread. The ultimate reason for the relative failure of the reformers, however, was simply that there were never enough monks who supported them. As internal quarrels became more bitter, it was useless to place small groups of strict observants in each monastery and hope they would convert the others. Instead, houses had to be cleared of their recalcitrant monks and refilled, a process which rapidly absorbed the available manpower, while displacing hundreds of people in the most public manner.

By 1643 it was evident that the sledgehammer approach had failed, and the inevitable reaction set in; it was much aided by one drawback of Richelieu's method of concentrating power, the numerous vacancies left by his death. It may be an exaggeration to describe this as a 'Fronde of the monks', but dramatic and violent events did occur in several places, all tending to reverse recent trends.[26] The Prince de Condé, for so long Richelieu's faithful henchman, played a leading part in undoing his work. He secured the election of his thirteen-year-old son Conti as abbot of Cluny, and supported *premier président* Bouchu of Dijon in the patently illegal coup which made the latter's half-brother Claude Vaussin abbot of Cîteaux. Soon the secular power was intervening, with armed men

[26] The phrase is taken from J. H. Elliott, *Richelieu and Olivares* (Cambridge, 1982), 75.

expelling the reformed Benedictines from the collège de Cluny and Saint-Martin-des-Champs. In October 1644 Cluny and Saint-Maur agreed to separate, while an interminable lawsuit between the strict and common observances of the Cistercians began before the Paris *parlement*. In 1660 the *parlement* finally ruled that la Rochefoucauld's old reform plan should be applied, an illusory triumph for the reformers; Louis XIV was far more interested in the foreign policy issues, which made him anxious to conciliate the foreign houses and the pope. His formal approval of the decision was nullified by the simultaneous permission for the Cistercian common observance to appeal to the pope, a move which ensured their final victory. Such reverses did not mean that the various reform movements disappeared, but they became virtually separate orders, with no real prospect of taking over houses which had resisted them. The rest of the century saw no general royal initiatives, merely piecemeal interventions when disorder or financial problems became apparent. The failure of Richelieu's policy is a salutary reminder of the legal, political, and human limitations on the crown's power to force the clergy into new patterns. In theory the king's right to appoint abbots and abbesses in all but a handful of special cases (effectively those great houses whose heads were also superiors of an order or congregation) might have been employed consistently to advance reform; this did not happen because of the widespread abuse whereby these posts were held *in commendam* by absentees who simply drew the revenues. This was such a valuable source of patronage that there was no real chance for its being forgone, and indeed one of its uses was to compensate bishops in the less remunerative sees. The renewed decay of many regular houses by the eighteenth century should not be primarily attributed to royal policies, however, since reformed congregations were far from immune to a process which was ultimately part of the *Zeitgeist*. The catholic reform was itself a major factor here, seeking as it did to abolish many of the frontiers between sacred and profane, and to propagate a vision of the laity as equally obligated to live a holy life. The notion of monks as a holy élite, compensating for the sins of both living and dead, tended to become very marginal in such a scheme of things. Richelieu, Colbert,

Louis XIV, and many others demonstrated a certain hostility towards institutions they considered to lack any direct utility to the state; there was a marked general preference for those new orders which served in more tangible ways.[27] Loménie de Brienne and his 1766 *commission des reguliers* might still lie a long way ahead, but the attitudes they would express were already taking shape.[28] Two other points should be made about the crown and the regular clergy. Firstly, the reformed congregations, with their strong emphasis on theological and historical studies, were often sympathetic to Jansenist ideas, so that the failure to establish stronger control over them was a considerable hindrance to later attempts to enforce orthodoxy.[29] Secondly, the feminine houses (excepting the very special case of Port-Royal) never came under comparable pressure, perhaps because an exclusively masculine government always felt at something of a loss when trying to deal with them.

Another long-established institution which gave the crown problems, and suffered a decline, was the Paris Theology Faculty. The intellectual prestige of the University of Paris had reinforced the European status of its theologians throughout the later middle ages, and their decisions had become another potential weapon for the kings to exploit. Whether it was a matter of anathematizing heretics, of deciding the legitimacy of a royal marriage, or of frustrating papal claims, their decision on a carefully framed question was a valuable card to play. In dealing with any controversial issue the government could ring the changes between the *parlement*, the assembly of the clergy, and the Faculty, each possessing its distinct traditions and prejudices. The latter had a special significance because of the curious situation whereby the Paris printers were supposedly under the supervision of the University, so that the approbation of two doctors of the Faculty was

[27] For Richelieu's views, see *Les Papiers de Richelieu*, i. 245, 256; for Colbert, C. W. Cole, *Colbert and a Century of French Mercantilism* (2 vols., New York, 1939), ii. 466; for Louis XIV, J. Lognon (ed.), *Mémoires de Louis XIV* (Paris, 1927), 220–1.

[28] This episode is well described in P. Chevallier, *Loménie de Brienne et l'ordre monastique (1766–1789)* (Paris, 1959).

[29] This emerges particularly well from R. Taveneaux, *Le Jansénisme en Lorraine, 1640–1789* (Paris, 1960).

required for the publication of any book on religious matters. Despite the notorious inadequacy of this system, the censuring of books by the Faculty had become one of the prime ways in which orthodoxy was defined, albeit in a largely negative fashion.[30]

The possibilities were eagerly exploited by both sides during the Jansenist controversy, which also brought about a major crisis in the Faculty's affairs, with acrimonious debates leading to the expulsion of Antoine Arnauld in 1657. These proved so damaging to the Faculty's unity and its prestige that after the 1660s it played a substantially reduced role in general ecclesiastical affairs; doubtless the development of the Assemblies as docile agents of the royal will during the same period contributed to this relative decline.

Royal actions could prove double-edged, for the need to guard against Jansenism generated an ultramontane tendency among the doctors, which emerged in an embarrassing fashion over the Four Articles of 1682. The Faculty withstood intense royal pressure for some time, refusing to accept any but the first Article; its resistance emphasized the true weakness of the royal position, in a way its final climb-down could not efface.[31] By the end of the century the Faculty had lost much of its old influence, with censorship much more firmly in the hands of the crown, and the great doctrinal issues handled elsewhere. There can have been only marginal compensation in Louis XIV's policy of recruiting bishops exclusively from among doctors of the Faculty, which rather emphasized the fact that the institution was now primarily limited to a teaching function.[32]

Such close delimitation of spheres of activity and influence was very characteristic of the personal rule of Louis XIV. The role of the bishops themselves was also made more precise; their authority over their subordinates was increased, but they lost many of their wider political capabilities. Even where they

[30] P. Féret, *La Faculté de théologie de Paris et ses docteurs les plus célèbres: époque moderne* (5 vols.; Paris, 1900–7); H.-J. Martin, *Livre, pouvoirs et société à Paris au 17ᵉ siècle (1598–1701)* (Geneva, 1969), esp. 440–4, 460–6, 695–8; A. Soman, 'Press, pulpit, and censorship in France before Richelieu', *Proceedings of the American Philosophical Society*, 120 (1976), 439–63.

[31] Martimort, *Le Gallicanisme de Bossuet*, 502–5.

[32] Peronnet, 'Les Évêques', 462–3.

continued to dominate local estates, as in Languedoc, possibilities for independent action were sharply reduced, not least because of a tightening of the royal grip on the estates themselves. The change was not sudden, for Cardinal Bonzi, as archbishop of Narbonne and president of the Estates, still held great power in the Languedoc of the 1670s; after his disgrace, however, the primacy of the intendant was never to be challenged again.[33] The pattern of episcopal appointments became far more routine, with middle-ranking noble families from south of the Loire predominating, after a *cursus honorum* through the Paris seminaries and the faculty of theology.[34] This was another policy which was largely determined by concern over Jansenism, for the government seems to have judged, probably quite rightly, that these *méridionaux* would incline towards a less austere style of religious practice than their northern cousins (notable among them the episcopal Colberts and Le Telliers). It is of course most improbable that there was any conscious decision to have a general policy; what happened presumably resulted from individual decisions about the orthodoxy of candidates, combined with a certain loss of interest among grandee and ministerial families as bishops became mere cogs in the machine, tied to their role as clerical administrators. A set of unwritten conventions established themselves, gradually hardening into a pattern which would last as long as the *ancien régime* itself. A subjective impression is that while the general standard of competence and assiduity probably rose, there were less exceptional figures among the bishops appointed after about 1680, and this had something to do with an excessive stress on conformity.

In some sense conformity and good order must always have been the central objective of the crown, not least because the management of such a complex structure required constant effort. Just how the kings formulated their ecclesiastical policy, and who can really be said to have run the church, is in fact a very difficult question. No primacy equivalent to that of Canterbury had ever developed within the church; the archbishop of Lyon claimed to be *primat des Gaules,* but his colleagues never seem to have taken much notice of this

[33] Beik, *Absolutism and Society,* 241–3.

[34] Peronnet, 'Les Évêques', 29–31, 462–5, 668.

formal title. In consequence a very fluid situation developed, with a loose group of bishops and other clerics who enjoyed court favour acting as brokers and advisors. Until 1629 the issues were largely seen in the context of the need to combat protestantism while avoiding catholic extremism; these problems were so central that they cannot be separated from the general formulation of policy by king or regent, ministers, and council. Henri IV naturally placed considerable reliance on those *politique* bishops who had stage-managed his conversion, such as Archbishop de Beaune of Bourges and Cardinal du Perron. Royal confessors had always been influential, and became more so when Father Cotton inaugurated the long series of Jesuit holders of the office. As in so many areas, the long domination of Cardinal Richelieu saw crucial developments. Uniquely well-placed to combine spiritual and temporal authority, Richelieu brought a new assertiveness and interventionism to bear, simultaneously pushing forward catholic reform and expecting the church to serve the king at the expense of old traditions and liberties. The cardinal sought not merely to manage the church, but to subject it to his will, as a means to modernize and rationalize it. In the reform plans he prepared at the beginning of his ministry it was natural that religious matters should come first, for there is no doubt that he shared the common opinion of the time, seeing the rebuilding of church and state virtually as one. While he naturally had to work through allies, some of the most notable being Léonor d'Estampes, bishop of Chartres then archbishop of Reims, Cardinal Sourdis, archbishop of Bordeaux, and of course Father Joseph du Tremblay, all crucial matters were kept firmly under his personal control. Earlier monarchs had regularly used the church as a source of both money and power, but Richelieu's approach was far more sophisticated and all-embracing; the Bourbon kings and their servants were quick to learn the lesson, which they would follow until it became absorbed into the absolutist style of Louis XIV. From now on there had to be men around the king who specialized in handling ecclesiastical affairs; for all his expertise in papal diplomacy, Mazarin lacked the knowledge to perform the task himself, and his ministry is notable for the emergence of the *conseil de conscience*. In its original form, between 1644 and 1652, this had Vincent de Paul as secretary, with a variable membership including the Regent, Mazarin, Condé, Séguier, and Hugues de Lionne, with as ecclesiastical members bishops Potier of Beauvais, Cospéan of Lisieux, and La Fayette of Limoges. The young Louis XIV would rely on a council which included his confessor, Father Annat, Archbishop Marca of Toulouse, and Bishops Péréfixe of Rodez and La Mothe Houdancourt of Rennes; by around 1700 the equivalent body would include Father La Chaize, Archbishop Noailles of Paris and Bishop Godet des Marais of Chartres, with Mme de Maintenon wielding considerable influence through the last two. Alongside its prime function of proposing suitable nominations to vacant sees the *conseil de conscience* could also advise the king more generally on questions of policy, although final decisions on important matters rested with the inner council, on which no ecclesiastic sat after Mazarin until the time of Dubois and Fleury.[35]

The *conseil de conscience* was only an informal organization, destined to be replaced in the eighteenth century by the *feuille des bénéfices*, placed in the hands of a single minister or cleric. Under Louix XIV it had already known a lengthy period of eclipse, between the 1660s and 1690s, for this was the age of the most impressive purely clerical politician of the century, Archbishop François de Harlay of Paris. For close on three decades Harlay was the king's man *par excellence* in the church, inevitably made president of every assembly, in constant attendance at court. It seemed as if the archbishop of Paris, only raised to that status in 1622, was now achieving a position of unquestioned dominance in the French church, underlined by the king's action of 1674, when he attached the *duché-pairie* of Saint-Cloud to the see so that its holder joined the six traditional clerical peers.[36] The only serious competitor for the king's ear was his confessor Father La Chaize but this well-bred Jesuit was generally content to play a secondary role, probably conscious that his political skills were no match for Harlay's. La Chaize was most influential in recommending candidates for promotion to bishoprics, although this was far from being a power of appointment, with the king taking advice elsewhere and making the final decisions himself. The

[35] For the *conseil de conscience,* see Peronnet, 'Les Évêques', pp. 515–16, 520, and Guitton, *La Chaize*, i. 177–81, 193–201, ii. 149–53.

[36] Peronnet, 'Les Evêques', 640.

two men together nominally formed the *conseil de conscience,* with Harlay very much the dominating figure on matters of policy.[37] Even if the archbishop's critics, led by the Jansenists, loved to pour scorn on his worldliness, his pre-eminence reflected both great ability and assiduity. It is remarkable that so important a figure has never received serious scholarly attention, leaving an enormous gap in our knowledge of the ecclesiastical politics of the second half of the century. As his secretary the abbé Le Gendre put it, in his time the archbishop of Paris came to be virtually 'le pape d'en deça des monts'.[38]

Harlay excelled as a negotiator and mediator, so that Le Gendre could call him 'l'oracle de tout le clergé'; twice a week there were sessions at his palace where the archbishop sat with Father La Chaize to hear cases submitted from all over the country.[39] For important matters of state there were also extraordinary meetings, with a *maître des requêtes as rapporteur* and the archbishop assisted by *conseillers d'état.* Whenever the assembly of the clergy met it was Harlay who interceded with the king to obtain some face-saving concessions for his colleagues, a role which created increasing difficulties for him as the financial demands of the crown grew disproportionately in his last years. As early as 1675 both the nuncio and his colleagues had suspected him of failing to press the case for the suppression of the new offices of *banquiers expéditionnaires en la cour de Rome;* this was probably one of those occasions when he knew the king would prove inflexible.[40] In 1690 he made no secret of his reluctance before he carried a protest to the king over the tax of one third imposed on all major benefices for which bulls had not been obtained, in order to provide funds for the *nouveaux convertis,* a tax which had become onerous when the dispute over the *régale* stopped the issue of any papal bulls after 1682. One could not blame Harlay if he inspired the stinging reply that those who were dissatisfied had only to resign their benefices, characterized by Le Gendre as a

[37] See note 35 above for references to Guitton, *La Chaize.*

[38] M. Rous (ed.), *Mémoires de Le Gendre* (Paris, 1863), 194. There is an excellent brief summary of Harlay's role in J. Orcibal, *Louis XIV contre Innocent XI* (Paris, 1949), 54–8.

[39] *Mémoires de Le Gendre,* 24–6.

[40] Blet, *Les Assemblées du clergé et Louis XIV,* 107.

'réponse despotique qui fit cesser tout à coup les clameurs indécentes de la plupart des députés'.[41] It seems clear that the archbishop made ruthless use of his position as president to keep control of all negotiations between the clergy and the king, and that although his colleagues distrusted him and murmured among themselves, no one dared risk royal disfavour by an open attack. In 1695 the pent-up resentments finally overflowed when Harlay negotiated clerical exemption from the new *capitation* tax against an annual payment of 4 million livres; Bishop Coislin of Orléans told his friends of a conversation with the king, from which he had learned that Louis would have accepted 2 million livres, had not Harlay informed him that the clergy could pay more. Since Coislin was an honest man, and no enemy of Harlay in general, it appears certain that the king did commit this indiscretion, which led to the archbishop being treated as a virtual traitor, contradicted and shunned in the assembly. When the session ended he retreated to his country house at Conflans to nurse his wounds, and died of an apoplexy on 6 August 1695.[42] Whether or not Louix XIV had intended to bring it about, he had effectively destroyed Harlay's position, so that his disappearance gave a timely opportunity to seek a replacement. The king, who never fully trusted any servant, had in any case been careful to allow considerable influence to other individuals and groups, with himself as final arbiter, so in his eyes and those of contemporaries it probably seemed a routine enough shuffle of the ecclesiastical cards. Such prominent figures as Bishop Bossuet of Méaux and Archbishop Le Tellier of Reims had long been feuding with Harlay behind the scenes, with the invaluable assistance of Mme de Maintenon. Every new nomination to a see or court position could become a power struggle;

[41] *Mémoires de La Gendre*, 112–14, where he also notes that Harlay was suspected of having been two-faced over an increase in the *portion congrue;* the chancellor Boucherat had raised this to 300 livres, and the bishops protested. See also Blet, *Les Assemblées du clergé et Louis XIV*, 514–16, who quotes a report that the archbishop of Albi stood to lose 14,000 livres a year on account of the *portion congrue* increase (which also included a payment of 150 livres for each *vicaire)*.

[42] *Mémoires de Le Gendre*, 198–200; Cans, *La Contribution*, 40. Harlay's sudden death, without receiving the sacraments, was regarded as a divine judgement by his numerous enemies, a punishment for both his 'tyranny' in the church and his 'scandalous' intimacy with Mme de Lesdiguières.

this emerges very clearly from Fénelon's experience. At a time when Fénelon was closely linked with Bossuet, Harlay succeeded in preventing his appointment as bishop of Poitiers in 1685, at least in part by suggesting that he was tinged with Jansenism (not so absurd as it seems in the light of later events). He could not prevent Fénelon's far more significant choice as preceptor of the princes in 1689, partly because he had less standing in the matter, partly because he was becoming more isolated and losing influence, his supposedly improper private life having brought him a semi-disgrace in 1687.[43]

Unfortunately Harlay had created a semi-institutionalized situation built around his own particular talents, in a manner characteristic of the period. It was almost inevitable not just that his successor would find it difficult to cope, but that the reaction against the supple and insinuating style he had embodied would bring in a prelate of a very different stamp. Probably no candidate in the same style was available, for Harlay was looking increasingly a man of an earlier age, a survivor of the church and court of the young Louix XIV, out of place among the *dévots* of the 1690s. The choice of Louis Antoine de Noailles, bishop of Châlons, as the new archbishop was a disastrous one on most counts. Although well-intentioned, Noailles simply lacked the stature or intelligence for the job; when he was promoted to the cardinalate *président* Harlay of the *parlement* observed that it was a very big hat for such a small head.[44] Much as he was despised for his worldliness, Harlay's lack of commitment to any ideological line had been ideal for the position he held, so that he had provided a vital element of realism and moderation at the centre of affairs. Noailles, welcomed with uncritical enthusiasm by the *dévots* as the man who would cleanse the Augean stables, blundered around creating crises everywhere. By 1700 the 'Clementine peace' of 1669, which had kept the Jansenist controversy under reasonable control, had collapsed, as Noailles was trapped in a series of untenable positions by unwise Jansenists

[43] For Fénelon, Hillenaar, *Fénelon et les Jésuites,* 41–2. Harlay's disgrace is described by Guitton, *La Chaize,* i. 195, but for a more cautious view see Orcibal, *Louis XIV contre Innocent XI,* 57.

[44] Quoted by A. Adam, *Du mysticisme à la révolte: les jansénistes du 17ᵉ siècle* (Paris, 1968), 296.

who hoped to force him into giving them open support. A doctrinal version of Pandora's Box had been opened, releasing deep and difficult questions which embroiled the whole French church in conflicts that would continue for decades. For Noailles himself the road would lead, through the bull *Unigenitus* and his appeal to a General Council of the church, to a long disgrace terminated only by his death in 1729. The thirty-four years of his tenure destroyed the special position of the archbishopric of Paris for the remaining years of the *ancien régime,* terminating what had seemed a logical and serviceable development. This had only become apparent at the end of Louis XIV's reign, however; for a long time Noailles continued to try and fill Harlay's shoes. He acted as president of the assemblies of the clergy until 1715, having been called in to replace Le Tellier, the original royal choice, in 1700. This was the result of a curious episode which saw Bossuet trying to pursue a violent anti-Jesuit campaign, with Le Tellier's connivance; the latter also turned out to lack the necessary personal skills for managing the assembly and succeeded only in raising hackles all round. Noailles allowed the deputies more freedom of speech, and seems to have been relatively popular in consequence, although he himself regarded presiding over the assemblies as a *corvée* he would have been glad to be excused.[45] When he was trying to justify himself to the king in 1712 he wrote of his role in the assemblies, drawing attention to the difficulties he might have raised over the 'dons immenses' they made. As he put it:

au fond, ne s'agissant que d'un bien temporel qui n'est pas essentiel à la religion et qu'on peut donner dans une pressante nécessité, j'ai surmonté mon scrupule et travaillé le premier à la ruine du clergé pour sauver votre État et pour soutenir votre trône.

[ultimately, since it was only a question of temporal possessions which are not essential to religion and can be given in case of pressing necessity, I overcame my scruples and took the lead in working for the ruin of the clergy to save your state and support your throne.][46]

[45] Martimort, *Le Gallicanisme de Bossuet,* 686–7; Cans, *L'Organisation financière,* 78–82.

[46] Quoted by Cans, *La Contribution,* 87–8.

The honest but maladroit Noailles expressed the dilemma facing a scrupulous leader of the church in explicit and uncourtly terms which his wily predecessor would never have used. The problems which appeared after 1695, with factional struggles doubling the doctrinal ones, make it plain how relevant Harlay's skills had been. His much decried opportunism had been invaluable in preserving some balance between the various extremisms, in an area where the king himself was at his least competent; once he was gone royal policy began to veer around abruptly and unpredictably as Jansenists, *dévots*, and Jesuits grappled for the tiller, and only succeeded in putting the ship on the rocks.

The evident weaknesses in the decision-making machinery must ultimately be laid at the door of Louix XIV himself. Very jealous of his authority, he made sure that even the most powerful clerics remained totally dependent on his support, and that they always had to contend with rivals. For the church there were actually disadvantages in the change from the young king who allowed the performance of *Tartuffe* and abstained from the sacraments because of his double adultery with Mme de Montespan to the elderly *dévot* of Versailles. The latter demanded more and interfered more, rarely to good effect. The great decisions of the reign, the declaration of the Four Articles, the revocation of the Edict of Nantes, the new purge of Jansenists after 1703, were essentially personal acts by the king. Since we know nothing of the proceedings of the council, it is impossible to judge how far he was manipulated into them by his close advisers, although it is plain that he did not always have full information. In practice there was great pressure on everyone concerned to guess what the king already wanted, then find the best arguments for agreeing. This emerged over the Four Articles, which quickly came to be seen as an error which all those concerned would have preferred to disavow. Bossuet claimed that Harlay did no more than 'flatter la cour, écouter les ministres et suivre à l'aveugle leurs volontés comme un valet', with the implication that it was the archbishop's patron Colbert who was the prime mover, whereas at the time Bossuet and the Le Telliers had been anxious to claim leading roles.[47] In retrospect it is

[47] Martimort, *Le Gallicanisme de Bossuet,* 380–91, 444–60, argues that Bossuet and the Le Telliers favoured a compromise position throughout,

certainly hard to understand why the government allowed the relatively minor question of the *régale* to become the occasion for a major conflict with Rome, opening up issues which were much better left in decent obscurity. The explanation would seem to lie in a combination of the king's own authoritarian temperament with the widespread resentment of Rome among the French secular clergy, and particularly the bishops. Royal determination never to give up a legal claim, however dubious, gave the Gallicans an opportunity they most unwisely tried to exploit to the full, hoping for a definitive statement which would assure their cherished independence in perpetuity. They had failed to realize that if the pope proved obstinate the manœuvre must threaten their own position, since there would eventually be more important problems on which the crown needed papal support, leading inevitably to a disavowal of the Articles. The attitude of the bishops towards the pope throughout the century, and not just in this affair, was a curious mixture of sincere respect and deep suspicion, largely sustained by the convenient notion that the Curia was fundamentally corrupt and ignorant, so that the Holy Father was regularly misinformed. Roman theologians and administrators were believed to be capable of almost any crime, while the French seem to have felt an unshakeable confidence in their own intellectual superiority. When in 1648 Father Combefis demonstrated that Pope Honorius really had been condemned by the sixth general council, so that Bellarmine and others had been wrong in claiming the relevant acts to be forgeries, his book was placed on the Roman index.[48] In France no more respect was paid to this than to the condemnation of Galileo; it was simply taken as evidence that Roman theologians disregarded historical truth as readily as they did scientific proof. For their part, the popes not unreasonably considered the French bishops to be the mere instruments of the king who appointed them.[49]

The Four Articles also had very serious consequences for

but Blet, *Les Assemblées de Clergé et Louis XIV,* 348–62, brings forward convincing evidence against this view. For Bossuet's comment on Harlay (made in 1700), *Journal de l'Abbé Ledieu* (Paris, 1856), i. 8.

[48] On this affair see Martimort, *Le Gallicanisme de Bossuet,* 172–3.

[49] Pope Alexander VIII shrewdly remarked to the Cardinal de Bouillon that he knew how great the king's authority was in the French church, so

Louix XIV himself, because they created an impasse between Rome and Versailles, to which other problems became attached. Innocent XI refused to issue bulls to the new bishops appointed after 1682, so their dioceses had to be administered through *vicaires-généraux*. The nadir of relations with the papacy was reached in 1687, with the absurd conflict over the *franchises* of the royal ambassador in Rome. The pope excommunicated the king, although no one in France took any direct notice of this drastic act. The royal response was to notify an appeal to a future council of the church, a step which the royal jurists alleged should nullify all censures until the appeal had actually been heard; Fénelon rightly foresaw that the violent 'plaidoyers des gens du Roi passeraient à la postérité comme des monuments publics des maximes de France'. Despite the compromise of 1693, which settled the immediate conflict, a most alarming precedent had indeed been set, which would be exploited by the opponents of *Unigenitus*. There were much more rapid consequences in terms of international politics, for the pope took the side of the emperor in the dispute over the election to the archbishopric of Cologne which was the flashpoint for the outbreak of general European war in 1688, and which played a vital part in encouraging William of Orange to attempt the overthrow of James II.[50] As Jean Orcibal demonstrated long ago, the breach with Rome in 1682 was also crucial, alongside considerations of international policy, in bringing forward the Revocation of the Edict of Nantes three years later, intended partly as a demonstration of the true depth of Louis XIV's catholicism.[51] This

that the bishops 'n'y auraient d'autres sentimens et d'autre religion que celle du Roi', and would support either a schism or papal infallibility as they were ordered. The exchange is quoted by Orcibal, *Louis XIV contre Innocent XI*, p. 60, n. 285.

[50] These developments are brilliantly analysed by Orcibal in *Louis XIV contre Innocent XI*. The comment by Fénelon is quoted on p. 75.

[51] J. Orcibal, *Louis XIV et les protestants* (Paris, 1951); for a recent account on the same lines, with some additional material, J.-R. Armogathe, *Croire en liberté: l'église catholique et la révocation de l'Édit de Nantes* (Paris, 1985). The outstanding modern account of the Revocation in its wider context is that by E. Labrousse, *'Une foi, ne loi, un roi?': erssai sur la révocation de l'Édit de Nantes* (Paris, 1985), which demonstrates the internal logic of royal policies over many decades, but does not give much space to the questions discussed here.

was essentially an act of state, decided and carried out by laymen, as is emphasized by the fact that the clergy neither formally requested the revocation, nor were consulted about its consequences.[52] Harlay has again been extensively blamed for his part in encouraging the king, generally without any understanding of the gap between his intentions and the outcome, or an awareness that he was personally dismayed by the Edict of Fontainebleau, which he did everything in his power to delay.[53] With hindsight the royal policy appears inexorable, consistent, and often hypocritical, whereas in reality crucial decisions were being taken at the very last moment, so that the course of events might have been very different. French catholics were naturally delighted to see the end of formal toleration for the protestant religion as such, and there is no reason to question the general enthusiasm for the revocation. This did not prevent, however, a deep undercurrent of concern on the part of many bishops and royal administrators, who had to apply the policy in practice. Harlay was merely the most prominent (and perhaps the least sincere) among many who remained faithful to the principles of Louis XIII and Richelieu in deploring conversions produced by the outright use of force. They were not merely hypocritical in believing that it might be permissible to use methods ranging from bribes to the billeting of troops in order to pressurize heretics into rejoining the church, while objecting to their compulsory incorporation *en masse*. What they required, in order to avoid both widespread sacrilege and subsequent obduracy, was that there should remain at least some element of free choice, allowing room for a conscious individual decision to convert.[54] This view, which seems both theologically and psychologically sound, was also that taken

[52] The assemblies of the clergy did of course press repeatedly for measures against the protestants. For some of their interventions, see J. Garrisson, *L'Édit de Nantes et sa révocation: histoire d'une intolérance* (Paris, 1985), 146–50, 190–2: Orcibal, *Louis XIV et les protestants*, 23; Blet, *Les Assemblées du Clergé et Louis XIV*, 423–44, 467–76.

[53] For Harlay's position, Armogathe, *Croire en liberté*, 84–6.

[54] For these episcopal doubts, see Orcibal, *Louis XIV et les protestants* 128–32, Armogathe, *Croire en liberté*, 105–26.

by the pope, whose lack of enthusiasm for the Revocation rapidly became obvious.[55]

Lurking behind the disastrous course that events actually took there was a crucial missed opportunity, for Harlay and many of his colleagues had actually wanted to make serious concessions to the protestants on matters of doctrine and religious practice in order to secure a general and peaceful reconciliation of the churches. The archbishop of Paris had gone so far as to sponsor a draft declaration of faith, which went further than Bossuet's celebrated *Exposition* in accommodating protestant views, only to be forced into disavowing it.[56] This was particularly unfortunate because many protestants were awaiting such a move, which would have made it far easier for them to go along with a reunion of the churches. On the intellectual level, at least, there had been substantial moves on both sides which made reunification seem a real possibility, as the areas of disagreement were narrowed down.[57] The idea was certainly worth trying; even if large groups of protestants would certainly have remained obdurate, it would probably have produced a considerable number of genuine conversions. For all the obvious advantages of such a policy, however, it was always extremely unlikely that it could actually be implemented, for two crucial reasons. Firstly, it would have needed something like a diplomatic miracle to get the various factions within the French church to sink their differences on such a question, then secure papal approval, all within a very short time. Secondly, the endemic theological wrangling of the period probably made it impossible for the document to be drafted in a form acceptable to all catholics, even before protestant susceptibilities were allowed for. If the scheme were to have any chance at all the king needed to appoint a small group of bishops with clearly delegated authority to carry it through; as it was even Harlay was kept out of the crucial discussions, something about which he complained bitterly. His enemies

[55] The papal reaction is described by Orcibal, *Louis XIV et les protestants*, 139–46, and Armogathe, *Croire en liberté*, 127–52.

[56] For these moves, Orcibal, *Louis XIV et les protestants*, 81–8, and Armogathe, *Croire en liberté*, 75–83.

[57] This background is most sensitively examined by Orcibal, *Louis XIV et les protestants*, 7–20, 29–38.

had no intention of allowing him to gain undue credit, while the papal nuncio Ranuzzi feared that any declaration of the essential points of the faith would deepen the breach with Rome and lead to a schism.[58] The Le Tellier family and Bossuet backed the nuncio's protests, so that the king irritably ordered the scheme to be dropped, then lost patience and tried to smash his way through the whole complex problem. It very soon became apparent that whatever the Revocation had done, it had not brought any kind of solution. Among the crown's lay officials even such determined enforcers as the intendant Bâville in Languedoc became increasingly disillusioned; for their part, the bishops were so unenthusiastic in applying royal policy that the infuriated king took control of the missionary effort away from them, entrusting it to the intendants.[59] The whole business appears as a classic example of bad decision-making, in which events ran away from people's control, and the autocratic power of king and council was used to push aside the highly justifiable reservations among the clergy.

For the king and his councillors clerical matters had become so integrated with other affairs of state that they were expected to respond to the same treatment, and policy might be changed periodically according to the same considerations of temporary advantage. This was to try and play by the rules of the wrong game, for if a degree of machiavellianism was *de facto* accepted in politics, this was certainly not the case in religion. Even if higher motives did not really obtain, it was necessary to pretend they did, and the local and secular inspiration of royal policy was all too clear. When this tendency was combined with the king's well-known ignorance in the religious sphere, the effects could be disastrous, as the doctrinal quarrels which beset the church from the 1690s onwards demonstrated. The king's new-found piety did not make the religious observances of Versailles any less pompous or empty, nor lessen his aversion to movements for religious reform, inevitably seen as forms of political conspiracy. One might well argue that in view of the almost total capture of the

[58] See n. 56 above.

[59] Oricibal, *Louis XIV et les protestants*, 128–9, Armogathe, *Croire en liberté*, 91–8 (for the use of the lay power), 98–104, 120–4 (the disillusionment of Bâville and other intendants).

church by the state, any challenge to the *status quo* automatically became a kind of rebellion. The Jesuits were foremost in supporting this *étatiste* view, for the Society had now lost much of its earlier dynamism, and had always relied heavily on royal protection; several decades of embattled resistance to their critics had turned its members into firm defenders of the established order.[60] There was a natural tendency for the king to see them as his principal allies, which was particularly unfortunate when nearly every doctrinal dispute evolved into another round of the endless contest between Jesuits and Augustinians. Step by step the crown, primarily concerned with good order and conformity, was drawn into supporting what was in reality a partisan and divisive position, fanning the flames it believed itself to be extinguishing. The process was powerfully aided by the curious state of hypertension which had built up within the church, with supporters of all camps eager to find heresy behind every bush. While royal stress on orthodoxy made this an obvious technique for discrediting rivals, it also reflected a complex process of cumulative discord. Theological differences had become much more tightly associated with institutional rivalries, each dispute leaving a new layer of resentments, so that the principle of mortmain seemed to apply to ideas as much as to property. A training which stressed the techniques of disputation, alongside the pervasive influence of legalism and a rather naïve belief that proper use of the authorities could produce final answers, exacerbated the situation. The quite numerous moderates who saw that the points at issue were often impossibly difficult, and preferred to accept the inscrutability of divine intentions, faced a long uphill struggle to get their views taken seriously.

The original persecution of Jansenism, from Richelieu's arrest of Saint-Cyran in 1638 to the Clementine peace of 1668, had given ample warnings of the risks involved in an attempt to define and enforce strict orthodoxy. An affair which had begun in an oblique, almost accidental fashion had become entangled with central religious and political issues, to the point that a fair amount of diplomatic duplicity was needed in

[60] This evolution is very well described by Hillenaar, *Fénelon et les Jésuites*, 17–23.

order to secure a fragile truce. The deep mutual mistrust between the parties, and the conviction that vital questions were at stake, made it easy for more extreme counsels to prevail. On the one side stood Antoine Arnauld, the real creator of Jansenism as a coherent position, ready to dispute every point interminably with all his legal and patristic skills, pre-eminent among those 'défenseurs de la grâce' whom Racine would have wished 'un peu moins attachés aux règles étroites de leur dialectique'.[61] On the other the Jesuits, with Father Annat as first their controversialist, then their man in power as confessor to the king after 1653, determined to seize on every opening to discredit the supporters of strict Augustinianism. To provide a suitable pretext there was Archbishop Gondrin of Sens, rashly protesting that the rights of the bishops had been disregarded by the bull *Cum occasione,* and that the decision should have lain with a council; an error quickly exploited by the ambitious and intriguing Pierre de Marca, archbishop of Toulouse, determined to exhibit his zeal in the service of both pope and king.[62] In fact both these powers were being dragged into factual quarrels they could not hope to settle, something the popes generally understood rather better than did the royal government. The trick was to manœuvre one's opponents into a position where they appeared to be challenging authority, then use this to discredit broader aspects of their position. The Jesuits and their allies were very successful in this, but at a heavy cost, for the Jansenists riposted by appealing to public opinion, much helped by the widespread (and rather inaccurate) belief that they were the victims of a conspiracy directed from Rome. It was very dangerous for the crown to allow itself to be used in this way, and risk sharing the resulting discredit; growing awareness of this, and of the impossibility of applying coercion successfully, underlay the compromise of 1668.

Once the government had pulled back in this way it paradoxically found it much easier to persecute small groups of extremists on particular grounds. So long as there was no

[61] Cited by Adam, *Du mysticisme à la révolte,* 252.

[62] Blet, *Le Clergé de France,* ii. 181–4; on Marca, see F. Gaquère, *Pierre de Marca* (Paris, 1932), although his favourable view of the archbishop has found few supporters.

general campaign to identify dissidents, these occasional victims of police action attracted little sympathy, for it was generally apparent that they had behaved provocatively. The relatively small number of outright Jansenists were gradually silenced or driven into exile, so that when Arnauld died in 1694 it seemed as though his party might disappear with him. Although both Harlay and La Chaize were naturally hostile to Jansenism, they were too shrewd or cautious to employ it as a weapon to discredit others, no doubt partly because they had no need to recommend themselves to the king in such ways. The surviving Jansenists also forfeited much support, notably from the *parlements,* when they courageously took the side of the pope against the king over the *régale,* incidentally confirming Louix XIV in his opinion of them as inveterate troublemakers. In their place there now stood a far larger and more powerful group of *dévots,* Gallican clerics who espoused a rigorous Augustinian moral theology, distrusted the Jesuits, and hoped to root out superstition and vice from French society. The great majority of these men were firmly orthodox on the Five Propositions which had been the original touchstone for Jansenism, but they tended to share an interest in education, purified forms of worship, and the repression of popular amusements, all of which brought them very close to the wider aspects of the movement. The powerful support of Mme de Maintenon ensured that they captured a number of key positions in the church, with the appointment of Noailles to Paris as crucial evidence of their growing strength. There was a fundamental incoherence in royal policy at this point, for Louis XIV did not share his consort's views, yet allowed her a degree of influence which made no sense unless he was prepared to back her consistently. The complexities of the situation had already become apparent in the tortuous affair of Mme Guyon and 'le pur amour', which had startlingly wide consequences. Mme de Maintenon had taken this very dubious mystic under her protection in 1689, only to find by 1693 that she risked being compromised in accusations of Quietism; she and most of her associates then changed tack abruptly, mounting a campaign against her previous associate. The political overtones were widely recognized, and when Pontchartrain originally alerted the king it was on the grounds that a party was forming at court which was 'rédoutable à la religion, pernicieux aux bonnes mœurs, et capable d'introduire un fanatisme aussi fatal à l'Église qu'à l'État'.[63] It is not difficult to see how charges of this kind could have lead to a catastrophic disgrace for Mme de Maintenon, so that she effectively ran the same risks as all other leaders of religious groups; when Noailles was later tarred with Jansenism she would again have to mark her distance from him.[64]

The Quietist affair developed into a lengthy crisis, partly because Fénelon broke with Mme de Maintenon, mounting a very effective defence of the mystical position. This produced the famous controversy between Fénelon and Bossuet in which the latter not only came off rather the worse in the argument, but lost all sense of proportion. Quietism was in fact no threat to anyone by this time, and if left alone would have attracted only marginal groups of the pious, so it was absurd to see it in the exaggerated terms Bossuet used to Noailles in 1698 about a Burgundian *curé* who had been propagating Mme Guyon's writings:

Beaucoup de confesseurs me font avertir que l'erreur se répand sourdement à Dijon, elle ne fait que couver sous la cendre. Vous savez la correspondance du curé de Seurre avec Madame Gyon. Enfin l'Eglise est terriblement menacée.

[Many confessors have warned me that the error is spreading secretly at Dijon, where it is only smouldering under the ashes. You know of the correspondence between the *curé* of Seurre and Madame Guyon. In the end the church is terribly threatened.][65]

The same way of thinking, regrettably characteristic of the period, led Bossuet to strain every nerve in order to secure the condemnation of Fénelon's *Maximes des saints*. When the issue at Rome seemed doubtful he finally turned to the king, whose repeated bullying interventions produced the desired result at

[63] The bibliography on Quietism is enormous; a good modern account is that by L. Cognet, *Crépuscule des mystiques* (Tournai, 1958). Pontchartrain's comment is quoted (from Aguesseau's memoirs) by Hillenaar, *Fénelon et les Jésuites*, p. 57, n. 1.

[64] Adam, *Du mysticisme à la révolte*, 320.

[65] Cited by Hillenaar, *Fénelon et les Jésuites*, 105.

last. The reluctant papal condemnation of 1699 was a Pyrrhic victory, however, which carried the seeds of future trouble.[66] Beneath his affectation of humble submission Fénelon was bitter and determined on revenge, in seeking which he could now count on the support of the Jesuits, who had fought for him to the last. The Society felt itself threatened by the renewed campaign against Quietism, for although most Jesuits had little personal sympathy for this kind of mysticism, they knew that the *dévots* were also hostile to them, and that a new offensive was being planned. Ever since 1682 Bossuet had been trying to use the assemblies of the clergy to secure a fresh condemnation of laxist morality, only to be frustrated repeatedly by Harlay. In 1700 there was no longer any check of this kind in an assembly dominated by Noailles, Le Tellier, and himself, and a vast list of 127 laxist propositions, mostly drawn from the writings of Jesuits, were duly stigmatized.[67] Beyond this, the *dévots* were quick to take advantage of the embarrassing question of the Chinese rites in order to accuse the Jesuits of tolerating idolatry and superstition. These were pointless and foolish moves, whose chief effects were to unite the Jesuits in self-defence, and to make it easy for them to claim that there was a conspiracy against them. Behind this they could readily identify the Jansenists, whose intrigues at Rome had been exposed with the seizure of Quesnel's papers in 1703. In truth these intrigues had been very marginal, and the prime opponents of the company were not Jansenists at all, but guilt by association was very hard to disprove. Once the *dévots* had been identified with the heresy, the king could be turned against them just as he had been against Fénelon, and the latter's revenge would be complete.

This is the crucial background to the last great religious crisis of the reign, that over the bull *Unigenitus* of 1713. Once again a king quite lacking in any real understanding of the issues forced the pope into a definition of orthodoxy that the latter would much rather have avoided. Learning nothing from past mistakes Louix XIV was still allowing himself to be

[66] This long conflict is analysed in detail by Hillenaar, *Fénelon et les Jésuites,* 69–189.

[67] For Bossuet's conduct in the assembly of 1700, see Martimort, *Le Gallicanisme,* 685–7, and Hillenaar, *Fénelon et les Jésuites,* 219–22.

used as a tool by the contending parties, determined to impose their version of the truth, and humiliate their opponents, with little regard for the wider interests of either church or crown. The appointment of the extremist Father Le Tellier as the king's confessor in 1709, which remains something of a mystery, would have been very unlikely if the Jesuits had not been previously driven so much onto the defensive; the normally powerful group which favoured more cautious policies had temporarily lost much of its influence.[68] Aided by the errors of Noailles, Le Tellier became the pivotal figure in the ecclesiastical politics of the last years of the reign. His task was certainly facilitated by the general nature of the prescription he and Fénelon offered, with the idea of an authoritarian, hierarchical church, in which the king would enforce orthodoxy as defined by the pope. The reforms demanded by the Jansenists, and by the far more numerous *dévots,* would be definitively removed from the agenda, so that an obedient peace might settle on the church at last. Louis XIV might well think it worth sacrificing his remaining Gallican instincts for such a reward, failing to realize that he was being offered a mirage, which would vanish in the effort to bring it about. *Unigenitus* emphasized in the most disastrous fashion the shortcomings of the method of censures as a means to establishing orthodoxy, since in many cases there was room for interminable argument about the precise sense in which a proposition had been found objectionable. No amount of ingenuity, however, could obscure the fact that the bull represented simultaneous assaults on the Gallican liberties, on the *dévot* plans for the reform of religious practice, and on almost any kind of positive involvement by the laity. When in addition it was seen as an act of despotism perpetrated by a highly unpopular government, there could be little doubt that it would arouse a storm of protest. Louis XIV could rely on the support of the Jesuits and the great majority of the bishops, with a few embarrassing exceptions led by Noailles; elsewhere he was faced with bitter hostility from the *parlement,* the theology faculty, many religious orders, and important groups among the parish clergy. Further drastic measures were envisaged, notably the convocation of a national council

[68] For this appointment, see Hillenaar, *Fénelon et les Jésuites,* 254–5.

of the French church, a plan which was disrupted by the death of its architect Fénelon in the first days of 1715, and by the reluctance of the pope to sanction such a dangerous precedent. By the summer the king had decided to press ahead regardless, dispensing with papal approval; informed that the *parlement* would not register the royal declaration setting up the council, he was about to force it through by a *lit de justice* when his fatal illness intervened.[69]

Louis XIV's belief that doctrinal orthodoxy was vital if the church were to prove a reliable adjunct to the state had ultimately proved highly counter-productive. Men and women were prepared to resist the king over their religious beliefs as on no other issue, backed up by the knowledge that Christian history was full of examples of similar conduct which had been retrospectively justified. They were also well aware that Louis, personally ignorant in such matters, acted on the advice of unrepresentative cliques. Merely to appear a victim of such groups was to attract immediate sympathy, as Fénelon found during the controversy over the *Maximes des saints*, and despite their often rebarbative views the Jansenists won significant popular support precisely because they represented opposition to the régime. Conversely, royal ministers and bishops became substantially discredited as timeserving placemen, who had aided and abetted persecution and injustice. More seriously still, a new cynicism spread about doctrinal disputes themselves, which was to find mordant expression in the writings of two of the most notable children of this age of misguided conflict, Montesquieu (b. 1689) and Voltaire (b. 1694). It had been a fundamental error for the state to become so heavily involved in controversies which could never really be settled by forcible means, and to turn its power against groups which, left undisturbed, would have remained relatively insignificant. The attempt to improve the mechanisms for ensuring orthodoxy created new tensions, to leave simmering discontents which would plague the monarchy during its final decades. Clashes with the *parlement* of Paris over the refusal to offer the sacraments to elderly Jansenists verged on the grotesque, yet they seriously

[69] There is a graphic description of this final crisis in Adam, *Du mysticisme à la révolte*, 315–30.

weakened the crown's authority in the middle of the eighteenth century. More serious still, the resentment *curés* felt towards remote and arrogant bishops led to the electoral revolt of 1789, which saw the *curés* take more than two-thirds of the seats in the First Order for the Estates General, then carry through the crucial vote of 19 June to join with the third estate and form the National Assembly. Although this dénouement naturally reflected many new factors which had emerged over the intervening decades, the church in which it ocurred was still essentially that created under Louis XIV. The attempt to eradicate protestantism also left an unfortunate legacy, most strikingly with the bitter and long-lasting revolt of the Camisards in the Cévennes, more permanently with the problem of *nouveaux convertis* whom the church never managed to assimilate, so that a limited degree of toleration was actually restored in 1787. Contemporaries already recognized that the forcible conversion of the protestants had been the occasion for a significant increase in scepticism and irreligion.[70] It might also be argued that the removal of official protestantism had unexpected drawbacks for the catholic church, which had actually obtained considerable stimulus from the rivalry during the period after 1598.

Whatever the negative aspects of Louis XIV's obsession with orthodoxy, there can be little doubt that the church did act as a vital adjunct to the state in maintaining an ordered, hierarchical society. This was partly a matter of ideology, with divine right monarchy and the duty of obedience to superiors at every level taken for granted by almost all clerical writers. Religious rituals naturally formed the central symbolic expression of these same values. In both cases there is a classic problem in deciding how far repetition is effective in implanting fixed ideas, how far it may become an empty show performed with tongue in cheek. The propaganda campaign of Louis XIV's time now seems perilously close to vacuous bombast, and it is possible that by 1715 it had imperceptibly gone over the top; certainly neither of his pre-revolutionary

[70] The point is very well made and illustrated by Orcibal, *Louis XIV et les protestants*, 159–67. Among others, he quotes the prophetic comment by Leibniz, 'qu'il faudrait maintenant s'appliquer bien plus à combattre l'Athéisme ou le Déisme, que non pas l'hérésie' (167).

successors found it appropriate to continue with the same style. For the church it was less easy to change course, so that the more traditional eighteenth-century clerics found some difficulty in evolving arguments suitable for the century of the Enlightenment. There was another level, however, at which church and state also worked very much in harmony in the seventeenth century in the interests of order and good behaviour among the population at large. The catholic reform movement, several of whose aspects are discussed elsewhere in this volume, can be characterized, with only slight exaggeration, as one of the greatest repressive enterprises in European history. Educated clerics had become convinced that France was a 'pays de mission', many of whose inhabitants knew nothing of true Christianity, and practised a religion based on paganism and superstition. Although the underlying vision may have been a positive one, of a truly Christian commonwealth, the practical expression was largely negative. There was an extensive attack on popular culture, with the suppression of pilgrimages and healing shrines, prohibitions on dancing and the *veillées,* and the elimination of many magical and protective rituals. The *dévots* would have liked to turn the *curés* into moral policemen, keeping *fiches* on the behaviour and attitudes of every family, and using their influence over the distribution of charity to reinforce other forms of pressure.[71] This campaign has obvious parallels with the simultaneous efforts by the state to eliminate political disobedience and reduce the importance of any competitors for authority. It is no surprise, therefore, to find that the secular power was repeatedly invoked to support such prohibitions, which royal administrators could easily see as an extension of their own preoccupations.[72] National values were to prevail over communal and local ones, pluralism to give way before linear power structures.

Given the formidable obstacles it faced, the relative failure

[71] For the keeping of individual dossiers, see L. Michard and G. Couton, 'Les Livres d'états des âmes: une source à collecter et à exploiter' *RHEF* 67, (1981), 261–75, and A. Croix, *La Bretagne aux 16e et 17e siècles* (Paris, 1981), 1210–11, 1407–10.

[72] These interventions are discussed and illustrated by Y.-M. Bercé in *Fête et révolte* (Paris, 1976), esp. 127–62.

of this enterprise was probably inevitable; popular beliefs and practices proved immensely resilient, the combined power of church and state very weak at parish level. It was most effective precisely in securing the kind of outward conformity which suited the crown fairly well, while frustrating the more profound ambitions of the reformers. The improvement in clerical education and standards, which in the long run produced the articulate and dissatisfied *curés* of 1789, had earlier tended to make the priest more of an outsider in relation to the community, as a representative of alien values. This increasing gap is surely one of the principal reasons why clerics ceased to feature as leaders of popular revolt after the middle of the century, after being very prominent in the major disturbances of the time of Richelieu and Mazarin.[73] This did not mean that they had become totally subservient, however, and beneath the deceptively smooth surface there were some dangerous cross-currents. The austere moral theology now taught in most seminaries contained much implicit criticism of a highly unequal society, dominated by privilege and disfigured by economic exploitation. Many severe clerical writers were at least as critical of the hedonism and selfishness of the ruling classes as they were of the 'idolatry' of the peasants, emphasizing how difficult it was for the rich to be saved from damnation.[74] It was dangerously easy to extend this particular view to the gross inequalities within the church, and it was hardly surprising if *curés* became increasingly resentful of the wealth of bishops, chapters, and monasteries, when their own qualifications and responsibilities had increased so markedly. Louis XIV had once thought of requiring that potential bishops should have experience of parochial and missionary work, but this was hardly practical, and it remained inconceivable for a *curé* to aspire to any of the privileged places in the church.[75] Although the young postulants to the episcopate did serve a kind of apprenticeship, as

[73] This point is made by E. Le Roy Ladurie in 'Révoltes et contestations rurales en France de 1675 à 1788', *Annales ESC* 29 (1974), 6–26, esp. 8–9, although I would not agree with his stress on the revocation of the Edict of Nantes.

[74] See Chapter 7 for a development of these points.

[75] Peronnet, 'Les Évêques', 468 (Louis XIV's ideas), 742 (position of *curés*).

vicaires-généraux learning the ropes of ecclesiastical administration, this provided no common ground between them and the ordinary clergy. Indeed, the functions of these posts, together with the expectation of future preferment, might be expected to encourage arrogance and self-importance.

Separated by social origins, income level, and practical experience, officers and other ranks within the church may also have developed ideological differences. As the crown became hypersensitive to any suspicion of Jansenism, particularly as defined by *Unigenitus*, the bishops chosen were commonly supporters of a more 'externalized' religion, with little enthusiasm for moral teaching which implied a challenge to the established order. So far as these representatives of privilege were concerned, there was an uncomfortable sense in which the main thrust of the catholic reform, as it had developed by the later seventeenth century, led to unpalatable conclusions, and therefore had to be aborted. This was more a matter of attitudes than of doctrines, and there could be no question of offering a direct challenge to the austere orthodoxy enshrined in the standard works of moral and pastoral theology, and taught in the seminaries.[76] While the *curés* themselves may have been glad not to come under much pressure to apply these rather unrealistic teachings, many of them probably felt some guilt on this score, and can have entertained few illusions about the motives of their superiors. The famous *curé* Meslier, with his atheistical communism, may well have been unique in his generation, but it is fascinating to note how many of his views were developed from the rigorist morality which was so common among his contemporaries.[77] They did not need to draw his extreme conclusion that religion was no more than a fraud designed to keep the people in subjection in order to feel some disquiet about the way the church operated. One might rationalize the attitude

[76] For this orthodoxy, see J. Guerber, *La Ralliement du clergé français à la morale Liguorienne: l'abbé Gousset et ses précurseurs, 1785–1832*, Analecta Gregoriana, 193 (Rome, 1973).

[77] On Meslier, see M. Dommanget, *Le Curé Meslier* (Paris, 1965); for his use of ideas debated in the *conférences ecclésiastiques*, D. Julia and D. McKee, 'Les confrères de Jean Meslier: culture et spiritualité du clergé champenois au 17ᵉ siècle', *RHEF* 69 (1983), 78–81.

of the bishops by suggesting that the need to preserve the social hierarchy led to the stifling of originality; in consequence the church lost touch with the development of ideas, and actually became a less effective guarantor of order in the long run. The situation was vastly more complex than this, of course; there had always been deep inconsistencies within the church's own ideology, and the resulting doctrinal wrangling had become sterile and counterproductive. As has already been explained, the increasingly repressive and reactionary line taken by the crown was often provoked by the endemic factionalism among the clerics themselves.

Whatever happened in the next century, there can be no doubt that the first three Bourbon kings were generally successful in exploiting the alliance between throne and altar in their own interests. The church was remarkably loyal, at times positively fawning on the monarchs who made use of it. Even those groups whom the crown persecuted—and this included the protestants after 1629—were always emphasizing their total dependence on the king. The church unfailingly supported the king against the pope, and contributed handsomely to the cost of his wars. Its individual members served the state in many capacities, for which few of them received any direct payment. Although the immediate effect of the tridentine reforms was to emphasize the role and status of the bishops, in the longer run it was the *curés* whose position changed most. By the eighteenth century this was emphasized by the literary commonplace of *le bon prêtre,* a reflection of the high public esteem now enjoyed by the secular clergy as a whole.[78] It was somewhat ironic that this esteem proceeded largely from the more secular functions of the *curé* as a local mediator, poor relief manager, and intermediary between community and government. By using the church in this way, the crown helped to desacralize it, so that it threatened to become just another institutional buttress of the *ancien régime,* sharing its increasing vulnerability. In this respect, as in virtually all others, the relationship between church and state was deeply ambivalent. The common sense of purpose which had marked the earlier decades of the period had been founded on

[78] P. Sage, *Le Bon Prêtre dans la littérature française d'Amadis de Gaule au Génie du Christianisme* (Paris, 1951).

an analogous sense of enemies to be overcome: popular disorder, licentiousness, heresy, and disrespect for legitimate authority. As the most obvious aspects of these evils were overcome, so more complex problems appeared on the agenda, to which official responses were uncertain, inconsistent, and often divided. Louis XIV's very success in asserting divine right monarchy made it natural that his reign should see the high water mark of co-operation between church and state, but may also help to explain why the tide had perceptibly turned well before his death. From the 1680s onwards a destructive interaction between heavy-handed royal authoritarianism and feuding within the church itself was set up, with grave consequences for the future of both institutions.

6

The church and family in seventeenth-century France

Marriage is a great sacrament, I say in Jesus Christ and in his Church, it is honourable to all, for all, and in everything, that is to say in all its parts. To all, for even virgins should honour it with humility; for all, since it is as holy among the poor as among the rich; in everything, for its origin, its end, its uses, its form, and its matter are holy. It is the seed-bed of Christendom, which fills the earth with the faithful to complete the number of the elect in heaven . . .[1]

THE famous opening words of François de Sales's chapter 'Avis pour les gens mariés', in the *Introduction à la vie dévote,* begin the seventeenth century on an encouraging note, followed as they are by exhortations to mutual love and tenderness. Not that even François de Sales questioned the inferiority of the married state to that of chastity, or believed that it should be normal for widows to remarry; nevertheless his humane and realistic approach might be thought to mark a new and promising departure.[2] Occasional echoes of his views can be heard in the later years of the century, but the dominant tone of clerical writers was very different. In his enormous *Pratique du catéchisme Romain* of 1630 Thuet, a doctor of the Sorbonne, expressed the general opinion:

Thus it was the commission of sin which has deprived and stripped man, and all his descendants, of the rich gifts of the original justice, from which has followed that great rebellion of the flesh against the spirit from which even the saints have not been exempt. For those who do not feel themselves strong and powerful enough to fight that

[1] F. de Sales, *Œuvres,* ed. A. Ravier (Paris, 1969), 233.

[2] Ibid., 244–9.

furious beast, the concupiscence of the flesh, so that they shall not fall from the justice and innocence which they have recovered in baptism through the Passion of the redeemer, God gives them marriage as the remedy for this sickness . . .[3]

Thuet takes his readers firmly back to the great authorities on the subject, St Paul and St Augustine, without even a nod towards the more positive approach exemplified by François de Sales. The austere Augustinian stance is found among clerics of all schools, Jesuits as well as Jansenists, and its hold seems to strengthen as the century wears on. Probably those responsible would not have objected to being told that it was more a regression than a development, since their position was ultimately based on the premiss that the primitive church was the essential model, to which the catholic reform should aspire to return.

The more detailed developments of this attitude, not to mention its wider implications, remain rather obscure. Despite much valuable recent work we are as yet some way from an adequate historical understanding of the family itself, and of the customs and sentiments which surrounded it. Historians still take widely differing views on many aspects of the subject in general, while in all the admirable recent research devoted to the catholic reform in its pastoral aspects the family and marriage make only the most fleeting appearances. No doubt historians should have done more, but the seventeenth century itself has left them somewhat starved of both evidence and fresh ideas. In France at least there is a startling lack of original contributions to the theology of marriage and family relations, which secular literature does little to remedy for any but the highest social levels, and which helps to explain the endless references to the *Introduction à la vie dévote*. A search amongst the dense brushwood of episcopal ordinances, handbooks for priests, works of devotion, and sermons yields only a meagre collection of repeated prohibitions and traditional commonplaces. One of the few historians to have perceived this strange void, and tried to make sense of it, John Bossy, was led to the striking conclusion that it was in

[3] E. Thuet, *Pratique du catéchisme Romain suivant le decret du Concile de Trente* (Paris, 1630), 1011.

its inability to integrate the nuclear family that the Tridentine church 'most damagingly failed'.[4] It is hard to dissent, although one must be aware of the extreme difficulties of proof in this case. One primary task must be to account for this failure, since Bossy's brief invocation of a reaction against protestantism can be only a very partial explanation. At the same time the dominance of tradition in this area does not mean that the catholic reform had no implications for the family; almost every aspect of its pastoral activity had the potential capacity to affect relationships both inside families and between them and the church.

One obvious difficulty is that both 'the church' and 'the family' are conventional unities and terms, masking wide divergences. It is difficult enough to ascribe general views to clerics, let alone to judge how effectively they were enforced on or even conveyed to the laity. As for the family, there is a fundamental ambiguity between the household and the lineage, which can never be tidily resolved without deforming the reality. One must also recognize the extreme diversity of both family types and legal systems in France, which constantly threatens to undermine all generalizations. Personal behaviour has always been so various that individual examples prove nothing, yet no satisfactory mode of quantification seems possible. The dangers inherent in taking theoretical treatises as evidence for social facts hardly need stressing, particularly if we look back to the Middle Ages. Indeed, the enormous gap between doctrine and practice in so many fields during the preceding period must be a crucial element in explaining the nature of both protestant and catholic reform, and it could hardly have been wider than it was in questions relating to marriage and sexuality. Most people presumably bridged the gulf by the widespread technique of compartmentalizing the world into the sacred and the profane; moral reformers of all denominations were to labour mightily to break down this barrier, and to assert the right of the sacred to invade all areas of life. This attitude was trenchantly summarized, with partial reference to

[4] J. Bossy, 'The counter-reformation and the people of Catholic Europe'. *Past and Present,* 47 (1970), 51–70. See also Bossy's later comments in *Christianity in the West, 1400–1700* (Oxford, 1985), 19–26, 122–5.

marriage, by the late seventeenth-century Jansenist Jean-Jacques Duguet:

There is nothing purely human, nothing purely political, in a Christian woman; religion is everything, enters everywhere, has control over everything; it is religion that should rule everything, sacrifice everything, ennoble everything. Salvation is not only the most important business, but the only one. One must work towards it independently of everything else, and only apply oneself to other matters with reference to that great purpose. Everything must be adjusted to it, everything respond to it; but it must never be adapted to fit our other purposes. A husband, children, friends, and all just associations, are only for salvation.[5]

The widespread adoption of such views led to an attempt to replace or purify a whole complex of entrenched beliefs and practices, an attempt to impose a new moral order which represents one of the most dramatic breaks in European social history. There have been some extreme views of the effect on the family, with defenders of women and children portraying their past experience in terms of a prolonged assault on their psyches and persons, while *bien-pensant* historians have tended to assume that there was no incompatibility between religious and domestic virtues. Although the more depressing portrayals of personal relations as brutal and lacking in affect have now been effectively criticized, the typical behaviour of spouses and parents remains very elusive.[6] What is far clearer, and in striking contrast to modern assumptions about catholicism and family values, is the very cool and distant attitude the French church took towards the family, at least in its official doctrines.

The limitations of clerical thought may be partly accounted for by the manner in which moral theology was organized around conventional structures, built around the seven deadly sins, the ten commandments, and the seven sacraments. Authors fell gratefully into the patterns these provided, with extensive use of compilation from their predecessors, in a

[5] J-J. Duguet, *Conduite d'une dame chrétienne pour vivre saintement dans le monde* (Paris, 1724), 316–17. This work was probably written several decades before its publication date.

[6] For differing versions of the 'black legend' of early modern emotional life, see works by L. de Mause, E. Shorter, L. Stone, and E. Badinter; recent correctives include R. A. Houlbrooke, *The English Family 1450–1700* (London, 1984), L. A. Pollock, *Forgotten Children: Parent-Child Relations from 1500 to 1900* (Cambridge, 1983), and F. Mount, *The Subversive Family: An Alternative History of Love and Marriage* (London, 1982).

context of respect for tradition rather than for innovation. The one important shift within these categories between the thirteenth and sixteenth centuries saw the rise of the commandments at the expense of the seven deadly sins; this meant more emphasis on specific prohibitions, resulting in less rather than more flexibility.[7] There was a partial escape route in the form of casuistry, the increasingly popular genre which employed sample cases to define moral law, but this too was negative by its very nature, and could only provide an indirect opening for any new theological approach. Despite these obstacles, some daring innovations concerning marital and sexual morality were propounded, notably by the Parisian doctor Martin le Maistre in his *Quaestiones morales* of 1490, and by the Spanish Jesuit Thomas Sanchez in *De sancto matrimonii sacramento* of 1602.[8] These authors, with a reasonable number of others, tried hard to combat the view that sexual pleasure was inherently sinful; had their arguments been generally accepted they would have made it much easier for the church to evolve a realistic and practical attitude to marriage. Unfortunately these novelties, prudently published in Latin, were only apparent to readers well versed in the details of traditional teaching. To have had much effect they would have needed not only a more favourable reception than they actually received, but also much development into a positive theology and a new pastoral technique. They had formed part of a movement away from Augustinian pessimism towards a more balanced view based on such authorities as Aquinas; quite strong in the later sixteenth and early seventeenth centuries, this ran into an Augustinian backlash later, which proved notably strong in France. It is no surprise, therefore, to find that most seventeenth-century catechisms state traditional views in an uncompromising fashion. Their characteristic tone can be heard in Bossuet's catechism for his diocese of Méaux:

To what purpose should one make use of marriage?
For the purpose of multiplying the children of God.
What other purpose can one have?

[7] Bossy, *Christianity in the West*, 116, and an unpublished seminar paper on 'Moral arithmetic', given in Oxford in 1986.

[8] For these innovations, see the admirable discussion in J. T. Noonan, *Contraception: A History of its Treatment by the Catholic Theologians and Canonists* (Cambridge, Mass., 1966), 303–40.

That of remedying the disorders of concupiscence.
What are the obligations of marriage?
To unite with one another, and support one another through charity; to bear mutually with one another and all the pains of marriage with patience; and to obtain salvation by the holy education given to one's children.

In the final section he includes:

Tell me the evil which must be avoided in the use of marriage?
It is to refuse conjugal rights unjustly; to make use of marriage to satisfy sensuality; to avoid having children, which is an abominable crime.[9]

Encapsulated in these dry questions and answers we can see the main issues debated by the theologians, resolved to their own satisfaction. How much enlightenment they brought to ordinary believers seems more doubtful, even when they were told that the husband represented Christ and the wife the church: 'How in particular should the husband represent Jesus Christ? In loving his wife cordially as the Son of God has loved the church, seeking the service of the church, and not his personal interests.'[10] Clerics resorted far too readily to this overworked commonplace, expressed in the kind of language which substituted a sonorous evasiveness for the refreshing directness of François de Sales. This habit would have been less significant if there had been a good and abundant secondary literature of guides for laymen and homilies to be read by priests. Such things do exist in small numbers, but few of them possess the slightest merit or conviction.

Among the general guides for secular clerics perhaps the best known was the *Instruction sur le manuel* by Matthieu Beuvelet, one of the priests attached to Bourdoise's parish of Saint-Nicolas-du-Chardonnet. This work for seminarists was popular from its appearance in 1654 for its practical, direct descriptions of the priest's duties, and was commonly recommended by bishops. Beuvelet included a series of 'Exhortations' for use during the various ceremonies, which could also serve as models for the *curé*. Those relating the ceremony of *fiançailles*

[9] J.-B. Bossuet, *Catéchisme du diocèse de Meaux* (Paris, 1687), 183–5.
[10] Ibid., 185.

or betrothal offer a series of warnings; God must be given his proper part, this is a contract on which there is no going back, the fiancés must not engage in 'aucune liberté ensemble', nor conceal any legal impediment to their marriage.[11] The exhortations for use during the ceremony of marriage itself emphasize the importance of the sacrament and the benefits God will offer those who follow him, but combine this with a series of pessimistic statements about general practice:

the grandeur and the dignity of marriage are not well understood by all Christians, and the dispositions in which most present themselves for it make all too clear to us the little esteem in which they hold it. . . . but at the same time as we speak of happiness in marriage, we are astonished to see so much unhappiness there. Do you wish to know the reason? It is that most people do not receive the grace of marriage. It is an article of faith that there are particular helps attached to the state and condition of marriage, to live in a holy manner therein, to support its charges and burdens, and to bring up children according to God: but how many are there, who receive these graces! The person from whom one should expect them is banished from marriage: the spirit of impurity, vanity, and evil intentions form obstacles to that divine light, so can one be astonished if people live in the shadows, in ignorance and forgetfulness of the affairs of their salvation? Pay attention to this, for today once past, it will no longer be time, this grace is only given at the moment of marriage, and to those who come to it properly prepared. This is what will decide all your happiness or your unhappiness.[12]

This was the tone which came most easily to the moralists of the period; here it is associated with the idea that the power of the sacrament is only efficacious if conditions are right at the time of its application. Quite apart from the crudely 'supernatural' element in such a doctrine, it is hard to see the pastoral merits of telling the participants that they may already have missed the bus. There is a strong suspicion here that many clerics thought of marriage as a regrettable necessity, and were (probably unconsciously) unwilling to allow it any positive role in the pursuit of salvation; they may also have been reacting against the protestant dismissal of celibacy as a holy state. They therefore made only the most passing

[11] M. Beuvelet, *Instruction sur le manuel* (7th edn., Paris, 1669), pt. i. 310–22.

[12] Ibid., pt. ii. 398, 403.

references to the possibility that the sacrament might bestow fuller participation in sanctifying grace, emphasizing instead the mutual aid and support the partners could offer one another against the tribulations of marriage.

Beuvelet's notion of the way the sacrament operated was clearly one he shared with his colleagues at Saint-Nicolas-du-Chardonnet, for in their own collection of treatises for the guidance of priests the same themes are hammered home. The sin of marrying without proper care or preparation is likely to be visited on the guilty even in this life, while the grace which could have been obtained with the sacrament is lost for good.

> . . . it is of the greatest importance, for the church and the state, for the families, and for individuals, that people should be instructed in a solid and Christian way of God's intentions in this sacrament . . .
>
> . . . but however abundant tears may be, they cannot wash away the sin one has committed in marrying against the rules of the scripture . . . their ignorance cannot excuse them before God for the unfortunate and perpetual obligations into which they have thrown themselves . . .[13]

According to Bourdoise and his disciples there were four basic rules of preparedness for marriage. Expressed in terms of the man, these were that he should feel certain of his capacity to govern and instruct a family, while giving a good example to his spouse. He must feel assured that his chosen spouse would help in every way possible in this task. He must be determined to put the commandments into effect among his children and servants. Finally, he must feel an interior spiritual inclination to sanctify his spouse and children, preparing them for the kingdom of heaven. To set such elevated standards was to ensure that most people would fall far short, and to justify in advance the widely held view that most marriages, lacking divine approval, were bound to prove unhappy.[14] A Jesuit,

[13] These quotations are taken from C. B. Paris, *Marriage in Seventeenth-Century Catholicism* (Paris and Montreal, 1975), 71, 79. This is an excellent and painstaking survey of the literature, which I have found to be very reliable wherever I have checked it, and on which I have drawn when I did not have access to some of the original texts.

[14] This advice is analysed by Paris, *Marriage*, 91, who understandably expresses amazement that anyone could have thought such standards attainable.

Father Le Blanc, wrote a treatise on *La Direction et la consolation des personnes mariées* in which, after briefly admitting that marriage is necessary for social life and the propagation of the species, he concerned himself entirely with unhappy marriages. This was no pioneer work of marriage guidance, for the consolations proffered were exclusively spiritual and other-worldly, making much use of the redemptive power of suffering.[15] When faced with such examples one wonders whether marriage would not have been better off without the status of a sacrament, if this involved the setting of absolute moral standards from which no deviation could be allowed. The parallel with the union of Christ and the church may have dignified marriage, but it also provided a quite inappropriate model of other-worldly perfection. Even the leading follower of François de Sales, Bishop Camus, providing model allocations for the *curés,* was swept along with the tide:

> . . . they mostly receive and contract this sacrament with a conscience charged with mortal sins . . . they have provoked malediction and it will come upon them . . . whereas those who have the fear of God . . . will see everything prosper in their families . . .
> Those who intend to marry should think of that covenant as a sacred good, which they should take great care not to contract in a state of malediction and disgrace, unless they wish to extend the resulting misfortunes even to their descendants . . .[16]

Clerical and popular views alike interpreted marriage as a *rite de passage* during which the partners were in a liminal state, exposed to peculiar dangers. One of the handful of clerics to write with some sympathy about marriage, the Jesuit Claude Maillard, synthesized these fears in a fascinating passage:

> The devil does not only make war on the newly married himself, immediately pushing them into disorder and excesses, because of his fear that they will bring blessed children into the world, who will fear God, serve him well, and finally fill up heaven; but he also makes use of his agents, who are the witches, magicians, and enchanters, who by their charms, knots, spells, ligatures, cause so many misfortunes in marriages, but whose charms and diabolical inventions are often prevented or dissolved by the blessing of the

[15] T. Le Blanc, *La Direction et la consolation des personnes mariées, ou les moyens infaillibles de faire un mariage heureux, d'un qui seroit malheureux* (Paris, 1664).

[16] Cited by Paris, *Marriage,* 71, 80.

priests, as these damned instruments of the devil have often confessed. Note that the devil's hostility to these young plants is particularly directed to preventing them having children, by his own work or that of his agents . . . When everything is joyful at the wedding, the devil seeks to trouble the feast, and convert the blessing into a curse, therefore it is necessary to make use of the ecclesiastical blessing against his design, and those of his fiends. If we bless the meat we eat to drive away any curse, and to ruin the enterprises of the devil, who seeks means everywhere to ruin mankind, one has good reason to bless the newly weds, against whom this sworn enemy of our nature will spare nothing to make the marriage unhappy.[17]

The widespread belief in the 'nouement d'aiguillette' as a means of causing impotence had led to a good deal of difficulty over clandestine marriages, the use of odd hours for the ceremony, and an array of superstitious counter-magical practices, but Maillard ignores these awkward issues to make it a major justification for the benediction.[18] In fact the priest blessed not only the couple, but the marriage bed itself. This final ceremony brought him into a still more exposed area, since he was entering into the family and neighbourhood celebrations. According to Beuvelet the purpose was twofold:

1. To chase away all foul spirits, and equip the newly weds against the malice of Satan, who busies himself by all means to trouble their repose and their salvation, to the point sometimes of preventing them having intercourse.
2. To damp down the heat of concupiscence, so that the married shall make use of marriage with a truly Christian modesty, and like children of the saints, shall make their bed without stain, as the scriptures say, and shall never dishonour such a holy union.

If, as sometimes occurred, the protection did not work: 'It is a punishment for their infidelity, or their past incontinences, or the brutal desires with which they have approached marriage.'[19] In one of the matching 'Exhortations' the priest explains that the benediction of the bed 'chases away the evil

[17] C. Maillard, *Le Bon Mariage, ou le moyen d'estre heureux et faire son salut en estat de mariage* (Douai, 1643), 206.

[18] For these problems see E. Le Roy Ladurie, 'L'aiguillette', in *Le Territoire de l'historien* (Paris, 1978), ii. 136–49.

[19] Beuvelet, *Instruction,* i. 369–70.

spirit, and disperses the traps and subtleties Satan may have set'. This seems appropriate enough, but it may not have been such a good idea for him to precede it with the suggestion 'think of this bed as a place where you will die one day', or to follow it with the story of Sara, whose first seven husbands were killed by the devil 'in getting into bed, because they did so with desires and thoughts entirely filled with immodesty'.[20] By no means all Beuvelet's remarks are unsympathetic, yet he never manages to suggest a reality behind his generalized invocations of holy and moderately happy matrimony. His general tone indicates that he would have found it easy to agree with the Oratorian Senault, in his sermon on marriage published in 1650: 'Do you wish to know why God has no part in your marriage? It is because you did not marry for love of him, you married by a passing fancy to seek your sensual pleasure.'[21]

These well-meaning clerics were caught within a network of constraints, social, intellectual, and political, of which they naturally remained largely unconscious. Within the community the family represented the chief focus of solidarity, yet it was one that clerics inevitably mistrusted, since they were excluded from it. They were much more at home with the community in its corporate manifestations. Sundays and feast-days, with their ceremonies and processions, gave the priest a defined and important place, not merely among his flock, but at their head. Pride, in all its manifestations, was a peculiarly insidious and widespread sin of the period, yet one which went relatively unnoticed by these severe clerical moralists, even in the form of *disputes de préséance.* It was all too easy to defend one's own position under colour of defending the prerogatives of the order to which one belonged, a more personal and more elusive form of *raison d'état,* and one to which the clergy were notably subject. Many of the internal disputes which played such havoc with the reform movement can be related to this endemic problem, as in the case of the quarrels between seculars and regulars, or the bitter rivalries between particular orders. In a less obvious form it also lay behind the emphasis clerics laid on the sacraments; these were

[20] Ibid., ii. 419. This story was regularly invoked by writers on marriage, particularly in sermons.

[21] Paris, *Marriage,* 76.

their professional speciality, the area in which only they could operate, and where their predominance was uncontested. The sacraments did extend into family life, since baptism, marriage, and extreme unction marked the *rites de passage,* but here the integration was less satisfactory. It was highly intermittent, while the priest was only one of numerous actors in a complex set of rituals and observances. The most far-reaching sacrament of all, that of penance, dealt essentially with individuals; this was probably one reason why it offered a more effective field of operation, and attracted so much more notice from clerical authors. A purely functional reason for its significance lay in the possibilities of the literature of casuistry, far more open to development than any other area of pastoral theology.

The preference for confronting individuals fitted in well with the tendency to stress paternal authority; the more this was done, the more the father could stand for the family as a whole. This opened the way for some notably heavy-handed advice, mostly concerned with the imposition of rules which were largely negative in character. The moralist Lordelot emphasized the central role of the fear of God:

> when at first fathers and mothers instil this salutary fear in the hearts of their children, it is a divine milk which nourishes them, and forms a solid substance for them which will maintain them in their duty for their whole lives. This first impression of fear is never lost, and if as a result of human fragility these children have just fallen into some sin, this divine seed, which seemed dead, immediately begins to appear; the fear of God appears in the mind; seizes the heart, and brings quick recovery when they fall. Apply yourselves therefore to bring them up well in this salutary fear.[22]

While it was commonly said that a father should ensure the saying of family prayers, with the participation of servants and children, no other communal religious exercises found any favour. This was perhaps less because of the taint of protestantism than because there was a general clerical reluctance to erect the family into an alternative centre of spiritual activity. One may compare this with the lukewarm and

[22] [Lordelot], *Les Devoirs de la vie domestique, par un père de famille* (Brussels, 1707), 42–3.

repressive attitude of bishops towards the *confréries,* whose mixture of profane and spiritual functions, and independence of control by the hierarchy, seem to have outweighed their obvious success in generating popular religious enthusiasm.[23] Nor do the rigorist clerics seem to have had much time for children, commonly taken to be obstinate embodiments of original sin; as the Jansenist Bishop Colbert of Montpellier put it: 'The Church believes that children coming into the world are slaves of the devil, that he holds them in his thrall, and that they are under his tyranny, which implies that original sin is a sin which kills the soul and merits death and eternal damnation.'[24] Parents themselves were held to sin when they failed to apply the necessary discipline. A particularly gloomy example of this appears in a sermon 'Du devoir des pères' by Father Lejeune, published in *Le missionaire de l'Oratoire* in 1670:

> How does it happen that so few children, even of worthy people, are virtuous and spiritual? It is that there are very few, and almost none, who are begotten and created for religious motives, and the love of God.
>
> If you do not punish your children they profane the churches . . . when they form the habit of swearing, lying, cursing . . . when they haunt meetings of girls . . . if you knew the evil which would result for them . . . you would prefer to break their arms and legs . . .[25]

Behind these dismal exhortations there lay a belief that God rewarded virtue on earth as well as in heaven; the godly family would prosper, the ungodly do badly. This was argued particularly strongly by Texier in his sermon on 'La famille heureuse', where one finds the following:

> the sin which is found in a household causes the interruption of prosperity and temporal blessings . . . the sin of a single person is

[23] For the *confréries* see G. Le Bras, 'Les confréries chrétiennes—problèmes et propositions', *Revue historique du droit français et étranger,* ive série, 19–20 (1940–1), 310–63; M. Agulhon, *Pénitents et francs-maçons de l'ancienne Provence* (Paris, 1968), and J.-P. Gutton, 'Confraternities, *curés* and communities in rural areas of the diocese of Lyons under the ancien régime', in *Religion and Society in Early Modern Europe,* ed. K. von Greyerz (London, 1984), 202–11.

[24] C. J. Colbert, *Œuvres* (3 vols., Cologne, 1740), ii. 585.

[25] Paris, *Marriage,* 78, 166.

enough to draw down the anger of God . . . Watch therefore, fathers of families, to see if sin, as a cursed obstacle to the blessings of heaven, is not hidden somewhere; it is enough sometimes that it is found in a child, in a girl, in a manservant or maidservant to be the downfall of the whole house. A family which is defiled by mortal sin . . . is out of that state which God wishes . . . there is an object in that household which draws down his anger and his malediction.

The contemporary obsession with the dichotomy between purity and pollution is very much in evidence here, helping to create a singularly unattractive doctrine. More sophisticated arguments were then necessary to account for the obvious fact that the ungodly often prospered in a most unfitting way:

God, whose conduct is always admirable, often takes pleasure in ruining and destroying the sinner completely by means of slight and short prosperities . . . so there are rich people who are so only in appearance, and who are fundamentally always poor, because they are always wicked, always starved, and their wealth serves only to torment them.[26]

The number of listeners or readers who were convinced by Texier's tortuous dialectic is perhaps open to doubt.

Parental authority was also crucially involved in the choice of marriage partners. Here the moralists constantly stigmatized parents who forced their children into unwelcome matches for purely worldly reasons, yet counselled strict obedience by the victims of such practices. It is not surprising to find the *précieuses* complaining about this aspect of marriage, and the occasional cleric gave them limited support.[27] In one of the most balanced treatises of the period Claude Maillard argued that children were not bound to marry as their parents ordered them to do, and even approved of marriage without parental consent when the parents were plainly unreasonable and paid no heed to the children's interests.[28] There was more agreement about the dangers of forcing children into the religious life without the necessary vocation, an abuse which was generally condemned. Whatever the formal position adopted over the choice of partners, a further difficulty arose

[26] Paris, *Marriage*, 175–7.

[27] For complaints by the *précieuses*, see I. W. F. Maclean, *Woman Triumphant* (Oxford, 1977), 115–16.

[28] Maillard, *Le Bon Mariage*, 203.

from the widespread fear of sin arising from contact between the two sexes. Segregation of children was the ideal, however unrealistic, advanced by clerics who alternated a deep fear of women with a much exaggerated trust in the purity and innocence of adolescent girls. Duguet advised his 'dame chrétienne': 'As the education of Mlle your daughter is a still more delicate affair, you must watch over it with care: she should never leave you, and you should never lose her from your sight.'[29] His main worry about this proposal was that it might inhibit the mother from praying and carrying out private acts of devotion. The superficial attractions of the policy of 'lock up your daughters' led its advocates to forget that girls brought up in this way would be even worse placed to exercise even a negative control over the choice of the husband with whom they might have to spend a lifetime. They were offered the occasional crumb of comfort, in the form of assurances that God and their parents would know best, but these were flatly contradicted by the insistent denunciations of parents for their irresponsibility in the matter. Again Maillard showed an unusual flash of insight when he commented: 'how is it that a young man will be able to know the morals of a girl, who is always closely guarded in the house of her mother and father?'[30] Unfortunately he was just citing this as another problem, in the context of the reasons why Augustine might reasonably have refused ever to advise anyone to marry, and did not develop the point.

Once married, and faced with the traumatic experience of childbirth, the young woman's only consolation—if she managed to discover it—was that her pains were a punishment for the original sin of Eve in the Garden of Eden. Suffering was seen as a redemptive act, as the hagiographical literature of the period shows, but there is something repellent about the treatment of childbirth by a male professional group which had so little real knowledge of it, and which was more interested in a midwife's religious knowledge than in her medical skills. Women were singled out for unfavourable comment in other respects too. The innumerable denunciations of extravagant and immodest clothing were directed

[29] Duguet, *Conduite d'une dame chrétienne*, 216–17.
[30] Maillard, *Le Bon Mariage*, 115.

exclusively against women, although this was an age of flamboyant male attire. It was constantly stressed that women profaned churches by displaying themselves in unsuitable dress, seeking to attract admirers. Lordelot complained:

> Today most women only come into the churches to see and be seen, and to display their luxury, and make an ostentatious parade of their vanities . . . One even sees some who are sufficiently dissolute to seek by their criminal glances and their indecent postures to draw away from God those who had come with a sincere intention of participating in the sacrifice at the altar, and who thus deprive him of all kinds of victories.[31]

Clerical puritanism in such matters is very understandable, but it could all too easily develop into a uncomprehending rejection of whole areas of life. Here perhaps one does see the grave disadvantages that could attach to a celibate clergy; the seventeenth century did not get very far towards overcoming these drawbacks, not least because they were never recognized as such. The clerical attitude to women was in any case only a version of much wider masculine fears. Although it would be far too simple to characterize this as a misogynistic age, without reference to the considerable amount of writing in favour of women, or their influence in social life at all levels, the numerous and forceful expressions of hostility towards them do suggest an unfortunate mixture of fear and aggression. A central aspect of this misogyny was the desire by men to repress feminine passions, regarded as a persistent threat to both husbands and society. Assumed to be inferior to men in the possession of reason, women were thought therefore less able to control the lower part of their nature, and their mysterious cycles emphasized how they were dominated by the womb. Defiled by the primary responsibility for original sin, their attraction for men was itself a cause of corruption and death; inconstant, loving only themselves, women were natural allies of the devil, who sought men's love only to betray them. These themes are to be found across the whole range of literature, from theology to the little books of the *bibliothèque bleue* sold by the pedlars.[32] It was easy to give a

[31] Lordelot, *Les Devoirs de la vie domestique*, 140–1.

[32] These themes are fascinatingly discussed by Arlette Farge in her introduction to *Le Miroir des femmes* (Paris, 1982), a selection of texts from the *bibliothèque bleue*.

double edge even to praise of female saints, as Senault did in his 'Panégyrique de S. Cecile': 'For beauty even in virgins is not exempt from this misfortune: and as if it were more in the devil's power than anything else in the world, he makes use of it to ruin men and make them into idolaters and lewd persons.'[33] The natural association between women, sin and temptation, and sexual pollution was regularly referred back to the figure of Eve, as the ultimate and undeniable source of these evils from which mankind could never escape. The struggle against feminine wiles was readily equated with that against the forbidden pleasures of the world at large.

Against this background it is no surprise to find a real note of panic in some clerical responses to women. One prolific author of pious works, H.-M. Boudon, archdeacon of Evreux, wrote to a *curé*:

> Distance from women, even devout women, is a precious form of grace. It is very difficult to preserve the purity of the soul, and not contract some pollution in conversation with them. I believe that one should only converse with them when it is absolutely necessary, that one should only say what is absolutely necessary; that even in restricting oneself to the absolutely necessary, one must be very much on one's guard, and have constant recourse to Our Lord and to his immaculate Mother; that one must avoid like the plague any amusement or useless talk with them.[34]

Boudon's desperate reiterations of 'le pur nécessaire' emphasize his intense fear of pollution from the slightest contact with women, while similar worries presumably caused Arnauld's pupil Antoine Paccori to recommend that the Christian should 'conceive a holy hatred of his flesh and of that of others.'[35] It is difficult not to think, however unfairly, of Dorine's retort to Tartuffe when he tells her to cover her breasts: 'Vous êtes donc bien tendre a la tentation, Et la chair sur vos sens fait grande impression?' ['You're mightly susceptible to temptation then! The flesh must make a great impression on you!'].[36] There is indeed a whole clerical literature directed against the exposure of the breasts, culminating in Jacques Boileau's *Abus des nudités de gorge* of 1675. For one of

[33] Paris, *Marriage*, 123. [34] H.-M. Boudon, *Lettres* (Paris, 1785), ii. 403.

[35] Cited in R. Pillorget, *La Tige et le rameau: familles anglaise et française 16ᵉ–18ᵉ siècle* (Paris, 1979), 74.

[36] Molière, *Tartuffe*, act iii, scene ii.

his predecessors, Father de Barry, 'la vanité de gorge' was: 'a portable plague and a venom which poisons at a distance when one looks on it or touches it.'[37] The language used by these clerics, with its pervasive metaphors of infection and corruption, reveals a great deal about their state of mind and the depth of their concern. In his much calmer style that most practical of seventeenth-century saints, Vincent de Paul, recognized that his Lazarists were much troubled by their contacts with women in the confessional; his sympathetic responses reveal that he envisaged this as a general problem.[38] When M. Tronson was corresponding with one of his Sulpiciens at the seminary of Le Puy in 1693 about the latter's forthcoming book he advised him:

> It would also seem that it would be suitable to delete the three examples which you cite to prove how much the purity of confessors is at risk, and how necessary it is for them to be exceptionally chaste. For these examples do not seem sufficiently important to run the risk of causing pain to chaste souls through the unfortunate ideas which they give.[39]

The general response to such dangers was to advocate the policy of the *cordon sanitaire,* insisting the missionaries and seminary teachers should keep to themselves, avoiding intimate contact with households. All the great sponsors of missionary activity, such as Bourdoise and Bourgoing, would have agreed with M. Vincent's instructions: 'We should make it our maxim never to enter houses, either in the town or the country, where we have no business, and to abstain from all visits, even when they would be useful to keep the good-will of certain people, unless they are necessary or are made to visit the sick or console an afflicted person, to whom one has been called.'[40] The ultimate effect of this tactic must have been to cut off members of the new orders from any experience of women and family life, and thereby to hamper the development of more realistic attitudes.

[37] P. de Barry, *La Mort de Paulin et d'Alexis* (Lyon, 1658), 86–7.

[38] V. de Paul, *Correspondance,* ed. P. Coste (14 vols., Paris, 1920–5), ii. 106–8. See also J. Eudes, *Le Bon Confesseur, ou avertissemens aux confesseurs* (Lyon, 1671), 120–2.

[39] L. Tronson, *Correspondance,* ed. L. Bertrand (3 vols., Paris, 1904), ii. 43.

[40] V. de Paul, *Correspondance,* iv. 125.

Writers naturally wanted to hold up the celibate life as an ideal; as in most religions, abstinence from sensual pleasures carried a corresponding charge of holiness. Even those authors who wanted to stress the sacred nature of marriage therefore tended to balance this with assertions about the superiority of virginity and the religious life, not only in theory but in practice. One of the few women to make her views known, Mme de Maintenon, did so in the course of a discussion with her *pensionnaires* at Saint-Cyr: 'There was then a long discussion of the hardships of marriage and the constraints on women . . . It seems to me, said a girl, that I have heard you say that the good families are not those where one suffers nothing, but those where one of the two suffers without saying anything. Yes, said Mme de Maintenon, or else when they possess enough virtue to put up with one another by turns.'[41] It is unclear whether she was reflecting on Scarron, Louis le Dieudonné, or both. Her spiritual adviser Fénelon sought to disabuse his hearers of the idea that children were a consolation; if they did not cause suffering by dying young, they would prove a disappointment in their abilities, their morals, or their lack of affection for their parents.[42] Texier preached an even harsher message: 'Furthermore, whatever name of gentleness, of tenderness and of friendship one can give to marriage, experience teaches us that it is a very heavy yoke: even if there is only the troublesome care of a family, the support of the children, who must be brought up and provided for, and who after that are so many dragons . . . so many vipers . . .'[43] This pessimistic view of worldly ties could extend to the priesthood itself. Hagiographers commonly emphasized the ways in which their subjects had been obliged to escape pressures from their families, while Bourdoise expressed his views with startling force: 'the cleric should keep himself apart from the world and from the house of his parents . . . I do not think that a cleric can go to paradise if he lives next to his parents . . . if you have taken the tonsure and your relatives

[41] *Recueil des instructions que Madame de Maintenon a données aux demoiselles de Saint-Cyr* (Paris, 1908), 27–9.

[42] Fénelon, 'Sur les avantages et les devoirs de la vie religieuse', in *Fénelon: pages nouvelles*, ed. M. Langlais (Paris, 1934), 224.

[43] Paris, *Marriage*, 143.

ask you to help with their affairs . . . say boldly to them: I am dead, I can do nothing for you . . . The worst air that a cleric can breathe is that of his native region.'[44] Tronson was more wistful, with a suggestion of bitter personal experience which might help explain these attitudes, when he wrote to a colleague in Canada:

> I am not surprised by what your mother orders you to do. She treats you as mothers are accustomed to treat their children in the church. They are not lacking in tenderness so long as one can be useful to the family, and can help to maintain its position in the world: but beyond that, one must expect nothing. For myself, I have always regarded this disposition of providence as a result of the special love which Our Lord bears to clerics, who would find it too painful to detach themselves from their relatives to the degree demanded by their position, if they perceived the same tenderness in them.[45]

Whether writing of priests or of the married, seventeenth-century authors give only the most cursory attention to the notion that families could work together to offer mutual help and encouragement, sometimes failing, sometimes succeeding. They seem only to recognize two absolute categories, the just and the wicked, the saved and the damned—and all of them imply that the latter are the great majority. Both individual and family have to keep the sinfulness of 'le siècle' constantly at bay, since once it touches them its corrosive power is seen as virtually irresistible. It is a dramatic vision, which clearly made a great appeal to the imaginations of its expositors: they loved nothing more than to evoke the powerful image of a just individual fighting against the surrounding world. The just could not afford the dangerous luxury of too much love for other people, which threatened the primacy due to the love of God. The normally moderate Nicolas Caussin wrote: 'if God does not give you children, do not love your partner any the less on that account, but persuade yourself that providence often permits sterility of the flesh to render the mind fertile to all kinds of virtue, by removing the need to care for children, and allowing the return to God of all the love one

[44] A. Bourdoise, *L'Idée d'un bon ecclésiastique* (Paris, 1667), 12–14.
[45] Tronson, *Correspondance*, ii. 261.

would have been obliged to bear them.'[46] Far from envisaging the possibility that love of other creatures might increase one's capacity to love God, Caussin apparently considers that one has only a finite amount of love to give, so that any devoted to others is lost to God. Some other writers added the curious argument that it was foolish to attach oneself unduly to other people, since their death would be felt so bitterly.[47] This appeal to demographic realities was forgotten when it came to second marriages; disapproval of these was expressed by omitting the blessing from the marriage service, while a man who remarried could not subsequently take orders, even if widowed for a second time. Writers stressed the domestic discords which they saw as the inevitable result of remarriages, yet provided no realistic alternative for those left with the care of a family. Even the most moderate writers treated second marriages as undesirable; as François de Grenaille put it: 'The church permits them without ordering them, and seems to punish them to some extent without however condemning them absolutely.'[48] In seventy pages on the subject the only argument he admits in favour of remarriages is the classic Pauline one that it is better to marry than to burn.[49] Maillard explained the absence of the benediction with reference to Aquinas; a second marriage could not be regarded as representing the union between Christ and the church, which could only happen once. This 'défectuosité' led to the rightful deprivation of the benediction, which also emphasized the church's preference for those who remained in the widowed state and did not remarry.[50]

The general tone of religious writers appears very badly calculated to bring familial and marital relations into any workable relationship with the church. Sweeping condemnations of popular practices, a ferocious rejection of virtually all emotional and sexual gratifications, and a deep suspicion of family loyalties are all characteristic of this literature. Even

[46] N. Caussin, *L'Année sainte ou la sagesse évangelique pour les sacrez entretiens de tous les dimanches et de plusieurs festes de l'année* (Paris, 1666), 212–13.

[47] As usual François de Sales was far more humane and realistic in his stress on the value of friendship;—see *Introduction à la vie dévote*, 185.

[48] F. de Grenaille, *L'Honneste Vefve* (Paris, 1640), 123.

[49] Ibid.; the section is pp. 123–92, and St Paul appears on p. 191.

[50] Maillard, *Le Bon Mariage*, 490–1.

those authors who were anxious to defend marriage often ended up giving a most depressing picture of its defects, partly because their anxiety not to discourage religious vocations led them to take away with one hand what they gave with the other. Sin and damnation generally dominated the scene, with spouses and parents seen as constantly failing in their near-impossible duties. This unfortunate evolution was one expression of a much wider difficulty which dogged the catholic reform movement, stemming from the lack of any effective mechanism for the positive formulation of doctrine and policy. Both the papacy and the assemblies of the French clergy operated by the cumbrous means of censuring unsound or heretical opinions. This could establish a trend, but it was overwhelmingly likely to be in the direction of narrowing and restricting; the more so because the crossfire between the Jansenists and their opponents was likely to kill off any innovation by either side, an inviting target for any competent theologian. The liberal views of Sanchez and other Jesuit casuists on sexuality in marriage were demolished with ease, but so were Jansenist efforts to use the vernacular bible and to involve the laity more directly in the church. The failure to receive the decrees of Trent in France went deeper than the political factors which were the immediate cause: many of the prudent compromises of the Council were rendered unworkable by the disputatious clerics who sought to dethrone or rethrone St Augustine and other Fathers. The everyday problems of ordinary parish priests were soon submerged beneath the paper wars waged by these specialists, whose superb command of the sources and techniques involved puts them into a different class from modern historians, let alone the great majority of their contemporaries. There is much justice in John Bossy's identification of 'the feeble hold of scholastic theologians, as opposed to canon lawyers, on the idea that the sacraments were social institutions'; in their remorseless pursuit of a true and logical religion based on authority these men could pose an unwitting threat to the more flexible bonds which held church and community together. Bossy is surely also right to suggest that the Tridentine decisions on marriage, making the reading of banns and the participation of the parish priest obligatory, 'transformed marriage from a social process which the church guaranteed in to an ecclesiastical process which it administered'.[51]

Many of the upper clergy were more adept in canon law and questions of church government than in the subtleties of theology; their prime concern would have been with the church's remaining jurisdiction in matrimonial questions. This was the only substantial area in which the diocesan *officialité* courts retained any authority over the laity, and accounted for a great deal of their business. The marked tendency towards endogamy in rural communities produced large numbers of applications for dispensation from the over-strict rules concerning the prohibited degrees of kinship. Women often brought suits against their seducers, seeking either compulsory marriage or financial compensation, while other litigants were hoping for declarations of nullity, separation, or divorce.[52] As far as the rich were concerned, the whole process could always be diverted by the procedure known as the *appel comme d'abus,* which transferred the case to the *parlement.* A series of royal edicts, from 1557 through to 1697, provided a separate secular law code for marriage, which allowed far closer parental control of children until they reached the age of majority in matrimonial affairs—thirty for men, twenty-five for women. The church insisted on the validity of the sacrament, even if the marriage were clandestine, while the monarchy, reflecting general opinion among the propertied class, encouraged its lawyers to dissolve marriages contracted against parental wishes. The most spectacular clash on this issue arose from the secret marriage between Gaston d'Orléans and Marguerite of Lorraine in 1632, which Richelieu and Louis XIII were determined to annul. In these exceptional circumstances the 1635 assembly of the clergy found sufficient doubtful points to declare the marriage invalid, but the pope was less accommodating, so that a stalemate arose, only resolved in 1643 when Louis XIII finally gave way, with a new marriage ceremony to save face.[53] In the

[51] *Christianity in the West,* 23, 25.

[52] The workings of the system are well illustrated in A. Lottin, *La Désunion du couple sous l'ancien régime: l'exemple du Nord* (Paris, 1975).

[53] For this episode see P. Blet, *Le Clergé de France et la monarchie* (2 vols., Rome, 1959), i. 404–45.

absence of any statistical information about proceedings before the various courts it is impossible to know exactly how the legislation was applied in everyday cases. For the poor at least, this was one area in which ecclesiastical authority was of genuine importance, and such evidence as there is suggests that the *officialités* were quite tough in enforcing the various prohibitions. One must suspect that there was widespread evasion of the rules, some of it arising from sheer ignorance, but there were factors working to assist the church in securing compliance. Apart from the genuine scruples many people must have felt, questions of legitimacy and inheritance were very important here; any suspicion of invalidity would have been an open invitation for greedy relatives to challenge the succession of property. Like so much else in *ancien régime* legal arrangements, the prohibited degrees were only really enforceable under the fear of denunciation by those close to one.

Bishops also tried to regulate the practices associated with marriages, by issuing ordinances and synodal statutes. Their most striking success was in changing the nature of the betrothal ceremony, the *fiançailles*. In the early sixteenth century this had been little more than a public—or sometimes private—contract between the parties, often followed by cohabitation before marriage. At first the bishops tried to integrate the ceremony within the other parts of the religious ritual, then in the seventeenth century they progressively reduced its role. In most dioceses in the Midi, where the practice seems to have enjoyed less popular support, the *fiançailles* were simply banned; in northern France the dangerous elements were neutralized by holding the ceremony in church only days before the wedding itself.[54] This was an interesting and rare success in changing popular habits, presumably achieved because there was some support for tighter control over marriages from within the village community, and because it took the form of deflecting rather than outlawing a common practice. It was a different story when reforming bishops tackled the wedding festivities

[54] A. Burgière, 'Le rituel de mariage en France: pratiques ecclésiastiques et pratiques populaires (16e–18e siècle)', *Annales: économies, sociétés, civilisations* 33 (1978), 637–49.

proper, as part of their assault on 'scandalous' and 'indecent' popular customs. The general view was summarized by Senault: 'for what is the purpose of these great feasts, these dances and other worldly shows: These are useless costs . . . they are sometimes the sparks and matches of disputes . . . This is the reason why the first day of your marriage is sometimes the last of your affection . . .'[55] In this sweeping repressive drive it seems as if everything not strictly comprehended within the religious ritual was now branded as superstition or paganism. The aim was a new kind of piety; silent, modest, controlled by authority, and threatening to cut off the church ceremony within a kind of circle of distrust—yet another *cordon sanitaire.* Many bishops issued ordinances like that of Mgr. de Clermont for the diocese of Laon in 1696:

> We desire that our diocese should be purged of this pomp and this profane display which the peasants are accustomed to use in marriages: and to this end, we forbid that the future spouses should process to the church to the sound of fiddles, either for the betrothal or for the marriage, or that the bells be rung for the betrothal, alongside all what are called welcomes, banquets, and similar displays which are redolent of the spirit of paganism.[56]

The way in which these relatively harmless customs were stigmatized as pagan underlines the hostility they now aroused, while it is none too clear on what basis the bishop could aspire to regulate activities outside the church. Such puritanical campaigns could hardly be anything but futile and alienating.

The bishop's reference to peasants also emphasizes an obvious conclusion which must strike anyone confronted with the whole range of clerical pronouncements about marriage and family life. The morality to which they endlessly refer is that of a social class, the educated urban élite, at least as its more devout members would have wished it to be. Virtually all the clerics concerned, from the aggressively reforming bishops through the doctors of the Sorbonne, the seminary

[55] Paris, *Marriage,* 103–4.

[56] Cited by Burgière, 'Le rituel de mariage', 646, on whose analysis this section is largely based. For other actions against wedding festivities in the 1640s and 1650s see R. Sauzet, *Les Visites pastorales dans le diocèse de Chartres dans la première moitié du 17ᵉ siècle* (Rome, 1975), 261–2.

teachers, and the members of new orders, came from this group. Much of the advice on such questions as the treatment of children, the ways to arrange marriages, and so forth only had meaning for the relatively well-to-do. Nowhere does one find any real comprehension of the problems and needs of either the urban poor or the great mass of the peasantry. This is not to say that this morality was actually adopted even at the upper end of society; many bourgeois households paid no more than lip-service to the more extreme advice, while the nobility largely remained stubbornly attached to its traditional licence. At this level it was very difficult to mount an open challenge to the new orthodoxy, partly because there was no obvious theoretical base from which to operate, even more because of the likelihood of severe punishment by a repressive state. The ninety years from the rise of Richelieu and the trial of Théophile de Viau to the death of Louis XIV saw church and state united by an aggressive and authoritarian attitude towards deviance of every kind. Great unseen dangers resulted for the church, since enforced conformity was wide open to foes which were much more formidable than public criticism; hypocrisy, evasion, ridicule, and contempt. Probably only a minority even of the respectable urban bourgeoisie aspired to the kind of religious life envisaged by the moralists. Town councils displayed a cool realism when dealing with proposals for the moral regulation of society or the establishment of new religious houses. The real *dévots* were to be found in such secret organizations as the *Compagnie du très-saint Sacrement,* whose clandestine nature reflected public hostility to these petty Tartuffes as much as the royal disapproval they also incurred. In 1677 the municipality of Dijon proclaimed that it had heard:

> that a sect was in existence under the name of the brothers of good works whose proceedings and actions, screened by the pretext of Christian charity, could produce no other fruit than to trouble the state and religion, because the inquisitions they conducted into the morals and the actions of people . . . of which they made reports in illegal assemblies, could only proceed from a wicked spirit . . . the Chamber . . . forbids all people of whatever rank to assemble to speak of, write about, or censure the morals of any inhabitants . . . this present deliberation shall be published from the pulpit at the

parish masses and to the sound of the horn at the crossroads where it shall subsequently be displayed . . .[57]

Paradoxically, the towns which were the driving force behind the catholic reform also presented the greatest difficulties in terms of pastoral control; parishes lacked the natural unity found in most rural areas, while alternative confessors and spiritual advisers might undercut the efforts of a rigorist *curé*. Despite this, it was certainly among the bourgeoisie that the reform was most securely established, notably in the upbringing of children. Since it is hard to imagine a set of doctrines better calculated to inspire adolescent rebellion, in the long run this could hardly avoid producing a reaction, or at least a polarization, which would become a major element in the widespread anti-clericalism and free thinking of the eighteenth century.

In the countryside the *curés* faced more straightforward but not necessarily more tractable problems. They were potentially isolated figures, who could only operate effectively if they maintained good relations with the majority of their parishioners. To the reformers the parish clergy themselves were a problem, in grave need of education and discipline, and in the long run a transformation was indeed secured. The *cahiers de doléances* of 1789 contain nothing like the 1614 outburst from the village of Rumilly-les-Vaudes in Champagne:

Pour ecclesiasticques sur les lieux sont encores plus que les gendarmes car ilz ne se contentent de prendre le bien du pauvre peuple mais ilz leurs oste leurs honneur au lieu de monstrer une belle et bonne excemple au peuple ilz leur montre a mal faire, il n'y a fille ny femme qui n'en soyent scandalysé par telle gens, aulcuns les tiennent en leurs maisons avec petits enffans . . .

[As for the local clerics, they are even worse than the soldiers, for they are not content merely to take the goods of the poor people, but deprive them of their honour; instead of showing a fine and good example to the people they show them how to do ill, there is no woman or girl who is not scandalized by such men, some of whom keep them in their houses with small children . . .][58]

[57] Y. Poutet, *Le 17e siècle et les origines Lasalliennes* (Rennes, 1970), ii. 304.

[58] *Cahiers de doléances des paroisses du bailliage de Troyes pour les États Généraux de 1614*, ed. Y. Durand (Paris, 1966), 286.

Sexual irregularity and concubinage had clearly been common for centuries, and were not at all easy for bishops to deal with. Estimates of moral probity based on visitation records are far from reliable, since the parishioners were rarely willing to denounce their *curé* to an outsider; where it is possible to find other records it appears that most offenders escaped adverse notice at the visitation.[59] Nevertheless, some episcopal *tournées* did produce startling results; when bishop Le Camus carried out a visitation of his diocese of Grenoble in 1672–3 he found that eighty-five out of the 142 *curés* for whom records survive were guilty of sexual misconduct. The *curé* of Entraigues dropped dead at news of the imminent visitation, no doubt repenting that he had been openly married in his own church 'devant le St. Sacrement et deux témoins, disant "que saint Pierre ayant esté marié il le pouvoit bien l'estre aussy" ' ['before the holy sacrament and two witnesses, saying that since St Peter had been married he could very well be so himself'].[60] After thirty-five years of resolute battle, Le Camus had still not eliminated such abuses at his death in 1707; while this mountain region was certainly backward, there was no shortage of similar troubles elsewhere. In the diocese of Reims at the end of the seventeenth century Archbishop Le Tellier recorded his opinions on 628 *curés*, of whom forty-six were censured for lack of chastity and thirty-seven for drunkenness.[61] It becomes easier to understand Tronson's remarks: 'The number of priests who are damned is far in excess of those who are saved . . . There are very few priests who can be saved . . . The majority throw themselves into the church without consulting God's will.'[62] By the last decades of the century, however, it does seem that the general introduction of seminary training, and more effective episcopal control, had brought about a major change in most dioceses. Scandalous clerics were now the exception, so that the parish clergy

[59] This point has been very clearly demonstrated by R. Sauzet, *Les Visites pastorales, passim*.

[60] J. Sole, 'La crise morale du clergé du diocèse de Grenoble au début de l'épiscopat de Le Camus', in *Le Cardinal des montagnes: Étienne Le Camus, évêque de Grenoble (1671–1707)*, ed. J. Godel (Grenoble, 1974), 186–7.

[61] D. Julia and D. McKee, 'Les confrères de Jean Meslier: culture et spiritualité du clergé champenois au 17e siècle', *RHEF* 69 (1983), 64–5.

[62] Cited in Poutet, *Les Origines Lasalliennes*, i. 325.

became much more plausible advocates of an austere morality, in which they had been instructed themselves.

A reform of the clergy was only the first step towards the ideal of a sober, dutiful, and religious populace. To forward this holy work the church had to confront the long-established practices of peasant culture; the attack on paganism and superstition must be extended to the household and the social life of the peasantry. A vast list of prohibitions emanated from the bishops and their advisers. Common sleeping arrangements, above all when these involved parents, children, and even servants sharing the same bed, were regularly stigmatized. As one Breton bishop put the matter in the sixteenth century: 'We forbid brothers and sisters or other relatives of different sexes to sleep together after the age of seven. This practice, which gives rise to an infinite number of horrible sins, as several confessors have reported to us, is therefore prohibited on pain of excommunication and a fine of 10 livres.'[63] While one must certainly suppose that incest was a problem, it was not necessarily shared beds that led to widespread immorality or impropriety. Their critics never recognized the need for warmth, the lack of means or space to provide additional beds, nor indeed the fact that multiple occupancy could be as much a safeguard as a threat. Those who were seriously tempted by such illicit pleasures were likely to find opportunities whatever sleeping arrangements were made. Another institution which resulted partly from the need for warmth was the *veillée,* the evening meeting of several families for needlework or spinning, story-telling, dancing, gossip, and flirtation. Numerous bishops tried to prohibit these, with predictable lack of success; a more promising idea was tried in the diocese of Châlons, in Lorraine, and in the Velay, where village schoolmistresses were encouraged to start up Christianized versions for women only, where pious exercises replaced the traditional activities.[64] This low-key challenge by competition does not appear to have caught on

[63] Statuts synodaux de Saint-Brieuc, 1507, cited in J.-L. Flandrin, *Les Amours paysannes (16ᵉ–19ᵉ siècle)* (Paris, 1975), 150.

[64] Poutet, *Les Origines Lasalliennes,* i. 479. For the institution in the Velay, see P. M. Jones, *Politics and Rural Society: The southern Massif Central* c. *1750–1880* (Cambridge, 1985), 134–5.

more widely, perhaps because its main claim was to be useful in itself, and it offered no serious prospect of destroying the traditional alternatives. Separation of the sexes was a great principle of all such schemes, and even more of the prohibitions, which naturally extended to village dances. Clerical hostility to dancing is an almost comical feature of the whole period; as Lordelot put it, dances were: 'shameful and unworthy extravagances . . . They only proceed from the depravity of the heart and the dissolute movement of the body.'[65] Whatever may have been the case in urban society, village dances seem so innocent as to make these claims absurd. They naturally formed part of the elaborate mechanisms whereby young people selected partners, but this hardly rendered them hotbeds of concupiscence. Here as elsewhere the reformers, like many nineteenth-century *folkloristes,* showed no understanding of the inner workings of peasant culture, which they saw from a thoroughly ethnocentric bourgeois standpoint.

It is important to realize that the majority of the French population did not make arranged marriages. Once below a certain level of wealth and status, the initiatives were normally taken by the young people themselves, with the aid of the various institutions of *la jeunesse.* The demographic patterns also meant that many would have lost both parents before reaching the late age of marriage characteristic of the period. While the well-born would then come under the authority of some kind of family guardianship, lower down the scale such arrangements were vestigial. On the other hand, the community in its various manifestations exercised a good deal of moral control, far more effective than the external efforts of the church. Low rates of illegitimacy and premarital conception are much more likely to have resulted from such communal pressures than from the direct effect of the catholic reform. The *curé*'s influence in his parish might be great, but it depended crucially on his success in integrating himself with the inhabitants. He was expected to perform his sacramental duties regularly, at times convenient to his parishioners,

[65] Lordelot, *Les Devoirs de la vie domestique,* 180–6. One should compare the much more moderate remarks by François de Sales, *Introduction à la vie dévote,* 74, 222–5.

organize local charity, mediate in disputes, and know the moment to withdraw. In practice the catholic reform was prevented from alienating the mass of the faithful only because the village proved an easy victor in the battle with higher authority for the soul of the priest. A brief training in the seminary, and the occasional visitation or mission, were quite outweighed by the steady everyday pressure of the parishioners among whom he had to live—and very often by his own good sense. The leading reformers had sought an impossible ideal, which would require the *curé* to be at once of the world and out of it; what they ultimately produced was an invaluable group of medical, legal, and agricultural advisers, instead of the austere proselytizers of whom they had dreamed.[66] This evolution was symbolized by the decline of some of the institutional expressions of reform, which simply lost impetus as the years passed. The monthly *conférences ecclésiastiques,* little refresher courses for the *curés* in each archdeaconry, usually disappear from the records after a decade or two of activity around the turn of the century.[67] The degeneration of the visitation to an empty formality in the eighteenth century is somehow symbolized by the appearance of printed forms on which the *curés,* in obvious bad faith, recorded negative answers to many questions about their parishioners' behaviour. Visitations depended very much on denunciations, of individual sinners by the *curé,* and of the *curé*'s faults by the parishioners; once some kind of *modus vivendi* had been reached, neither party was keen to disturb it in this way. Such attitudes are already apparent in the 1650s, when *curés* who lived with their concubines and children, but otherwise performed their duties to local satisfaction, escaped detection at the visitation.[68]

There were occasions when energetic bishops or missionaries were found using the possibilities of the confessional, delayed absolution, and close observation of

[66] I owe this point to Professor John McManners's 1986 Hensley Henson lectures.

[67] They seem to have had a very patchy existence in the diocese of La Rochelle; L. Pérouas, *Le Diocèse de la Rochelle de 1648 à 1724* (Paris, 1964), 256, 368, 377–8. For the diocese of Reims, Julia and McKee, 'Les confrères de Jean Meslier', 78–81.

[68] See Sauzet, *Les Visites pastorales,* 101–13.

individual families to carry the process of evangelization further. In Brittany the great Jesuit missionary Julien Maunoir recommended that the priests should visit each house, enquire about the number of persons, their age and status, their performance of their religious duties, then institute a system of *fiches*. Knowledge of the catechism, piety, and morals each had a column, and were to be classified on a scale from one to five; at least one Breton parish actually possesses a set of these *fiches* for 1668–72.[69] They can be seen as a development of the *livres des états des âmes*, pioneered by Borromeo and imitated in at least fifteen French dioceses during the seventeenth century.[70] Few examples of these confidential documents have survived, so that it is impossible to know how widely they were used, but the intention was clearly to enable the *curé* to make a much more critical assessment of the confession. Although organized by families, the assessments were in fact of individuals, in the characteristic manner of the catholic reform. When Claude Joly, bishop of Agen, wanted to make his *curés* keep such registers in 1673 the consuls and canons of Agen lodged an *appel comme d'abus* with the *parlement* of Toulouse on the grounds that the measure was contrary to 'le repos, la tranquillité et l'honneur des famillies'; even if the registers were kept, it seems unlikely that they could have been extensively employed without sparking off numerous hostile reactions of this kind.[71] Maunoir's evangelization of Brittany, a remarkable feat, was primarily based on other techniques, including great histrionic talents and a relentless exploitation of popular concern for the fate of the dead through hell-fire sermons. What he achieved was to take over a vital part of popular culture, the cult of dead ancestors, for the church, and employ it as a lever to secure the acceptance of the purified Tridentine faith. Not even Maunoir or the greatly improved parish clergy who continued his work could bring off the trick completely, yet they succeeded in placing the ordinary parishioner in a permanent dilemma. Now that

[69] A. Croix, *La Bretagne aux 16ᵉ et 17ᵉ siècles* (2 vols., Paris, 1981), ii. 1210, 1407–10.

[70] L. Michard and G. Couton, 'Les livres d'états des âmes: une source à collecter et à exploiter', *RHEF* 67 (1981), 261–75.

[71] Ibid., 267.

he knew his elementary catechism and the basic moral teachings of the church, there was an incompatibility between his deep concern for the fate of the dead members of his family and his continued adherence to a popular culture condemned as superstitious or sinful. Under the constant menace of hell-fire for his ancestors and himself, the believer took refuge in emotional and ostentatious devotional practices, which he hoped would compensate for the failings of his private life. Dances and cabarets, perhaps closed down for a few years in the aftermath of an intense mission, would soon reappear.[72] Similar characteristics have been remarked in other regions which remained strongholds of the faith in the nineteenth century, like the southern Massif Central, where the church can be seen as the agency of a kind of ancestor worship.[73] Perhaps in the end hell-fire was a weapon which burned the fingers of its users, creating new variants on the old deviations of mechanistic devotion and semi-Manicheanism. It certainly failed to burn out the normal forms of popular culture, which were the external expression of traditional family life among the peasantry. It was relatively easy to introduce new elements into religious practice, but virtually impossible to transform the nature of a faith, perfectly genuine in itself, which took the form of attitudes to be lived in a world where doctrine was largely meaningless. Missions were successful precisely because they fitted into this schema, setting up collective occasions which associated penitence with festival, and allowed people to escape from their everyday surroundings. They created an ephemeral society in which everyday divisions disappeared and the participants were united by emotion; the difficulty was to integrate these exceptional cathartic events with the everyday practice of a Christian life.

This problem was in practice rendered insoluble by the extreme and unrealistic vision of such a Christian life built up by the reformers. Their emphasis on sin, guilt, and punishment led to a religion based on prohibitions, so extensive in

[72] Croix, *La Bretagne*, ii. 1183–1246.

[73] P. M. Jones, 'Parish, seigneurie and the community of inhabitants in southern central France during the eighteenth and nineteenth centuries', *Past and Present*, 91 (1981), 74–108; also idem, *Politics and Rural Society*, 134–40.

their coverage that they were bound to be broken repeatedly. Within this context it was perfectly understandable that sexual behaviour should assume a very important place, as the most obvious case of the triumph of the inferior elements in man over the superior, of the senses over reason.[74] This commonplace did not fit very easily with the notion of procreation as the legitimate purpose of sexual relations within marriage, particularly since many theologians did not completely reject the Galenic notion of the double seed, and therefore insisted that both partners should achieve orgasm; it was for this reason that they permitted women to stimulate themselves when necessary, either during or after intercourse.[75] Not surprisingly, such advice was only published in Latin, usually in highly technical treatises; vernacular writers generally followed Maillard's line in *Le Bon Mariage,* where he said that 'la pudeur' obliged him to refer many matters to the works of the doctors or the advice of a wise confessor: 'rather than exposing to the common people things which might perhaps be read with more curiosity than utility, with more destruction than edification'.[76] The relatively liberal Maillard opposed the many authors who claimed that 'l'usage du mariage' (the common euphemism for marital sex) always constituted at least a venial sin, a view predictably taken by the severe Paccori: 'Marriage is in truth a remedy against incontinence; but it is one of those remedies which should only be used with caution and with fear. Its use is disordered, because it is always mixed with concupiscence, and concupiscence joined to consent becomes a sin.'[77] The debates on these issues may not have mattered very much in practice, since there is no evidence of any serious attempt to apply such moral teachings in ordinary pastoral work, but they do emphasize how difficult it was to find a logical yet workable approach to marital sexuality within counter-reformation Christianity. There was no disagreement about all other sexual activity, and here only an inveterate optimist could

[74] For a typical example of this argument, Maillard, *Le Bon Mariage,* 30.

[75] For this question see J.-L. Flandrin, *Le Sexe et l'occident* (Paris, 1981), 127–35, 'Homme et femme dans le lit conjugal'.

[76] Maillard, *Le Bon Mariage,* 74.

[77] Paccori, *Règles,* 71.

have expected much success in imposing total sexual continence on a population which married so late. If masturbation was stigmatized as a crime against nature, avoidance of which was one reason for making *le devoir conjugal* a mutual obligation, then the church was surely committed to repressing the irrepressible.[78] The same was true of the 'frequentations' between the young of both sexes, with their ever-present danger of encouraging 'la paillardise'. Fear of such occasions lay behind the endless attempts to separate the sexes, in schools, in the church (where families were therefore divided by sex), in religious processions, and so on. Worst of all was the period of Carnival, when 'an unbridled populace dances and rushes in the streets, as the Bacchantes once did, so that they give scandal to religion by their shameful licentiousness.'[79] An attempt to counter-attack was made from the 1570s, with the introduction of the Italian devotion of the 'prières des Quarante-heures', which were particularly favoured by the Capucins. Preachers were to denounce 'les vices et les malices de ce siecle universellement depravé en tous estats', and the faithful attended church with a candle to make *amende honorable* to God 'for those who offend him, and dishonour him in public with so much audacity during the unfortunate debauches of these days of Carnival'.[80] Originally this was on the night of Shrove Tuesday, but the clergy saw the night as the prime occasion for sin, and in the 1630s at Saint-Nicolas-du-Chardonnet it was provided that the prayers should be interrupted at night, so that the faithful could return home 'in order to avoid the evil encounters they might make in the streets in the evenings, and to close the church to avoid the disorders which might be committed there later'.[81] In 1686 this nocturnal suspension was formally approved by Rome, at the request of the French Capucins, on the grounds of the indecencies which the presence of women might cause; it does not seem to have occurred to anyone that this was tantamount to an admission of defeat.[82]

[78] Flandrin, *Le Sexe et l'occident,* 251–78, 'Mariage tardif et vie sexuelle'.

[79] Lordelot, *Les Devoirs de la vie domestique,* 181.

[80] B. Dompnier, 'Un aspect de la dévotion eucharistique dans la France du 17ᵉ siècle: les prières des Quarante-heures', *RHEF* 67 (1981), 5–31; quotation p. 23.

[81] Ibid., 24.

[82] Ibid., 24–5.

If numerous seventeenth-century clerics appear to have been obsessed with the pollution of the world by unbridled sexuality, this does not mean that all were, or that there was necessarily anything unhealthy about their concern. The issue does occupy a very large place in the pastoral literature of the time, but is far from excluding such other major problems as usury and avarice, hatreds and violence. Within the context of basic Christian teachings, and in the light of what we know about the behaviour of most contemporaries, it is not easy to quarrel with the assessment made by an anonymous brother of the Doctrine Chrétienne in 1649: 'lechery is the sin which should be feared most for several reasons, because it is the most common sin, which all kinds of people can commit at virtually any time; because the inclination to it is more violent; because it causes more people to be damned than the others'.[83] Rather than treating such comments as yet further examples of clerical paranoia, we might do better to recognize them as corresponding pretty well with the facts. There has been a tendency to use the relatively low level of illegitimate births and premarital conceptions, particularly in rural France, as evidence for the successful imposition of a more austere Christian morality. The demographic evidence also shows that there was fairly widespread obedience to the periods of prohibition, for marital sex as well as for weddings. Nevertheless, these facts do not tell us very much about the true level of sexual activity, nor can we assume a simple cause and effect relationship with the catholic reform movement. The low level of extramarital conceptions is more likely to have resulted from the use of various practices stopping short of full intercourse than from total abstinence; in any case the downward trend antedated the general application of the reform, and started upwards again in the eighteenth century. Straightforward illegitimacy does seem to have decreased, probably on account of the restrictions on the *fiançailles* and the pressure on offenders to make 'marriages of reparation', which were more moderate and workable policies. The number of premarital pregnancies was still quite high, however; in the

[83] Paris, *Marriage*, 36, citing *Instructions chrestiennes et familières pour apprandre aux pères et mères à élever leurs enfants dès leur jeunesse à connoistre aimer et servir Dieu*.

early eighteenth century between 5 per cent and 15 per cent of brides were already pregnant, depending on the region, and 10 per cent would seem a plausible national average. As such figures are bound to represent only a certain fraction of those who anticipated their marriage, they do not suggest more than a very partial success in imposing sexual continence.[84] The Jansenist Bishop Colbert of Montpellier was sufficiently alarmed to issue an ordinance in 1699 which began:

The visitation we have just made of our diocese has shown us that one of the greatest disorders which holds sway there is the indulgence in those shameful intercourses which only too often precede marriage; which after having dishonoured the holiness of the sacrament, cause further affliction to the church in the form of those unhappy fruits of incontinence which become apparent to the eyes of all, and it has seemed to us that public penitence, whose practice we found established in this diocese to serve as a barrier against these excesses, is one of the most effective means for preventing them.[85]

In the towns illegitimacy rates were increasing from the mid-seventeenth century, although they were still far from spectacular: the communal solidarities of the urban poor were generally weaker than those of the peasantry, and young girls in service were in a particularly vulnerable position. The overall impression is that the clerical campaign had only a marginal effect at best, and that fluctuations in the number of extramarital conceptions are better explained by social and economic changes. In this area too the main effect was probably on the educated minority; transgressions of the moral code may have remained just as numerous, but they became less open. More dangerously, a kind of sexual dialectic was created, with a much higher level of self-awareness, so that the forbidden acquired a new charge of excitement.[86]

Despite all the momentum the reform movement had built up, and the somewhat belated conversion of Louis XIV to a respectable life, the later years of the century did not bring any great surge of optimism among the devout. They rightly

[84] On this subject see Flandrin, *Les Amours paysannes*, 223–43.

[85] Colbert, *Œuvres*, ii. 833–4.

[86] This argument is developed, and perhaps rather overstressed, by M. Foucault, in *Histoire de la sexualité*, i: *La Volonté de savoir* (Paris, 1976).

perceived that they had failed to impose much more than an outward conformity, while the ruling élites themselves were often hostile towards the more far-reaching proposals for moral reform. The kind of principles advocated by some members of the *Compagnie du très-saint Sacrement* revealed that they did not see their own society as truly Christian at all; they hoped to make it so by inspiring indifference to riches and honours, and replacing the search for personal pleasure by a desire to serve others. Increasingly the leading advocates of such ideas turned to the education of children as the only hope of breaking down resistance; they thought this should be the primary duty of the parish priest, who could strike at both laziness and religious ignorance in this way.[87] Here too there were to be many deceptions, for priests showed little enthusiasm for taking up a secondary role as schoolteachers, given both the demands and the greater prestige of their sacramental functions, while teaching of the catechism was always liable to degenerate into parrot learning. It may be that one should take some complaints on this last point with a grain of salt, however; much of the catechism was probably meaningless to those who learned it, but the broad outlines of the faith were better understood than some visitors, interrogating alarmed and taciturn village children, perhaps realized. Any attempt to use education as a means to moral reform also had to expect some resistance, less on account of the content of the teaching than because there was lack of enthusiasm for education itself. On the one hand many members of the ruling classes thought it better for the poor to remain ignorant, on the other many parents saw no point in sending children to school when they could be usefully employed at home.

The most interesting and enterprising effort to overcome these obstacles was that of the Frères des Écoles Chrétiennes, founded by J. B. de la Salle in the 1680s. La Salle confined his operations to towns, where the need was greatest and the schools could be larger, using his scarce manpower to greatest effect. He expressed the ideas behind them very clearly:

All the disorders, of the artisans and the poor above all, derive ordinarily from the fact that they have been abandoned to their own

[87] Poutet, *Les Origines Lasalliennes*, i. 348–9.

guidance and very badly brought up in their early youth, something it is almost impossible to put right at a later age because the bad habits they have contracted are only abandoned with great difficulty and almost never completely.[88]

Whereas most *dévots* wanted schools to emphasize the teaching of religious knowledge, la Salle recognized that parents were 'peu touchés' by this; instead, they must be persuaded that the schools were useful in a practical way, giving their children the chance to find a worthwhile career. His schools were designed to provide a practical education, in French, which would leave pupils fully literate and able to use simple arithmetic. Latin was firmly excluded; la Salle thought it useless to poor children, while to teach it even to a low standard all other kinds of education had to be sacrificed. If French was taught, and used as the language of instruction, then the pupils would leave the schools reading fluently, so that they could make use of the abundant devotional literature available to them. If anything, the schools proved too successful; the poor were naturally attracted by this intelligently conceived free education, and began to be joined by children of better-off families. This aroused bitter opposition from the professional teachers whose livelihood was threatened, while many clerics were hostile because by excluding Latin la Salle had effectively prevented his pupils from aspiring to the priesthood. His foundation survived, but many of its schools, notably those in Paris, had to be abandoned, and its operations contracted.[89] This was particularly regrettable since education offered a most promising opportunity for co-operation between church and family. Alongside la Salle's foundation were the numerous feminine orders (many of them very local) providing education for girls, which can also be seen as embodying the most positive and hopeful side of the catholic reform. Unfortunately their prestige and popularity with ordinary people was not enough to counterbalance the reticences of the hierarchy and the hostility of those whose vested interests they disturbed; in consequence their influence was far less than it might have been.

[88] La Salle, 'Règles communes' (BM Avignon), cited by Poutet, *Les Origines Lasalliennes*, i. 527.

[89] For these points see Poutet, *Les Origines Lasalliennes, passim*.

The catholic reformers certainly do not deserve to be pilloried or ridiculed for the inadequacies of their vision of the family. They held high ideals, for which one might well feel more sympathy than for the hedonism and selfishness of other members of their class. Their failure perhaps represented a lack both of self-knowledge and of capacity to analyse the unfamiliar worlds of other social groups. They were saddled with a set of inherited rules and traditions which did little to help them, while a variety of pressures induced them to take a hard and rather unrewarding road. The Christian family of their dreams was a bourgeois family, spritualized and purified, and in the process so sterilized as to have lost most of its humanity. Even among the bourgeoisie this proved an unattainable ideal, which probably ended by generating a reaction. What made the policy even more misguided was that among the poor the family did exist as a far more real and profound force than most outside observers recognized; the efforts of the reformers, if seriously applied, would have damaged rather than enhanced it. The anxiety to cleanse the sacraments of all extraneous or worldly elements threatened to separate family and church just where they should have been most firmly linked. Fortunately most *curés* lacked the intellectual rigour, or had too much good sense, to follow the stony path indicated by their superiors, so that in reality the rites of passage continued to mix official and popular culture. As so often, the everyday pastoral practice of the church did not greatly resemble the formal doctrines which are now so much easier to recover. The over-ambitious nature of the reform therefore tended to make it only a secondary influence on changes in family life, except among those bourgeois groups from which it emanated. On the other hand John Bossy probably exaggerated in suggesting that this failure was a crucial obstacle to the success of the Tridentine reform, since it was only part of a much wider set of difficulties out of which it proceeded. The whole style of the reform movement was so antipathetic to popular culture that it was almost impossible to build bridges between them. In the terms most people understood at the time, the church was probably best served by such methods as those of Maunoir, which did involve the church, via the cult of the dead, with the family as both lineage and household. Whether this could possibly have been applied over the whole country is a different matter. The Breton cult of the dead was peculiarly well-developed; in regions where it was less fervent it might not have withstood the strain placed on it by such techniques of mass persuasion. In any case Maunoir was an exceptional figure, untypical of the educated clergy even in his own day, and prepared to exploit popular attitudes in a way many would have thought highly dubious. As ideas evolved, the clergy themselves would come to see such extreme forms of evangelization as improper and indecent; the rationality inherent in the new faith was in a sense self-limiting.

In many respects the family, alongside the local community within which it was the basic unit, represented a formidable challenge to the counter-reformation church. The culture and values of a traditional society were embodied in and transmitted through these immensely resilient and flexible agencies, which were therefore increasingly liable to be seen as centres of subversion. Just as the state devoted much effort to taming the community as a political force, so the church would have liked to penetrate and rebuild the family as part of its drive to create a godly society. In so far as this ambition found conscious expression, however, it came from 'moral absolutists' whose austere vision was hopelessly at odds with the life-style and attitudes of peasants or artisans. It is also apparent from their writings that most of them found it much easier to discuss the problems of prosperous households, and that they were incapable of engaging imaginatively with the situation of the mass of the population. A strongly negative message was commonly projected, in which one can detect deep fears of the family as an embodiment of corporate self-interest which prevented individuals from putting God first, and as the centre of the inescapable pollution associated with sexuality. In these ways particularly the family could be associated with the long tradition of 'le mépris du monde', to be seen as yet another form of worldliness obstructing the divine purpose. The drawbacks of this vision are evident, most notably that it could result in an attempt to destroy most of the positive and central features of the family, in the name of ideals which were essentially other-worldly. Fortunately there is usually an enormous gap between rhetoric and action in such cases, coupled with a tendency for extreme doctrines to lose support as their implications become apparent. The peculiar feature of this period was that social norms came to be extensively discussed as such, particularly within the context of pastoral theology, instead of evolving virtually unnoticed through social interaction. The immediate result was the rather unbalanced and unmanageable doctrines which have been illustrated here, with their tendency to carry prescriptive logic to extremes. In the longer run François de Sales and even Thomas Sanchez would prove more influential than the gloomy rigorists who had dominated the seventeenth and early eighteenth centuries, while the family itself would prove the toughest and most enduring phenomenon of all, taking what it wanted from Christian teaching, and largely ignoring the rest. The contest between church and family was a hopelessly unequal one, in which clerical discourse had finally to confront a stark alternative between drastic self-modification and total irrelevance. This choice was not faced by the writers of books, nor even by missionaries, but by the ordinary *curés*, who learned the hard way how people actually lived, and came to take a very different attitude. Their management of local charity, above all, gave them a crucial role in sustaining poor families.[90] It was as a great charitable organization, given fresh impetus by the committed clergy of the catholic reform, that the church made its positive contribution to the family life which its outward rhetoric threatened to repudiate.

[90] For a very clear description of these charitable activities, see O. H. Hufton, *The Poor of Eighteenth-Century France* (Oxford, 1974), 194–201.

7

The sins of the people

Auricular confession and the imposition of social norms

ONE of the first features of the medieval church to disappear almost everywhere once the reformed faith gained the upper hand was auricular confession, at least in the obligatory form laid down by the Fourth Lateran Council of 1215. It is not hard to see why this happened, despite the residual approval Luther and others still expressed for confession as a devotional practice. The formal requirement for annual confession, as a precondition for admission to the sacraments at Easter, had been an ambitious claim for clerical power, made at the height of the papal revival. It had taken a long time to secure even nominal compliance, while the nature of the obligation was always likely to make it unpopular with most of the laity. Here, in the eyes of its critics, was a classic example of the church arrogating power to itself, by an evolutionary process devoid of biblical or patristic authority. The whole penitential and disciplinary structure elaborated by Rome was seen as the nursery of a range of abuses. These included indulgences, alongside the stress on good works which was crucial to the debate over the doctrine of justification. A natural corollary was to connect confession with the 'quantification of salvation' and the doctrine of purgatory. The so-called 'power of the keys', which made formal absolution by a priest necessary for salvation, implied just that kind of separation between clergy and laity to which protestants objected.[1] It also interposed the priest between the believer and God in a fashion incompatible

[1] The most comprehensive treatment of the subject for the medieval period is still that by H. C. Lea, *A History of Auricular Confession and Indulgences in the Latin Church* (3 vols., Philadelphia and London, 1896). For the development of the theology, see B. Poschmann, *Penance and the Anointing of the Sick* (London, 1964). Popular hostility to confession is illustrated from the farces in *La Mort des pays de Cocagne*, ed. J. Delumeau (Paris, 1976), 36–7.

with the stress on faith and the need for personal communication with the deity. A fully predestinarian theology left no logical place for a system of satisfaction and absolution in any case, although some ingenious dialectic might get round this difficulty. The reformers themselves had no intention of diminishing moral control, and in France the consistories tried to impose disciplinary standards worthy of the early church. The creation of institutions which transferred a large element of moral control to the elders who represented the community was a bold move; if it had worked properly it would have made it possible to enforce norms with a ruthlessness no individual priest could have contemplated.[2] It is obvious that in most cases the results fell desperately short of the ambitions, but the hope of establishing some such religiously based normative system appears to have been widespread among the educated classes of early modern Europe, both protestant and catholic.

The Anglican church, as the product of a somewhat secularized compromise, was thought by many to have ended up with the worst of every world. The task of enforcing discipline was left to the church courts, institutions which appeared to lack both the personal touch of the priest and the moral authority of the community. Their shortcomings were probably much exaggerated, for there was much to be said for a formalized judicial mechanism, and in many areas they did become agencies of moral reform by the early seventeenth century, under the domination of the local élites.[3] This did not prevent nostalgia for the old situation, so that a fair number of conservative Englishmen might have agreed with John Aubrey's claim 'Then were the consciences of the people kept in so great awe by confession that just dealing and virtue

[2] The functioning of the consistories is very well discussed by J. Garrison-Estèbe, *Protestants du midi, 1559–1598* (Toulouse, 1980). There is an excellent example of a consistorial register in A. Soman and E. Labrousse, 'Registre du consistoire de Coutras, 1582–4', *Bulletin de la Société de l'histoire du protestantisme français,* 136 (1980), 202–28.

[3] For a guide to the recent literature, and a very telling analysis, see M. J. Ingram, 'Religion, communities and moral discipline in late sixteenth- and early seventeenth-century England: case studies', in *Religion and Society in Early Modern Europe*, ed. K. von Greyerz (London, 1984), 177–93.

was habitual.'[4] This egregious piece of wishful thinking bears little relation to what we know of pre-reformation social behaviour, but by Aubrey's own day reforming catholics were certainly trying to enforce 'just dealing and virtue' through confession. The fact that English travellers tended to be struck by the ease with which the system was subverted, rather than recognizing how much effort was being invested in it, although more significant than Aubrey's speculations, did not represent any very deep knowledge of the matter.[5] The failure of clerical moralists to suppress swearing, theft, fornication, and drunkenness can hardly surprise us much, nor indeed does it appear to have surprised that professionally lugubrious body of men themselves. There is no obvious way in which historians can test the degree of this failure, or in some cases relative success; counter-factual propositions cannot really extend to positing the economic or sexual behaviour of Europeans injected with varying doses of Christian ethics. It is quite likely that perceptible differences between countries in such matters as illegitimacy rates and rules about lending at interest are related to their possession, or lack, of a tightly ordered system of social discipline run by the clergy, but the casual links are so ambiguous and obscure that it is hard to get beyond mere speculation. It may be more helpful to consider the ways in which confession could have operated as a dynamic agency of change in catholic Europe, then examine the various obstacles to such a process, whose operation produced the kind of muddled results we would intuitively expect.

The church's own position was established (and sometimes confused) through a wide range of formal and less

[4] Cited by K. V. Thomas, *Religion and the Decline of Magic* (London, 1971), 155. The whole section on pp. 154–9 is particularly interesting on confession in England.

[5] For the general opinions of English visitors on catholicism in France (which do not include any specific reference to confession) see J. Lough, *France Observed in the Seventeenth Century by English Travellers* (London, 1984) and 'Two more British travellers in the France of Louis XIV', *The Seventeenth Century*, 1 (1986), 159–75. The subject is discussed directly by Sir Edwin Sandys, *Europae Speculum, or A View or Survey of the State of Religion in the Westerne parts of the World* (The Hague, 1629), 10–17; admittedly his unfavourable views date from the beginning of the seventeenth century.

formal pronouncements, ranging from papal and conciliar decisions, through the decrees issued by individual bishops, to the increasingly controversial literature of moral guidance and casuistry. This material has been subjected to a particularly acute analysis by John Bossy, who has identified a marked shift in its centre of gravity over the later medieval and early modern period.[6] The emphasis moved from a concern with objective social relations to a scheme of internalized discipline for the individual. Late medieval confession had been an annual settlement of social accounts, with the emphasis on restitution and reconciliation; those who were unwilling to settle their quarrels often stayed away. The theologians were already taking a more 'modern' view in 1215, when the object of the sacrament was clearly defined as reconciliation to God rather than to the community. This position was overwhelmingly endorsed by the Council of Trent, but over several centuries it only filtered slowly and imperfectly down to the local level. The process was considerably speeded up by the intervention of reforming bishops, with Borromeo at Milan as the heroic prototype. Borromeo was a crucial figure in propagating a new penitential style in northern Italy, which was symbolized by his introduction of the confessional box itself. This device gave far greater privacy, while revealing a fear that face-to-face confession was dangerous on sexual grounds; for several decades only women were obliged to use the boxes. It took more than a century for this fascinating piece of religious technology to spread across most of catholic Europe, symbolizing the triumph of the individual over the communal view of the sacrament, not least because it was a physical obstacle to the ritual laying on of hands by the priest which had traditionally stood for reconciliation to the community. Bossy sees these changes as being accompanied by shifts in both the 'economy of sin' and the 'moral arithmetic' which can be found in the literature of confession.[7] In the

[6] J. Bossy, 'The social history of confession in the age of the reformation', *Transactions of the Royal Historical Society*, 5th series, 25 (1975), 21–38.

[7] Apart from the article cited above, Bossy's views are expressed in his *Christianity in the West, 1400–1700* (Oxford, 1985), 45–50, 127–8, 134–5. I have also drawn on the arguments of his unpublished seminar paper on 'Moral arithmetic', given in Oxford in 1986.

first case, medieval practice had been based on the primary significance of hatred and its consequences, so that the primary objection to even adultery had been the hatred and dissension it caused, whereas later it came to be seen above all as a problem for the moral health of the individual. In the second, the seven deadly sins were largely replaced by the ten commandments as the basis for identifying and classifying sin. Here a flexible system which emphasized the need for charity gave way before a precise list of prohibitions, directed above all towards sins directly against God.

Between the thirteenth and sixteenth centuries, therefore, a linked group of intellectual changes had occurred among the élite of theologians who developed and defined the official position of the church. They were very slow to affect the attitudes or behaviour of parish priests and laity, mainly because the mechanisms for transmitting them down to this level were so inadequate. The decrees of the Council of Trent, implemented through the agency of reforming bishops, were the catalyst for their widespread implementation from the late sixteenth century through the seventeenth. In his first approach to the question Bossy asked whether this amounted to 'a Copernican revolution in sin', taking as the vital test the replacement of hatred by individual sexuality as 'the primary material of the sacrament'. An investigation of clerical teachings about masturbation and contraception suggested that changes in this direction, although discernible, fell well short of anything that could be described as revolutionary. This result was predictable, and perhaps indicates that the test itself was not quite correctly aligned. Even if one accepts the choice of the 1650s as the date by which the shift should have occurred, which certainly seems very early in the French context, the polarization between hatred and sexuality does not really pit like against like. Sexual pollution could not act as an organizing principle for the treatment of sin, since it is a specific case which forms part of a larger whole. The Copernican analogy, appropriate enough in many ways, also risks being misleading; sin is far more pliable and subjective than celestial mechanics, and has tolerated much higher levels of internal incoherence on the part of its expositors. It might be better to make a comparison with the new discoveries, and the addition of countries to an existing map in a way which eventually changed its whole aspect. When we also take into account respect for tradition and authority, with the prevailing habit of compilation from earlier sources, we would expect just the kind of muddle and inconsistency which is indeed to be found in much seventeenth-century writing on sin and penitence. Far from arguing against Bossy's general approach, I think that he was quite right to look for some kind of revolution in sin, and that the choice of an excessively severe test led him to understate the extent to which it took place. In his latest discussion of the question he adds the point that 'almost everything, from the revival of the Ten Commandments downwards, conspired to enforce the conception that the proper description of sin was disobedience—disobedience to God, church, king, to parents, teachers, and authorities in general.'[8] This is very convincing, but I would combine it with an emphasis on the hostility to sensual indulgence and concupiscence of every kind, not merely sexual, which was the hallmark of moral rigorists in the seventeenth century.

The obvious way to discover whether such changes occurred might seem to lie in a comparison of writings on moral theology from the later middle ages with those of the counter-reformation period. For reasons already hinted at, this is in fact very unhelpful; at first glance we would have to conclude that an internalized, individualized morality already held sway in the fourteenth century, if not earlier.[9] Whereas evidence for the theory of confession is abundant, that for actual practice is inevitably scarce and ambiguous. Historians of the later medieval church generally assume a substantial gap between the two, particularly in rural society, and Bossy's argument depends explicitly on the existence of such a time-lag. The great bulk of the technical literature was in Latin, which would have rendered it inaccessible even to many clerics, while general diffusion was clearly impossible before the age of the printed book. The sixteenth- and seventeenth-century catholic reformers certainly saw themselves as having

[8] *Christianity in the West*, 135.

[9] This emerges very clearly from T. F. Tentler, *Sin and Confession on the Eve of the Reformation* (Princeton, NJ, 1977).

to confront widespread ignorance and abuse, both clerical and lay; they were prone to a degree of self-dramatization in such respects, but it is hard to find any serious evidence on the other side. It is also important to understand that religious rituals have multiple meanings, so that the sophisticated clerical conception of the sacrament, even supposing it to have been shared by the mass of the parish clergy, need not have found much response among a laity deeply attached to older ways of thought. Since confession could really only function with the collaboration of the mass of penitents, their attitudes were bound to remain as important as the theories of the intellectuals. In the light of these factors, this chapter will be based on a (necessarily somewhat hypothetical) model of change:

(*1*) That there was an 'official' doctrine of confession, dating back to at least the twelfth century, conceived in basically individual terms, which stressed the reconciliation of the sinner to God, after an investigation of how well he had kept the moral code.

(*2*) That this was balanced by a 'popular' understanding of the sacrament in essentially communal terms, as a device for reconciling enemies through the mediation of the priest, and maintaining communal harmony, which saw no need to probe for hidden sins lacking any public dimension.

(*3*) That only with the implementation of the Tridentine decrees did this latent conflict impinge on the mass of the population, as rising levels of education, information, and discipline swung over the majority of the lower clergy from the 'popular' to the 'official' view.

Such a picture would certainly fit with many of the observed facts; the great expansion in the literature of advice for both confessors and penitents, much of it now in the vernacular, the place given to confession in seminary teaching, its central role in missionary activity, the bitter disputes over 'laxism' and 'rigorism', and the limited but significant evidence of popular reticence towards the new style. Fuller consideration of such questions may help us to build up a more precise picture of the pivotal role confession clearly played in the relationship between church and society in the France of the catholic reform.

The shifts in values involved were relative rather than

absolute, for an emphasis on individual sinfulness did not imply the automatic relegation of offences against the community to some subordinate category. Public sinners continued to receive special attention from the moralists, although the primary point was now the bad example they set others rather than the hostility they created. Both styles of thought centred on conceptions of purity and danger, with pollution as the constant threat to the Christian commonwealth; for the individualist school breaches of communal norms were only one aspect of man's failure to mould his life in God's image. Such a view naturally tended towards a kind of moral imperialism, seeking to bring the whole of human life under scrutiny, and to extend the sphere of the sacred until it swallowed up the secular. There was a sense in which popular attitudes also repudiated any division of the world between sacred and secular, but the primary movement here was to adapt religion to communal needs, within a flexible system which was very tolerant of inconsistencies. The new moralists took the opposite stance, spiring to the taut, logical coherence which had been the hallmark of the scholastic philosophers who had played a crucial part in initiating the whole process. Inevitably their conceptions of sin and pollution were formulated within the context of a faculty psychology which was essentially scholastic. It was by no means an agreed and stable theory, for a lively debate was in progress over the nature and status of the passions, with notable contributions from Montaigne, Charron, Descartes, and others, but most clerical moralists subscribed to a broadly similar, slightly vulgarized version of it.[10] The outlines are admirably stated by the Capucin Jean François de Reims, writing in the 1630s:

> One must therefore understand that there are in us, or rather in our souls, two completely opposed parts. The one is called inferior, and contains the concupiscent and irascible passions, which when excited by the objects which present themselves to the exterior senses, and to the imagination, carry the soul towards those objects which suit them, no more and no less than in the beasts: the passions of the concupiscent appetite move it towards objects of delight; and those of the irascible appetite, towards the good which is useful or delightful, but can only be secured with difficulty; and it comes so

[10] There is a very helpful analysis of these debates in A. Levi, *French Moralists: The Theory of the Passions, 1585 to 1649* (Oxford, 1964).

naturally to our passions to move the soul towards the said objects, without considering whether they are according to the pleasure of God or not, that if God had not given us the reason and the will to command these passions, we would differ in no way from the beasts. We therefore have another part of the soul, which directly combats the first, a part which is called superior, because it is moved towards celestial things, and towards God's pleasure, and this is nothing but the reason or the will: for it is always in our power, by reason of the free-will which is in us, to subject this superior part to God: but it is not always in our power to subject that other part, and make it follow what is good, but often it opposes both God and reason. This is why God has indeed obliged us to hold this [superior] part of the mind subject to his pleasure, but we are not obliged to hold the inferior part subject in the same way, since this is impossible. Note this distinction well, for it is very necessary to know how to detect when there is sin or not, within thoughts and internal movements.[11]

Although formally the passions were classified as morally neutral, it is evident here that they readily became the villains of the piece, the gateway by which the devil constantly made his entry. It was dangerously easy for vulgarization to produce a version of dualism, with the superior faculties of the soul required to exercise a repressive role towards its baser components.

The ambiguities and tensions within this form of psychological philosophy were indeed very great, and played their part in the furious disputes over confessional practice which marked the seventeenth century. The quarrels between Jesuits and Jansenists, and their respective allies, ranged from the most abstruse theological issues to everyday pastoral methods. The real starting-point was probably the 'new theology' of Molina, who was concerned with questions of free will and predestination. These arguments fitted neatly with the ambition of encouraging all men towards self-amendment, by denying the existence of any firm dividing line between the elect and the reprobate. They also had implications for the handling of confession, in which area they could be combined with the Jesuit expertise in the science of casuistry, with its technique of examining test cases to establish boundaries between mortal sin, venial sin, and no sin at all. Both Jesuits

[11] J.-F. de Reims, *Le Directeur pacifique des consciences* (4th edn., Paris, 1645), 48–9.

and Jansenists were committed to enlarging the role of confession, but there were significant differences in their approaches. For the Jesuits confession provided an ideal opportunity to coax men and women into godliness, with the confessor fitting his prescription to the individual case. Moderate penances and quick absolution encouraged the penitent to return often, gathering spiritual strength, and securing divine aid by taking communion frequently. This method did not have any necessary connection with a laxist moral theology concerning the nature of sin itself, so was relatively unaffected by the various condemnations of Jesuit casuistry. It is quite implausible to suggest that the Jesuits were deliberately setting out to provide a moral system 'within which the middle classes could live with untroubled consciences'.[12] Whatever their enemies alleged, their laxism should not be exaggerated, while such a theory of intentionality is patently anachronistic. The Jesuits aimed to win souls for Christ, and believed that their methods would lead to a more intense spiritual life. At the same time they did seek to become the confessors of the rich and powerful, from the king downwards; the royal confessors were Jesuits throughout the seventeenth century. It was hardly surprising that the Society's enemies should have seen its use of confession as summing up most of their objections, while the writings of the Jesuit casuists provided Pascal with the materials for one of the most devastating satires of all time. Readers of the *Provincial Letters* will hardly need reminding how the Jesuit father is presented, with the greatest subtlety, as a well-meaning *bonhomme* who repeatedly fails to see the implications of his own remarks, as he revels in the ingenuity of his order's casuists. This is both very funny and very perceptive; until the 1660s the Jesuits seem to have been so persuaded of the rightness of their approach that they stirred up an unnecessarily large hornets' nest.

It would certainly be most unwise to take Pascal and the other Jansenist polemicists as fair guides to what the Jesuits taught and practised. Laurence Brockliss, drawing on his

[12] M. Hepworth and B. S. Turner, *Confession: Studies in Deviance and Religion* (London, 1982), 47. This book contains an interesting sociological analysis of confession, but its historical sections are often questionable.

exceptional knowledge of faculty and college teaching in the period, has recently argued that the *Provincial Letters* were a calumny on Jesuit moral theology.[13] He points out that of some thirty casuists cited by Pascal only one, Etienne Bauny, was a Frenchman, and that his views had been condemned in both Rome and Paris in 1640–1; moreover the numerous enemies of the Jesuits could only produce a tiny group of accusations about their many courses in moral theology. These arguments are powerful, although not conclusive. Since the enormous row over tyrannicide, given such sharp focus by the assassinations of Henri III and Henri IV, the French Jesuits had been keenly aware of their vulnerability to charges of lax moral teaching, so it is not surprising if they observed greater caution in this area.[14] While Brockliss is surely right in claiming that their formal teaching rarely deviated from orthodoxy, this does not prove that they would not have liked to do so had they felt less constrained. Bauny may be a single example, but this was hardly Pascal's fault, since the only other relevant publication by a Jesuit teaching in France was the 1634 *Theologia Moralis* of an expatriate Scot, James Gordon, an unexceptionable work which seems to have passed almost unnoticed.[15] The *Somme des pechez,* by contrast, was something of a popular success, already in its fifth edition by 1638, while there were French editions of some of the other casuists attacked, including the prime target Escobar, a Latin edition of whose compendium, originally published in Spanish in 1630, appeared at Lyon in 1644. There is no mistaking the literary origins of Pascal's Jesuit as a stock pedant, part of the artifice of the polemicist; when one reads the maladroit counter-attacks of Fathers Nouet, Annat, and Pirot, however, he begins to take on a certain credibility outside the text.[16]

[13] L. W. B. Brockliss, 'The *Lettres provinciales* as a Jansenist calumny: Pascal and moral theology in mid-seventeenth century France', *Seventeenth-Century French Studies,* 8 (1986), 5–22.

[14] For the debates on tyrannicide, see R. Mousnier, *The Assassination of Henry IV* (London, 1973).

[15] Brockliss, 'The *Lettres provinciales*', 6–8.

[16] [J. Nouet], *Réponse aux lettres que les jansénistes publient contre les jésuites* (Paris, 1656); F. Annat, *La Bonne Foi des jansénistes en la citation des auteurs reconnues dans les lettres que le secrétaire de Port-Royal a fait courir depuis Pâques* (Paris, 1657); [G. Pirot], *L'Apologie pour les casuistes contre les calomnies des*

Their methods of defence were rather less eccentric than that adopted by the Spanish Jesuit Mateo Moya, under the pseudonym Amadeo Guimenius, who collected outrageous opinions from other casuists to show that his order did not deserve to be singled out.[17] Nevertheless, the French Jesuits did not simply assert that they had been traduced; their spokesmen tried to defend several of the positions Pascal mocked, and plainly reasserted the case for a more flexible and pragmatic approach to human sinfulness.[18] There is no real doubt that members of the Society had been 'selling easy devotion to gentlewomen', something they would continue to do.[19] Pascal was not so far from the truth when he had his Jesuit say that his colleagues would have liked to impose the strict evangelical morals they personally followed, but recognizing the corruption of mankind, 'le dessein capital que notre Société a pris pour le bien de la religion est de ne rebuter qui que ce soit, pour ne pas désespérer le monde' ['the essential design our Society has formed for the good of religion is not to rebuff anyone at all, so that the world shall not despair'].[20]

Although this passage highlighted a genuine tendency, it was patently unfair in suggesting that there was any deliberate policy or conspiracy to such an end, as Roger Duchêne has pointed out in his recent reformulation of the case against

jansénistes (Paris, 1657). These responses are discussed by R. M. Golden, 'Jesuit Refutations of Pascal's *Lettres provinciales*', in *Church, State, and Society under the Bourbon Kings*, ed. R. M. Golden (Lawrence, Kansas, 1982), 83–124; Golden rightly emphasizes that they made many good points, but also shows how Pirot in particular gravely embarrassed the Society as a whole.

[17] A. Guimenius, *Adversus quorumdam expostulationes contra nonnullas jesuitarum opiniones morales* (Bamberg, 1657). A French edition published at Lyon in 1664 was censured by the Sorbonne in 1665, leading to a complex series of disputes involving the pope, the king, the *parlement* of Paris, and the Assembly of the clergy. P. Blet, *Le Clergé de France et la monarchie* (2 vols., Rome, 1959), ii. 319–25; A.-G. Martimort, *Le Gallicanisme de Bossuet* (Paris, 1952), 240–73.

[18] This was particularly true of Pirot; see below, pp. 350–1 for some examples.

[19] The phrase is used by Brockliss, 'The *Lettres provinciales*', 6.

[20] B. Pascal, *Les Provinciales*, ed. L. Cognet (Paris, 1965), 103. Pascal may have had in mind the reply made to Arnauld by Nicolas Caussin during the earlier controversy, *Réponse au libelle intitulé la theologie morale des Jesuites* (Paris, 1644), where it is remarked 'si la question estoit prise dans cette rigueur, on desespereroit tout le monde' (p. 42).

Pascal.[21] Duchêne shows how the *Provincial Letters* use literary devices, quotation out of context, subtly misleading translations of Latin technical terms, and such techniques as guilt by association to create a seductive fiction, a travesty of the real opinions of the Jesuits. As he says, if one actually reads Escobar or Bauny it is to discover that these casuists were generally austere and serious moralists, whose attempts to match Christian ethics with everyday life sometimes led them into awkward or imprudent formulations. There is no need to suppose that Pascal was insincere, for he detected two dangers to the church which were of supreme importance in his eyes and those of his friends, and which led naturally into exaggerated visions of conspiracy. The attack on Jansen's *Augustinus* was implicitly also one on St Augustine himself; this probably happened far more by accident than design, since all parties were interpreting (or misinterpreting) his doctrine to suit their own purposes. Secondly, the casuists and devotional writers were the exponents of a practical Christianity for ordinary men and women, who recognized the difficulties of the world, and were prepared to compromise with them. Port Royal and the *solitaires* were the embodiment of an alternative view which recognized only moral absolutes, required total commitment as a sign of grace, and dealt with moral dilemmas by retreat and self-abnegation.[22] The Jansenist view held obvious advantages as a debating position, reinforced by the need to uphold an illusory continuity and coherence in the past doctrines of the catholic church, as well as by Pascal's supreme literary gifts. It is rather too easy for modern readers to exaggerate the resulting polemical success, for Pascal's tone of voice, which now seems so sympathetic, shocked many contemporaries; in any case the Jesuits were too powerful and influential to be defeated in the crucial struggles on specific issues.[23] Arnauld was expelled from the Sorbonne, and it was probably only

[21] R. Duchêne, *L'Imposture littéraire dans les Provinciales de Pascal* (Aix-en-Provence, 1985).

[22] This point is very well made by Duchêne, ibid., 157–8.

[23] There is of course good contemporary evidence for the effect of the *Provinciales*, and the critics were probably concentrated among the erudite clergy and magistrates. For the reactions of the fashionable world, see the opinions of Gabriel Daniel (p. 350 below) and F. Hébert, *Mémoires du curé de Versailles, François Hébert (1686–1704)*, ed. G. Girard (Paris, 1927), 138–40. Daniel and Hébert both attribute enormous influence to Pascal's work.

when they realized that this was inevitable that the Jansenists decided to appeal to public opinion, a tactic which was certain to alienate many moderate clerics. The immediate result was a redoubled persecution, which had few serious opponents even among the *parlementaires* whom Pascal must have hoped to win over. The majority of seminaries and colleges remained in the hands of the Jesuits themselves or of other orders who taught a Molinist theology of grace.[24] Jesuit confessors, protected by the secrecy attached to the sacrament, continued to practice 'la dévotion facile' without needing to assert any systematic moral theology to justify their application of individual judgement. Yet the debate did have a very significant effect on the formal position of the French church on questions of casuistry and confession. 'La morale relâchée' became a routine *bête-noire*, pursued in successive condemnations by the Assembly of the clergy and the papacy. The elements of formalism and hypocrisy involved did not escape the eye of contemporaries; in 1700 Archbishop Le Tellier was offering lavish hospitality to the members of the Assembly while they condemned 127 laxist propositions. A placard was circulated in Paris, showing on one side Jesuit missionaries being flayed and burned under the title 'Morale relâchée', on the other various plump bishops and *abbés* sitting down to a feast under the title 'Morale sévère'.[25]

By 1700 the attempt to smear the Jesuits with charges of laxism had become dangerously counter-productive, so that the 'triumvirate' of Bossuet, Noailles, and Le Tellier succeeded only in arousing public sympathy for their opponents.[26] The episode does make explicit the motivation behind so many of these clashes; they were struggles for power and esteem, in which doctrine was as much a weapon as an end. Neither the seven deadly sins nor the ten commandments really dealt adequately with this endemic clerical vice, which permeated the church from top to bottom. Perhaps men who had renounced so many other kinds of gratification were more

[24] For this point see Brockliss, 'The *Lettres provinciales*', p. 17, and *French Higher Education in the Seventeenth and Eighteenth Centuries: a Cultural History* (Oxford, 1987), 247–66.

[25] The placard is described by Hébert, *Mémoires du curé de Versailles*, 309.

[26] See above, p. 226.

vulnerable on this front in consequence, particularly when they could persuade even themselves that they acted to support either lofty principles or the dignity of their own order or position. In the case of confession, such feelings were very evident in the bitter running battle over the privileges of the regular orders, whose unpopularity with bishops and *curés* helps to explain why the Assemblies were so ready to sniff out laxism. There was a second level at which it operated here, for these questions of jurisdiction were more than a matter of prestige; the sacrament of penitence was a peculiarly direct assertion of clerical power over the laity, and was valued accordingly. All this helps to account for the decision of the 'great Assembly' of 1655–7, in general so severe against the Jansenists, to promote the cause of moral rigorism through the circulation of Borromeo's *Instructions to Confessors*.[27] At the official level the French church now took a consistently hard line very close to that advocated by the Jansenists, shorn only of a few extreme positions adopted by individuals. Far from inviting penitents to communicate often, confessors were encouraged to preserve the dignity of the sacrament by admitting only those who were worthy, excluding habitual or public sinners until they gave evidence that they had mended their ways. Moral rules became increasingly rigid, with only very small areas of discretion left on such matters as Sunday observance. The formal teaching offered in colleges and seminaries across the whole country, including the numerous Jesuit establishments, seems to have been couched in such terms, so in principle these doctrines should have penetrated extensively down to parish level by the end of the seventeenth century. Such apparent unanimity is certainly suspicious, particularly when it is partly motivated by a fear of denunciation, but that it could be reached at all does rather confirm the view that the Jesuits were much nearer to rigorist standards than their critics thought.

Some of the most heated arguments took place over issues which are unlikely to have had much significance for the average layman. Probabilism, the casuistical principle by which a minority opinion could justify absolution, and which the

[27] A. Degert, 'S. Charles Borromée et le clergé français', *Bulletin de littérature ecclésiastique*, 4 (1912), 145–59, 193–213.

Jansenists saw as a Trojan horse for almost any turpitude, looks like a case in point. So does the tangled debate over the nature and necessity of contrition and attrition; was it sufficient for the sinner to repent out of fear, or must he attain a true love of God before he could purge his sin? To a large extent these subtleties belong to the world of the lecture theatre and the controversial tract, where they were applied to a classic group of test cases, the stock-in-trade of casuistical writing and teaching. Such cases were chosen more for their theoretical interest than for their practical relevance. Over the vexatious issue of duelling, for example, there was considerable discussion of obsolete practices, such as a duel between two condemned men ordered by a judge.[28] Another much debated problem was how the confessor should proceed when faced by a married woman who said that one of her children had not been fathered by her husband; should he require her to inform her husband, so that the illegitimate child could be excluded from the inheritance?[29] The standard answer was that he should not, since the injustice done to the legitimate children was less serious than the breakdown of marital relations and family honour which was likely to result from such a revelation, leading to further disorders and sins. Since those devout women whose scrupulous consciences were seen as something of a danger would hardly commit such sins, and the less moral ones who did—and managed to keep them secret—were unlikely to volunteer confessions, it is doubtful whether many priests found opportunities to apply this sensible advice. Casuistry was drawn by its very nature to pick out such problems, which helps to explain why it could easily give rise to scandal, when its tendency was always to spotlight the points at which moral norms conflicted. The Jansenist fear

[28] F. Billaçois, *Le Duel dans la société française des 16ᵉ–17ᵉ siècles: essai de psychosociologie historique* (Paris, 1986), 176–7. The whole section on pp. 175–82 dealing with the casuists is of great interest, showing them struggling with the problem of the loss of honour and reputation which might result from declining a challenge.

[29] For a full discussion of this case, see J. Pontas, *Dictionnaire des cas de conscience* (3 vols., Paris, 1734), iii. cols. 628–30 (restitution, case 155). The problem is also mentioned as a difficult one by J. Benedicti, *La Somme des pechez et le remede d'iceux* (Paris, 1602), pp. 116–20 (for this work see below pp. 293–7).

that this involved walking on quicksands is easy to understand, particularly when one considers such authors as Sanchez, who used the technique in a constructive way to evolve a more humane, realistic, and flexible approach to sexual morality.[30] The ultimate crime of the casuists was to demonstrate that a simple Christian morality was impractical, and that the real world required the endless use of human judgement to interpret and nuance the commandments of God. Unfortunately this approach, which had great potential for the creation of a more realistic moral theology, implied the kind of sensitive individual application by the confessor which aroused the darkest suspicions, and could not survive exposure to public discussion in the religious climate of seventeenth century France.

The result was to leave a gulf between the theoretical discussions and the everyday problems which faced the ordinary priest. In order to approach the latter more closely we need to turn to less polemical literature, beginning with the manuals written for confessors, as distinct from treatises on moral theology. Then there are similar, if shorter, sections in the *Rituels* and *Statuts synodaux* of individual dioceses, and in the more general handbooks for priests. As might be expected, the tone of this literature changes significantly around the middle decades of the century, as authors, no doubt acting on some combination of conviction and self-preservation, adopted a more austere, essentially rigorist posture. Earlier writers were less inhibited from venturing personal opinions on vexed points, or from trying to reconcile the demands of the godly life and the human world. The most impressive and encyclopedic work produced by a French cleric was *La Somme des pechez et le remede d'iceux,* by the Franciscan theology professor Jean Benedicti. First published in 1584, this was a massive folio volume of over 700 pages, which displayed a great range of subject matter, a knowledge of recent debates and of the Council of Trent, and a serious yet flexible view of the confessor's job. The book often strikes an unexpectedly sympathetic note, and on many issues Benedicti evidently strives to find a middle position; he regards attrition as sufficient, 'tante petite soit-elle', being turned to contrition by the sacrament, for 'il

[30] For Sanchez see above, p. 239.

ne faut donc pas Lutheraniser, disant que telle repentance est hypocrite'.[31] On Sunday observance a long list of prohibitions is followed by a large number of exceptions, including the argument that the really poor should be allowed to work, provided that it is a matter of absolute necessity, and they avoid scandal by doing it in secret.[32] Dancing is the subject of a very complex argument, for while Benedicti does not condemn it outright, he only permits it under six conditions. It must not take place during periods of ecclesiastical prohibitions, the participants must be decent and well-behaved, and must sing no indecent songs; clerics must not participate, and the dance must not be organized with the intention of encouraging luxury. Finally

Il faut danser honnestement, et non point à la façon du jourd'huy, lors qu'on fait faire *la volte et madrigalle impudamment, impudiques, furieuses, et paillardes,* a ses dames et damoiselles, qui bien souvent y sont descouvertes ignominieusement et frauduleusement, chose du tout indecente a personnes d'honneur . . . Mais quoy? Voulez-vous dire que ces danses qu'on fait ordinairement aux festes et Dimanches par ces villes et bourgades soient peché: Il se peut faire quelquefois que n'enny, jaçoit que difficilement, quand on les hante seulement pour se recréer, et non point pour mauvaise intention: cela estant toleré à cause de l'infirmité du simple peuple souvent à demy brutal.

[The dancing must be proper, and not in the modern fashion, when people dance *the volta and the madrigal shamelessly, which are scandalous, furious, and lascivious,* for the women and girls are often exposed ignominiously and fraudulently, something honourable people must regard as totally indecent . . . But so what? Is one to say that the dances which are held on Sundays and feast-days in towns and villages are sinful? It may sometimes be that they are not, although this is difficult, provided people only attend for recreation, and without evil intentions; they are tolerated in this way on account of the weakness of the common people, often half brutal.][33]

As he twists and turns, in his mind rather than on the village square, one can sense his desire to reconcile stern principles with practicality, even to the extent of showing a certain wary tolerance of popular feeling.

[31] Benedicti, *Somme* (1602 edn.), 626.
[32] Ibid., 79–80. [33] Ibid., 365–6.

As his worries about new dance styles reveal, Benedicti was very conscious of sexual problems, and his book follows its medieval predecessors in giving them plenty of space. There is a substantial and uncomfortable section on masturbation, which leads into the difficult case of 'wet dreams', which he obviously sees as a particular problem for clerics.[34] While he is reluctant to write too much in French, he insists this is necessary when many priests are weak in Latin; as he sensibly observes, 'it is not books which teach how to sin, it is our nature, alas so deeply flawed, which teaches sin all too well'.[35] He declares in the orthodox manner that a husband must not seek excessive sexual pleasure with his wife, and that only the so-called missionary position is allowed, although so long as conception is still possible irregularities will mostly be only venial sins.[36] Contraception is of course a mortal sin, whether by taking some potion or by those who 'were content to satisfy themselves in their marriage by caresses, from which pollution outside the natural place might occur, without intercourse . . .'[37] Benedicti is conscious of the implications of this position, claiming that one should never fear to have too many children, since God will provide the means to nourish them, as he does for all his creatures. If poverty does result, this is the cross which Jesus himself carried, and in any case the Scriptures indicate that it is the mark of the truly predestined.[38] He would also allow kissing, caressing, and loving words between husband and wife, as a prelude to intercourse, for 'I do not want to be such a rigorous censor on this point, as to blame such things between husband and wife, putting sin where there is none.'[39] Despite these sections, and others dealing with such matters as lascivious songs, it would be quite misleading to suggest that Benedicti gives disproportionate space to sexual problems. His remarks reflect his manifest tendency to place the emphasis, in the manner which had become current during the later middle ages, on the individual, his intentions, and his personal salvation. What one does not find is any suggestion that sexual offences are any

[34] Ibid., 138–45. [35] Ibid., 144–5. [36] Ibid., 148.
[37] Ibid., 148. [38] Ibid., 149.
[39] Ibid., 148. Benedicti was also willing to permit 'honest' touching and kissing between fiancés (p. 186).

worse than the other sins to which he devotes attention, such as drunkenness, a vast range of superstitions, or the many varieties of avarice.[40] The one occasion when he does relate lack of chastity to other sins is in discussion of a particular act of *maléfice*, the infliction of impotence by the 'nouement d'aiguillette'; granted his basic assumptions, it is not unreasonable for him to treat this as a divine punishment for the number of marriages contracted primarily to satisfy carnal desires.[41]

In some respects the most striking feature of the book is the prominence given to 'economic' sins, notably in the form of a tremendous list, occupying nearly thirty pages under forty-eight headings, of offences calling for restitution.[42] These include usury, simony, theft, clipping and coining, using false measures, and more specialized offences such as keeping too many pigeons. For Benedicti all professions have their characteristic crimes, whether these involve innkeepers who adulterate their wine, apothecaries whose drugs are old or intended for evil purposes, or butchers who sell bad meat. Whether one takes it as a catalogue of actual misdoing, or as a list of popular prejudices, it adds up to a dismal commentary on the reputation of French tradesmen. More significantly, it implies an immensely ambitious view of the confessor's task, although Benedicti is so aware of the difficulties arising from ignorance and established habits that he cannot have envisaged this as lying within the immediate grasp of the church. His discussion of the relations between priest and penitent looks back to 'the rigour of the solemn penitence observed in the primitive church. If ecclesiastical discipline were observed in our kingdom, we would not admit the good and the bad pell-mell to communion in this way, like Judas among the apostles.'[43] The world being as it is, he has to recommend

[40] Nevertheless he clearly faced some criticism for his frank treatment of the subject, for the dedicatory letter of the second edition of 1586, to Bishop Pierre de Gondi of Paris, remarks: 'Et touchant le 6. commandement, duquel on a tant voulu quaqueter par l'advis de quelques grands personnages, et mesmes Docteurs de la Sorbonne, j'en ay adoucy et mesme changé quelques mots et sentences pour contenter un chacun, ce qui est toutesfois bien difficile.' Unfortunately I have not been able to compare the relevant editions to see exactly which changes he made.

[41] Benedicti, *Somme*, 463. [42] Ibid., 671–97. [43] Ibid., 225.

prudence to the confessor, who must avoid giving scandal or revealing the secret sins of a penitent by refusing him communion. He also feels it necessary to insist that venial sins should be confessed as well as mortal, and for this reason 'one must not make a quick and rash judgement that someone who goes to confession has on every occasion committed a mortal sin, because one can very well confess having only committed venial sins.'[44] The ultimate effect of Benedicti's work is to develop the tradition of the late middle ages, taking topics individually in a quest for a workable balance between the ideal and the realistic. He offers no challenge to established approaches, although he tends to extend the range of confession very widely, while regarding it as an essentially individual sacrament. There is no reference to any communal function of confession, beyond the exclusion of public sinners from taking communion with the people they have scandalized, until they have done public penance or the *curé* has been able to announce they have repented.[45] The community appears primarily as a vector of activities which threaten the moral health of its members, whether these are dancing, the outrageous Sunday drunkenness found in much of France, or the superstitious practices which are 'les reliques du Paganisme'.[46] At the same time there is a genial realism here, which suggests almost an age of innocence by comparison with the taut, logical systems of a century later; Benedicti is concerned with the everyday sins he sees around him, and writes as much out of personal experience and feeling as out of theological expertise.

Something of the same quality is apparent in a work which appeared in the next decade, *La Vraye Guide des curez, vicaires et confesseurs* by the Benedictine Father Milhard. First published at Toulouse in 1597, this small handbook was censured by the Sorbonne in 1619, on grounds which included the suggestion that a confessor could speak to another priest about sins of which they had both heard in confession. Alongside some laxist propositions about simony the Faculty also singled out sections claiming that a marriage between two impotent people might be licit, and that a woman might refuse intercourse

[44] Ibid., 214.
[45] Ibid., 225.
[46] Ibid., 81 (drunkenness), 42–3 (paganism).

to a importunate husband, or when they already had too many children.[47] The whole tone of *La Vraye Guide* is certainly more accommodating than that of Benedicti; written confessions are held to be admissible in certain circumstances, while some rather suspicious exceptions to Sunday observance are allowed on grounds of custom, to such persons as lawyers and notaries.[48] When it comes to the sixth commandment Milhard's points are mostly very conventional, and indeed he is essentially a collector of opinions from earlier casuists. The choice he makes and the manner of presentation is however of some interest, as when he baldly states 'with regard to the immodest caresses of married people; if they say nothing of these, there is no need to enquire about them, since at the worst they are only venial sins'.[49] It is certainly hard to understand why he gives a whole paragraph to the question of feminine attire when his argument seems designed to get the worst of every world:

> If a woman adorns and paints herself, and exposes her bosom or breasts for an empty show and worldly glory it is a venial sin, but becomes mortal if she does it as a means to attract someone into concupiscence. But if these artifices and fine clothes were worn without vanity, and simply in order to conform with other people of her rank: or indeed to find a suitable marriage partner, matched to her own birth: and such other good ends, then there would be no sin at all: but since such good ends are among the rarest there are, the confessor ought to exaggerate and repress such dissolutions and vanities, which belong rather to paganism than to Christianity, without nevertheless finding mortal sin where it is no more than venial.[50]

On questions of rural usury Milhard is quite severe, listing

[47] The Faculty condemnation of 1619 is reported by P. Féret, *La Faculté de Théologie de Paris et ses docteurs les plus célèbres: époque modern* (5 vols., Paris, 1900–7), iii. 342–4. Brockliss indicates that there was another condemnation by the Faculty in 1627, which included the suggestion that it was permissible to kill false witnesses in defence of one's life, honour, and goods: 'The *Lettres provinciales*', 8. Since I have failed to find any such passage in my copy of the 1603 Paris edition, or indeed the points about *le devoir conjugal* censured in 1619, it seems almost certain that they were added in later editions, which do seem to have been considerably expanded.

[48] For written confessions, fos. 36ʳ, 41ᵛ; Sunday observance fo. 48ʳ.

[49] *La Vraye Guide*, fo. 60ʳ.

[50] Ibid., fo. 60ᵛ.

various loans of grain and sharecropping contracts for animals which are prohibited.[51] He goes on to complain that many people are so taken up with the desire of gain that they take no notice of clerical admonitions, but continue to be extortioners and usurers; in such cases, the only hope is to provide them with an alternative, but licit, route to enrichment. It is hard not to feel that the good father was a frustrated businessman, for he spends twenty pages listing twelve ways of increasing one's substance, by investment in land, *rentes,* trade of various kinds, and even a kind of bet on the life of a named person.[52] The final twelfth form of investment is the best of all: charity to the poor, which will not only store up heavenly treasure, but ensure the giver full granaries and wine-cellars.[53] On relations between parents and children Milhard adopts a moderate position, making the consent of both a condition for marriage, while emphasizing that those who defy their parents or maltreat them will suffer at their own children's hands.[54] The differences from Benedicti are subtle but interesting; whereas the former's exceptions were mostly in favour of the simple people, Milhard's confessor is invited to go easy on the privileged groups in society. He also enquires a good deal less earnestly into both the intentions of the individual and his private life, seeming more concerned with outward social conformity. His social preconceptions emerge in the clerical context when he warns his confessor:

> Not forgetting besides to advise the common people that they should never persuade their children to become priests, unless they have the means to take their studies as far as those of letters, having regard to the inconveniences and prejudices this has brought and is still bringing to the church of God, since these poor people have made use of this holy profession of the priesthood, as of a trade and mechanical estate, simply to earn their bread. In such a way that since their ends have been different, they only very rarely embrace the duty of their charge, whether in the service of God or their neighbour, the true and principal end towards which every cleric should direct his intentions.[55]

No doubt many other clerics would have used similar arguments to justify the widespread introduction of a

[51] Ibid., fos. 74r–76v. [52] Ibid., fos. 84v–95r. [53] Ibid., fos. 94v–95r.
[54] Ibid., fo. 114r. [55] Ibid., fo. 113r–v.

property qualification for ordinands in the later sixteenth century.

There are striking similarities betwen Milhard's moderate laxism and that of the much-vilified Jesuit Étienne Bauny, some of whose points sound like extensions of those made by his predecessor. Bauny argued, for example, that women might not incur blame if they allowed their faces and hands to be touched by those they knew to have improper designs on them, when this was a matter of common courtesy, or where refusal might harm their marriage prospects and expose them to ridicule.[56] In another passage, singled out by Pascal, he suggested that usury might be avoided through a form of contract by which profits and losses were shared with a merchant. 'Here, in my opinion, is the means whereby a quantity of people in the world, who provoke God's just indignation by their usuries, extortions, and illicit contracts, may obtain salvation, if instead of lending their money, they lay it out in the manner indicated below . . .'[57]

A long section dealt with the problem of those who did not pay their debts, giving seven reasons why they might at least be allowed to defer payment.[58] This began with sympathetic treatment for the poor who stole bread or firewood when they were in extreme need, held neither to commit mortal sin nor to be obliged to make restitution. The fourth reason, however, allowed deferment of payment

> when the person does not have the means to repay without reducing his status, as would occur for a judge, if in making restitution he became unable to meet the expenses necessary for the upkeep of his family, or for a gentleman, if he were to lack the means to appear in public according to his status, to have no horses, no arms, no clothes, nor servants suitable to his rank . . .[59]

Advice which would have been music to the ears of Jane Austen's Sir Walter Elliott and many others. Some of Bauny's worst difficulties resulted from his well-meaning attempts to distinguish levels of moral responsibility, as in the famous

[56] E. Bauny, *Somme des pechez qui se commettent en tous estats* (Paris, 1635 edn.), 125–6.

[57] Ibid., 258–64. For Pascal's criticism, *Les Provinciales*, 137–8.

[58] Ibid., 165–79.

[59] Ibid., 170–1.

passage where he claimed that the man who induced a soldier to set fire to his neighbour's barn should not be liable to make restitution.[60] Despite such instances of misplaced ingenuity, the tone of his work is very far from justifying the jokes that he would have eliminated sin by making so many exceptions. Many rigorist clerics would have agreed with him that 'worthy men, those who love honour with all their soul, flee from the sight of women, from which they can only suffer censure . . .[61] His general views on confession and absolution were hardly outrageous, for he held that the former was invalid without true repentance, and that this applied to

those of usurers, and others of similar humour and inclination, as are all those who knowingly do not wish to abandon the occasions of proximity which tempt them to evil, such as wicked practices with girls and women whom they have around them night and day, the reading of dirty and disreputable books, continued usurious lending: for how can the heart possessed by the love of such things be contrite? The sadness which is part of the sacrament detaches the soul from earthly attachments to turn it towards God, but these things of which we speak hold it down, and place it in the state of sin, with which grace can no more coexist than night with day, or hot with cold.[62]

As for absolution, Bauny ended his book with thirty pages on the question of when it should be deferred, whose opening section set the tone:

When someone has often confessed to having the property of another, and nevertheless has not restored it as he had promised, although he has the means, when he presents himself once more for confession he should be sent away without being absolved until he has given satisfaction: the same is true of those who keep concubines, and of all those others who see themselves within those proximate and unquestionable occasions of offending . . . the reason one gives for this, is that there is a danger that the excessive softness of the confessor will be their downfall, and that in the hope of always obtaining absolution for the sin from him, they will remain in it; they

[60] This passage is in fact missing from the 1635 edition, but the principle involved is explained on pp. 224–6. Cognet, referring to the first edition, says it is correctly cited by Pascal, *Les Provinciales*, 142–3.

[61] Bauny, *Somme*, 137.

[62] Ibid., 810.

must therefore be refused it, unless there is some special reason for thinking they may yet again be absolved that time . . .[63]

Pascal did manage to find a passage to ridicule in this section, where Bauny sought to prove that those whose business or social life brought them into contact with women who then caused them to sin could be in a state of grace without renouncing such frequentations. He cited de Bea, Azpilcueta and Sa to the effect that there could be legitimate reasons for them not to break off such contacts, which were not evil in themselves, 'such as would be the inability to excuse oneself, without causing people to talk about it, or that it would cause them some inconvenience . . .'. It was hardly honest, in this case, for Pascal to omit the qualification that they could only be absolved *dummodo firmiter proponent non peccare* (as long as they seriously resolve not to sin).[64] Bauny might have opened up a dangerous loophole, but he plainly had no thought of using it as a general escape route, and his overall position on absolution makes few concessions to hardened sinners. Since well over 90 per cent of his advice was the common currency of moral theology it is easy to understand why the unfortunate Jesuit was bewildered by the storm which blew up several years after the original publication of the *Somme des pechez*.[65] The book was censured by Rome and by the Paris Faculty of Theology; the latter picked out passages concerning hatred of one's neighbour, the seduction of a virgin, usury, restitution, defamation, and the right of regulars to absolve 'reserved cases'.[66] Chancellor Séguier prevented the publication of this censure, whose ultimate fate was still pending at Richelieu's death, but meanwhile the Assembly of the Clergy had joined in the hunt.[67] Even before the controversy which began with *La Fréquente Communion* in 1643, casuistry was drawing unwelcome attention to itself.

Another lengthy compilation of the same period was *Le Directeur pacifique des consciences*, by Capucin Jean-François

[63] Bauny, *Somme*, 871–2.

[64] Ibid., 878–80; Pascal, *Les Provinciales*, 180–1.

[65] The first edition is normally cited as 1630, but Cognet, *Les Provinciales*, 56, n.2, suggests that this is an error and that it appeared in 1634.

[66] For the condemnation, and Séguier's interference, see Féret, *La Faculté de Théologie*, iii. 344–9.

[67] P. Blet, *Le clergé*, i. 500–1.

de Reims, already quoted to illustrate the Faculty psychology of the day. This was a very conventional work, whose failure to explore difficult cases may have helped it to escape adverse comment; the author often pushed the discussion towards the specific problems of feminine religious houses, which provided relatively safe territory, away from the dangerous area of everyday conduct by the laity. Nevertheless his statements on economic matters, including almsgiving, usury, the concept of a fair price, and restitution, do have a generally laxist tone, so that had he been a Jesuit they could well have found a place in one of the Jansenist collections of scandalous opinions.[68] As with the other authors, this would have been to caricature his general outlook, for he emerges as a pious cleric whose ideals were those of an austere life of self-mortification and denial of sensual appetites. He devoted no fewer than eighty pages to sins against chastity, despite omitting actual fornication and adultery, from which the devout were supposed to be relatively safe, to concentrate on impure thoughts, gestures, and words.[69] No doubt this reflected his conviction that 'sensual or carnal love . . . is the greatest enemy we have, the most importune, the most subtle, and the most difficult to overcome.'[70] Like other moral theologians writing before the middle of the seventeenth century, Jean-François de Reims tended to subscribe, probably unconsciously, to something of a double standard. On the one hand stood the intense commitment of the truly virtuous few, mainly to be found among the clergy, on the other the more modest and compromised life-style which was the best that could be expected from those who lived in an inherently corrupt world. Even so severe a reformer as Bourdoise expressed a similar view, combined with his usual harshness towards clerical failings:

Les peuples seront sauvez, Dieu leur fera Misericorde, s'il luy plaist, car ils croyent, ils sont dociles, ils se laissent persuader: mais les Prestres, pensez-vous qu'il s'en sauve, pour moy je le crains fort, ils font trop les entendus. Quand il est question de corriger un Prestre, il vous apportera plus de repliques, et plus de raisons, qu'il n'y a a Paris des pavez, et de maisons, ils n'ont point d'humilité.

[68] J.-F. de Reims, *Le Directeur pacifique*, 533–7, 567–91, for the sections in question. [69] Ibid., 690–770. [70] Ibid., 652.

Ce que le monde sera damné, ce ne sera que par la faute des Prestres.

[The people will be saved, God will have mercy on them, if he is so pleased, for they believe, they are docile, they let themselves be persuaded: but do you think that the priests will be saved? For my part, I very much doubt it, they pretend to know too much about it. When it is necessary to reprove a priest, he will give you more replies, and more reasons, than there are paving stones and houses in Paris, for they quite lack humility.

If the world is damned, it will only be the fault of the priests.][71]

Such lacerating judgements, placing moral responsibility heavily on the clergy itself, were hardly likely to command assent among the mass of clerics, although they were by no means without their supporters. The most sweeping alternative vision was that of the Jansenists, with their insistence on one moral standard for all men, which inevitably led to the notion of 'le petit troupeau', whose lay members largely repudiated the world and its pleasures. At a less ambitious level, most reformers seem to have aimed at the simultaneous imposition of discipline on an unworthy clergy and an ignorant, recalcitrant laity. While in practice they recognized the necessity for differential standards, they shied away from any formal admission on such lines. It was always possible to call for more catechizing, frequent missions, and above all stricter application of penitence, while covering failure with the widespread assumption that God's will was for the reprobate to be far more numerous than the elect.

Whatever reservations people may have felt about the *Provinciales* themselves, from the 1650s onwards no author in the area of moral theology showed any inclination to risk being branded as a laxist. Rigorism was now firmly established as the official view, not least by the decision of the Assembly of the clergy in 1657 to adopt Borromeo's *Instructions aux confesseurs* as an officially sponsored text. The *procès-verbal* of the assembly makes it clear that this was seen as a counter to the excesses of casuistry, which might seem a surprising move from a body which had taken such a determined stand

[71] A. Bourdoise, *L'Idée d'un bon ecclésiastique, ou les sentences chrestiennes et clericales* (Paris, 1667), 70–1, 75.

against Jansenism.[72] Certainly the Jansenists were delighted by the step, and it was one of their sympathizers, Bishop Antoine Godeau of Vence, who drew up a circular letter recommending the book to all bishops. Most members of the Assembly, however, probably saw it as a deft way of averting trouble. Borromeo's prestige was such that neither the pope nor the Jesuits could object to the dissemination of his book, which at the same time gave only very limited or ambiguous support to specifically Jansenist positions on penitence. It is very far from being a comprehensive treatise; in the typical small format of my copy there are approximately 180 words to the page, seventy pages being sufficient for the main section of advice to Milanese confessors, a further fifteen for the additional guidance to *curés*, and twenty-seven for the specific instructions in connection with the jubilee.[73] Even within this exiguous compass there is considerable repetition, some sections reappearing virtually word for word. The work was so unfamiliar in Paris that a copy had to be obtained from Toulouse, in the 1648 translation by Archbishop Charles de Montchal, for the members of the Assembly to peruse it; this neglect is easier to understand when one recognizes how far Borromeo was concerned with a particular local situation. As a guide to confessors his book does not begin to compete with those already discussed; its main purpose is to lay down rules, with relatively little reference to examples or likely difficulties. The orthodoxy of Borromeo's position is still a matter of debate, with one recent scholar, Jean Guerber, denying that he was any kind of crypto-Jansenist, and portraying him as a moderate exponent of sound tradition, who was somehow hi-jacked by Arnauld and his supporters.[74] The case is well argued, but does not really convince, for it relies on the letter

[72] The fullest account is still that by A. Degert, 'S. Charles Borromée et le clergé français', but see also M. Bernos, 'S. Charles Borromée et ses "Instructions aux confesseurs": une lecture rigoriste par le clergé français (16e–19e siècle)', in Groupe de la Bussière, *Pratiques de la confession* (Paris, 1983), 185–200.

[73] *Instructions de S. Charles Borromée aux confesseurs de sa ville, et de son diocèse;* I have used a 1763 edition printed at Besançon.

[74] J. Guerber, *Le Ralliement du clergé français à la morale liguorienne: l'abbé Gousset et ses précurseurs (1785–1832)* (Rome, 1973), Analecta Gregoriana 193, esp. 309–27.

of the text to the almost total exclusion of the spirit. A much more balanced assessment is offered by another Jesuit historian, Paul Broutin, who rightly points out that Borromeo was opposed to Arnauld in advocating frequent participation in communion, but goes on to characterize him as 'an athlete of penitence', who wanted to turn all men, even great sinners, into saints.[75] What does seem generally agreed is that even if Borromeo was misinterpreted, the sudden enthusiasm for his work after 1657 gave impetus to the severe moral theology which was coming to dominate the Gallican church. It is disconcerting to note the response of the generally level-headed and charitable Vincent de Paul: 'The holy severity so often recommended by the holy canons of the church and renewed by St Charles, produces incomparably more fruit than excessive indulgence on whatever pretext.'[76]

One crucial point about Borromeo's little essays on the confessional is that they assert a single set of rules, mostly of a rigorous kind, with no concessions to any social group, whether on grounds of special privilege or of ignorance. Indeed he lists among the occasions of sin participation in a wide range of worldly activities, including the following:

Such commonly are for many through the corruption of our times war, commerce, the magistracy, the profession of barrister, of proctor, and other similar activities, in which the man who is accustomed to sin, often mortally, by blasphemies, robberies, injustices, hatreds, frauds, perjuries, and other similar crimes against God, knows that while he continues these same activities, he will find himself at the same risks . . .[77]

The same tone comes across in the list of matters he selected as being of special concern in a jubilee year, in which confessors had unusually wide powers; it is also a good guide to his work as a whole, since he was referring them back to his previous declarations:

Of the internal and external preparation of the penitents.
Of those who are negligent in learning Christian doctrine and morality, and in making their children and servants learn them.

[75] P. Broutin, *La Réforme pastorale en France au 17ᵉ siècle* (2 vols., Paris, 1956), ii. 379–98.
[76] V. de Paul, *Correspondance*, ed. P. Coste (14 vols., Paris, 1920–5), v. 324.
[77] Borromeo, *Instructions*, 47.

Of heads of households who fail to take the necessary spiritual care of the rest of their family.
Of the proud and dangerous displays and costumes of men and women.
Of those who are habitual sinners, and of those who, even though they confess yearly, nevertheless give no effective sign of amendment, nor of true penitence.
Of the obligation to withdraw from occasions leading to sin.
Of illicit contracts.
Of necessary restitutions, and of satisfactions to pious legacies and other debtors, and to the reputation of neighbours, which are necessary before receiving absolution.
Of public satisfaction and penances.
Of causing the penitents to make a general confession, and subsequently to confess and communicate often.
Of the necessary help to be given to pentitents after confession, to lead them into the way of their salvation, and to constant perseverance therein.
Under all these headings, and several others under which one is often engaged in sinning, without always knowing it, as when one keeps impure books, scandalous pictures and figures, or when one owns houses made ready for gambling, which are let to those who live an evil life; when in inns, houses, or rooms which are let out, or rented in some other manner, one holds or allows games, gambling with cards, prostitutes, or allows dances on feast days, and similar dissoluteness, and engages in forbidden practices. In the matter of trade, and especially moneychanging, the confessors should examine and interrogate the penitents with such precision, and such diligent discussion of every point, that they truly remove them from the shadows and the slavery of the Devil, and that they do not succumb with them.[78]

Borromeo goes on to warn that although the confessors have been given special powers to absolve in reserved cases, 'nevertheless they should not use them for the ruin of souls, but only for their edification.'[79] His prime reaction to the jubilee seems to be concern lest it sap the moral fibre of his diocese.

Apart from Borromeo's introduction of the confessional box, the constituent parts of his treatise may be traditional enough; the way they are combined, and the moral tone they create, leave a very different impression. His conception of the

[78] Ibid., 100–2. [79] Ibid., 105–6.

confessor is a dynamic and interventionist one, and the very first sentence of the book calls for them to administer the sacrament 'with the fruit which it can produce in souls, which is the true amendment of life.'[80] The result is not merely an internalization of the religious life, with greater weight on the individual, who must be properly reconciled to God through the operation of penitence. The personal relationship to God is essentially untestable, so institutional religion inevitably looks to external signs as the test. The result is the same kind of negative feedback as is observable in many forms of protestantism, with the call for personal spirituality rapidly transmuted into a general demand for the reformation of society. Sin becomes an offence against one's neighbours as well as against God, when it exposes them to bad examples or open temptation; the only real answers to such dangers are either mass conversion to the new standards, or their imposition by repressive means. The natural tendency of enthusiasts for Christian morality was to employ both techniques simultaneously, without recognizing that they were to some extent incompatible. Like many other theologians, Borromeo symbolized this by employing the metaphors of the judge and the doctor in writing about the relationship between the confessor and the souls of his penitents. After tough-minded discussions of the circumstances in which absolution should be deferred, and the imposition of public penitence on notorious sinners, he went on to give a positive image of the confessor in his role as a director of consciences, encouraging private prayer, self-examination, and frequent communion. There is no sign that he appreciated the difficulty of moving between these tasks, or how serious a barrier might be created by the harsh and authoritarian role assigned to the confessor in most of his treatise. This same tension is apparent in the application by French reformers of techniques which drew on both Augustine and Borromeo for justification; indeed it is probably inherent in any generalized system of confession. Although such problems cannot have any definitive solution, it is potentially disastrous if they are not recognized, so that a constant effort is made to strike a balance in practical terms. The pursuit of laxism effectively ruled out serious discussion

[80] Borromeo, *Instructions*, I.

of this central issue among the French clerical intelligentsia after the middle of the seventeenth century, leaving the ordinary clergy to strike what compromises they could between an idealistic, austere doctrine and worldly realities.

The consequences at the formal level of moral theology and seminary teaching have been very well described by Guerber; only in the nineteenth century would the French church accept the more humane and practical position established by Alphonse of Liguori. The Jesuits themselves took good care to fall into line, although there may have been an element of formalism in this. Father Antoine, whose *Theologia moralis universa* of 1726 was so severe that it disarmed critics of the order, seems to have participated in a mission a few years later which employed much more characteristic Jesuit techniques.[81] When Father Daniel tried to defend the casuists in his *Entretiens de Cléandre et d'Eudoxe* of 1694, the Jansenist Dom Mathieu Petitdidier was quick to use the book as evidence of duplicity, ironizing on the 'belles apparences' it had undermined:

> It has been noticed for some years that several young Jesuits proclaimed openly enough that they did not agree with the opinions of their elders, any more than those of their casuists, with regard to moral theology. It is known that Father Gaillart busied himself saying everywhere that the moral teachings of those Jesuits aged less than fifty were so different from the doctrine of those above that age, that there was what amounted to a wall separating them.[82]

Despite Petitdidier's jibe that 'le singe est toujours singe', Daniel's arguments were notably more cautious than those of Pirot and his contemporaries, and there is good reason to think that the French Jesuits had shifted their position considerably. Even if Antoine exaggerated his rigorism for tactical reasons, his book resembled virtually all others of its type in lacking any pastoral dimension or perspective; after the 1660s such considerations effectively disappeared from view, with

[81] For Antoine's work see R. Taveneaux, *Le Jansénisme en Lorraine* (Paris, 1960), 674–5, and Guerber, *Le Ralliement*, 64–6. His participation in a mission in 1731 is discussed by Guerber, pp. 68–9, and for the Jesuit missions in Lorraine and their laxist attitudes see Taveneaux, pp. 686–90.

[82] M. Petitdidier, *Apologie des lettres provinciales de Louis de Montalte contre la derniere reponse des PP. Jesuites, intitulée: Entretiens de Cléandre et d'Eudoxe* (Rouen, 1697), 13.

rigorist assumptions going unchallenged in the absence of serious debate. What is much less clear is the extent to which this strident moral theology affected actual practice. Concentrating on the use of delayed absolution as a method for compelling sinners to reform, Guerber writes as if theory and practice must have been virtually identical, yet the evidence for this is alarmingly scanty. Only a handful of Jansenist bishops and *curés* are known to have made systematic use of delayed absolution, whatever the manuals said, and it is hard to believe that the widespread employment of this technique would not have produced conspicuous and vocal protests. By far the most cogent evidence Guerber cites is taken from the correspondence of one of the strictest Augustinians among the bishops, Cardinal Le Camus of Grenoble, who explained repeatedly and eloquently why such an approach would be disastrous. Le Camus thought it would need two or three centuries, in a large diocese such as his, to administer the sacrament on such a basis, since:

these gentlemen find that tests lasting a whole year are necessary before absolution can be given, and that there must be no relaxation at all of the severity of the ancient canons with regard to giving satisfaction: a director who has only three penitents to conduct may, if he considers it appropriate, follow such rules; but in a diocese they are impracticable.[83]

In any case his *curés* had not been taught such rules, while if he had to rely on missionaries they would only cover twenty parishes in a lifetime. In earlier letters he had pointed out that Antoine Arnauld himself 'has agreed with me that in practice all these maxims must be softened in many ways: that in the best-ruled centuries they were much tempered',[84] while his own attitude showed admirable sense:

Believe me, let us never judge anyone nor despair of anyone, so long as the people retain their faith and have some fear of God's judgements. It is only the business of a very few heroic souls to perform canonical penitence, and to renounce all to follow the crucified Jesus. These are the graces of the study, and of the particular elect. But in the ordinary way, grace only destroys our temperament, our inclinations, and our bad habits little by little, in as imperceptible a manner as they became established; and in that case, purity does not

[83] Guerber, *Le Ralliement,* 90. [84] Ibid., 91.

appear at first either in our actions or in our heart, and all we do at the start is mixed with many impurities.[85]

To follow the literal prescriptions of the moralists would have required a combination of industry, saintliness, and tyranny very few bishops or confessors were likely to possess. The record of Le Camus's episcopate suggests that he scored highly on the first two, but was deficient in the last.[86]

Two early Jansenist bishops did come significantly closer to the ideal, egging one another on in their remote Pyrenean dioceses between the 1640s and the 1670s; Caulet of Pamiers and Pavillon of Alet.[87] Both sought to establish tight control over their clergy, approving only suitably qualified priests as confessors, while using newly founded seminaries and monthly *conférences ecclésiastiques* as instruments for clerical instruction. Apart from direct attacks on dancing, drinking, charivaris, and superstition, external phenomena which could be the object of prohibitions and police action, it was through the use of delayed absolution that Caulet expected to transform his people. For him the mark of a good confession was a change of life, while:

Experience shows that only too often, unless one makes sure of the sincerity of the promises penitents make by having them performed in advance, they always receive the sacraments and never leave their evil state . . . The practice of always absolving the penitents as soon as they have confessed is not that of the church, but that of lazy, compliant, and self-serving confessors.[88]

The same tone is found in Pavillon's *Rituel* of 1667, which was promptly condemned by the pope, but went into several editions, and had a significant influence on similar compilations for other dioceses.[89] The bishop harped almost obsessively on the need to withhold absolution from the unworthy,

[85] Ibid., 90–1.

[86] For Le Camus in general, see *Le Cardinal des Montagnes*, ed. J. Godel (Grenoble, 1974).

[87] For Caulet see J.-M. Vidal, *François-Étienne de Caulet, évêque de Pamiers (1610–80)* (Paris, 1939); for Pavillon see E. Dejean, *Un Prélat indépendent au 17e siècle, Nicolas Pavillon* (Paris, 1910), and Broutin, *La Réforme pastorale*, i. 199–214.

[88] Vidal, *François-Étienne de Caulet*, 285.

[89] *Rituel Romain du Pape Paul V à l'usage du diocèse d'Alet* (2 vols., Paris, 1667).

to which he devoted a whole section.[90] On Easter communion he accused the clergy of having lacked the necessary diligence:

> since it is universally seen that there is so little change, and so few true conversions among the people after the reception of the sacrament. This can only be attributed principally to the excessive facility, and weak compliance of the confessors in granting without fruit, and sometimes with sacrilege, the grace of absolution to their penitents, instead of withdrawing them from the state of sin in which they are wallowing, by deferring, or by refusing them absolution until they have produced worthy fruits of penitence, and they have effectively given up their sin, not merely in words.[91]

Elsewhere Pavillon recommended the keeping of 'registres de l'état des âmes', developing a Borromean model, in which the *curés* were to record the merits and faults of each parishioner, household by household, as far as they could be known from external observation: 'this is easy for them, since every year they visit every house to interrogate each inhabitant and neighbour, which gives them a more reliable knowledge of consciences than the confessions themselves, so that they cannot be deceived afterwards.'[92] While there does not seem to be any conclusive evidence that Caulet followed suit, it is hard to believe that he was left behind on this point. Whatever techniques his *curés* used, they clearly risked retaliation in kind, for the bishop claimed that four *curés* and a *vicaire* had at different times been falsely accused of incontinence 'in hatred of the way they have shown themselves zealous in the execution of ordinances against those who profane feast-days, against innkeepers, and against those who publicly keep concubines.'[93] On the other side, one of his opponents suggested that the bishop employed 'inspectors of confessions', who ensured that *curés* were rigorists, with those who were not approved being harassed until they resigned.[94]

The effects of this discipline seem to have borne out many of the warnings of more moderate clerics, for the number of communicants fell sharply in the dioceses of Alet and Pamiers.

[90] *Rituel Romain*, i. 114–21.
[91] Ibid., i. 83–4.
[92] This is taken from the 1669 report of the brothers Foreau, cited by Vidal, *Francois-Étienne de Caulet*, 288–9; the model is given in the *Rituel d'Alet*, i. 464–5. For these *registres* in general, see above, p. 266.
[93] Vidal, *François-Étienne de Caulet*, 132.
[94] Ibid., 289.

It appears that by 1669 there were people aged between twenty-five and thirty who had been unable or unwilling to acquire basic knowledge of the faith, who were therefore not being admitted to take their first communion.[95] In various parishes the *curés* reported that between 50 per cent and 80 per cent of their flocks did not take Easter communion, a situation they preferred to tolerate rather than force them into sacrilege. The extreme case seems to be a parish where only eighty out of 700 had taken communion, because the main profession was that of *flotteurs* working the river-boats, which made Sunday work obligatory.[96] This was just the sort of special case the despised casuists had tried to accommodate, and here rigorism does seem to have reached a point of absurdity. Profanation of the Sabbath was rather a special issue, since breaches of the rules were likely to be public, so that a tough-minded *curé* was in danger of being trapped by his own logic, unable to compromise without losing face. Most of the sins which agitated clerical moralists were more complex in their nature, combining elements of the public and private. Perhaps the best guide to assumptions in this area can be found in the short lists of sins picked out as examples by the writers of manuals, approximate though this technique must inevitably be. A typical passage occurs in the 1700 *Rituel* for the diocese of Toul: 'The most common public sinners in the parishes are the unchaste, the drunkards, the swearers and blasphemers, the married who live apart, those who have old hatreds, those who regularly fail to observe Sundays and feast-days, the innkeepers who normally receive the local people in their taverns on those days . . .'[97] These were public sins, those for which absolution could be refused openly without betraying the secret of the confessional, whereas private sins could only be sanctioned in concealed ways.[98] Pavillon's *Rituel d'Alet* provides another list, this time not confined to public sin, under the heading of the 'occasions prochaines' which lead to sin, and from which penitents must remove themselves. Those specified are (1) servants living in houses where someone is the cause of their falling into impurity; (2) those who frequent

[95] Ibid., 285.
[96] Ibid., 292.
[97] *Rituel de Toul* (for Henry de Thyard-Bissy) (Toul, 1700), 9.
[98] Ibid., 201–2.

houses or company where they fall into sin; (3) women who show their breasts; (4) gamblers who blaspheme and quarrel; (5) those who attend balls and plays, even if they do not commit the sin of impurity so common there; (6) those who read heretical or lascivious books; (7) those who have lascivious pictures; (8) those who profane Sundays and feast days.[99] Pavillon also stigmatized certain holders of public office—*curés*, judges, doctors, apothecaries, surgeons—who lacked competence, and were therefore liable to commit repeated errors and injustices.[100] In general he thought it was dangerous to give absolution to a penitent who was trying to cure himself, but had occasional lapses; such a person was not truly cured. If the lapses had been frequent he might be tried out for two or three months 'at the end of which if one recognizes a true amendment caused by the faithfulness of the penitent, and by the violence he has practised on himself, he may be absolved, since he will have given effective proofs of his conversion and his penitence.'[101] Scandal could be avoided, in the case of secret sinners, by saying prayers over the penitent in the same posture as used for giving absolution, while making it clear to him that he was not absolved.[102]

The lists cited above are typical of many to be found in the moralistic literature of the time. Their emphasis on sexual pollution should not perhaps be over-emphasized; the varied range of temptations to the sins of the flesh was liable to give them a privileged place, which did not need to represent an unhealthy obsession. Certainly Caulet was very concerned about sexual questions, as his long campaign against dancing demonstrated.[103] Like many other bishops, he ordered parents not to share their beds with children, and insisted on the separation of boys and girls, who were not even to cut wood or herd beasts together.[104] The stress on the sinfulness of tempting others inevitably brought sensuality in all its forms to the fore as a target for repression, since the rigorists had such a

[99] *Rituel d'Alet*, i. 115–16.
[100] Ibid., i. 116.
[101] Ibid., i. 117–18.
[102] Ibid., i. 118.
[103] Vidal, *Francois-Étienne de Caulet*, 149–51, describes this campaign and some of the acts of defiance it generated. On pp. 152–4 he discusses the associated attack on strolling players, carnival practices, and charivaris.
[104] Ibid., 157.

poor opinion of the capacities of fallen humanity for resistance. As the Jansenist Treuvé put it in 1676:

If you have the slightest difficulty over sins concerning chastity, do not fail to obtain instruction about them. Corrupted nature is subject to a thousand secret disorders, into which those persons consecrated to God may even fall more often than the worldly, if they do not take care. The majority are not well enough informed on the subject. Cardinal Tolet said with much truth of these faults that they damn the greatest part of the reprobate; to the extent that he thinks that of a hundred thousand Christians who are damned, more than 80,000 are so because of these secret sins.[105]

His concern about the ignorance and animality of the married led him to make a very dangerous suggestion, which would surely have placed many confessors in an unenviable position:

there are many things which married people believe to be permitted, which are not: this is why it is necessary for them to obtain instruction from a wise confessor about the sins one can commit in marriage.

It must also be observed, that when one has the slightest doubt on this matter, one should consult one's confessor; nothing is more common than to find elderly persons who have long wallowed in infamous sins, without knowing for sure that such actions were sins, although they had some doubts about this.[106]

A slightly later Jansenist manual, that by Louis Habert, advised the confessor 'that friendship with a wicked man is less dangerous for him than long and frequent interviews with a devout woman.'[107] Habert later devoted a full twenty pages

[105] S. M. Treuvé, *Instructions sur les dispositions qu'on doit apporter aux sacremens de penitence et d'eucharistie* (Paris, 1753; first edn. 1676), 235.

[106] Ibid., 291.

[107] L. Habert, *Pratique du sacrement de pénitence, ou méthode pour l'administrer utilement* (Paris, 1729; first edn. 1691), 62. The whole section on pp. 62–70 is concerned with the dangers of confessing women, and includes the remarkable statement 'On voit par experience que les hommes, sous une bonne direction, font de grands progrès dans la vertu, et qu'au contraire, plus un directeur donne de son temps à une femme, plus elle devient foible, parce qu'elle s'attache de plus en plus à sa personne.' (p. 64), see also above, p. 251. It would be possible to compile quite an anthology on the subject; good examples include J. Eudes, *Le Bon Confesseur, ou avertissemens aux confesseurs* (Lyon, 1671; first edn. Paris, 1666), 120–2, and J.-F. de Reims, *Le Directeur pacifique*, 121–4.

to an elaborate discussion of the sixth commandment, beginning with the caution: 'The confessor has need of an angelic chastity, if he is not to be polluted in turning over such dirty matters.[108] It would be possible to multiply these citations many times over, yet it would still be a distortion to claim that sexual problems are the central and dominant theme of such works. The language and the tone do indicate a greater degree of concern than that shown by Benedicti, for example, but it is integrated into a comprehensive view of sin which involves the separate and exhaustive treatment of nearly all its manifold forms. Not quite all, because the seven deadly sins and the ten commandments still left one or two curious gaps, most notably cruelty, which therefore never received specific attention.[109]

What can be said about both the practices demanded by rigorist bishops and the manuals they favoured is that they virtually abandon any notion of using confession and absolution as a means to console and support the faithful. They have become exclusively a means of social control, admittedly in the cause of godliness. The rigorists might not have been able to resurrect the primitive church, but they had certainly managed to strip away most of the more flexible and humane attitudes evolved during the middle ages. As Mgr. de Saint-Vallier, bishop of Quebec, put it in 1691, the confessor should perform 'la fonction de Père, de Juge et de Medecin', and he envisaged all these in their most repressive sense.[110] The medical analogy was perhaps less encouraging than at first appears, in view of the notorious unpleasantness of seventeenth-century treatment, but the bishop was running no risks of misunderstanding. 'Absolution is of the type of those great medicines which become mortal poisons to those who take them without being well prepared. Although it has the virtue of erasing all sorts of sins, it nevertheless becomes pernicious and fatal to those who receive it without the necessary dispositions.' [111] His stress on the need to ration absolution was clearly

[108] Habert, *Pratique du sacrement*, 345.

[109] This point is made by Delumeau, *Le Péché et la peur*, 271.

[110] G. Plante, *Le Rigorisme au 17ᵉ siècle: Mgr. de Saint-Vallier et le sacrement de pénitence* (Gembloux, 1971), 78; the quotation is from the bishop's circular letter of 1691.

[111] Ibid., 77, quoting the diocesan *Rituel*, ii. 135–6. The same analogy with poison is used by Habert, *Pratique du sacrement*, 448.

derived from his predecessors in France, notably Pavillon, numerous passages from whose *Rituel d'Alet* are followed very closely in his own similar work. Guy Plante has counted 197 mentions of absolution in his works, no fewer than 106 of them associated with the idea of refusal or delay, while he stated in 1691: 'it is laxity in giving absolution that has bred the disorders in the life of Christians.'[112] Yet as with his mentor Le Camus, it is not clear that Saint-Vallier was really prepared to apply his own doctrines beyond generalized exhortations and threats; in almost the only specific case recorded, in 1719, he merely threatened a settler and his concubine with the refusal of absolution.[113]

Another way of forming some general impression of the weight attached to particular sins is to use the works of the casuists, with their long lists of debated cases, and undertake some statistical analysis. This is not entirely reliable, since the amount of space devoted to a question may reflect its complexity rather than its importance, but in practice it seems to work quite well. For this purpose I have used an encyclopedic work in three folio volumes, the *Dictionnaire des cas de conscience* by Jean Pontas, first published in 1714. This is particularly suitable because it is almost entirely a compilation from the works of others, which displays no individual bias, but sums up the casuistry of the previous century, and above all because the 1736 edition has a very full index. There are twenty-three headings which have two full columns or more in this index; way out in front are *restitution*, with nine columns, and *empêchements de mariage*, with eight. To these we may add, respectively, *legs* (three), *vente* (two), *usure* (two), to give a total of sixteen columns for economic questions, and *mariage* (five), *devoir conjugal* (three), *adultère* (two), *dispense de mariage* (two) and *fiançailles* (two), producing twenty-two columns on marital and sexual matters. Apart from *juge* (two columns) all the other entries are connected with the church or confession itself: *confesseur* and *confession* (a total of five columns), *excommunication* (five), *irregularité* (five), *absolution* (three), *simonie* (three), and seven more entries of two columns under such headings as *curé* and *vœu*. A tripartite division therefore emerges, with economic abuses, marital problems, and the internal workings of the church itself as the main themes, and

[112] Ibid., 72–3. [113] Ibid., 85–6.

this is very much in line with my subjective impression of the literature as a whole.[114] The thirty-five columns devoted to internal clerical offences, such as simony, and the problems associated with benefices and the complex structures of the church, are a striking reminder of the natural priorities of many clerics. The role of the church as a career structure loomed very large in the eyes of its members, many of whom had entered it for rather worldly motives, while reformers very reasonably thought that clerical misbehaviour was the most unfortunate example of all. Simony might be the most scandalous aspect, but there were other much-debated issues such as the right to act as a confessor, the status of vows, and the obligation to reside. These were also legal questions, which remind us of the common parallel between confessor and judge; the same applied to many of the issues over marriage, where the *officialité* courts continued to exercise the one remaining ecclesiastical jurisdiction over the laity. The absence of the legal element makes the sixteen columns given to economic sins and the problem of restitution all the more impressive, and a wider survey of the literature only emphasizes how much attention they received from Benedicti and his successors. Habert stressed that the seventh commandment was the most difficult of all, because there were so many occasions for greed and misconduct in one's relations with others, and so many subtle ways of taking advantage of them.[115]

Pavillon's concern with such questions led him to consult the celebrated Parisian doctors Sainte-Beuve and Porcher in the early 1660s on a series of *cas de conscience* which are clearly based on his impression of what was happening in his own diocese. The replies were published by Pavillon in 1666, apparently without the knowledge or approval of the authors, although they were later reprinted in the posthumous edition of Sainte-Beuve's consultations by his brother.[116] There were

[114] It may also be compared with the analysis by Delumeau, *Le Péché et la peur*, 475–97, and especially with his table on p. 477, which is based on sermons. Since these were essentially addressed to the laity, clerical problems are missing, but the dominance of sexual and economic sins is again clear.

[115] Habert, *Pratique du sacrement*, 367–8.

[116] *Resolutions de plusieurs cas importans pour la morale et pour la discipline ecclésiastique* (Paris, 1666). Also in J. de Sainte-Beuve, *Resolutions de plusieurs*

twenty-nine cases in all, and eighteen of the first twenty were concerned with economic misbehaviour by the rich and powerful, especially the local nobility.[117] Gentlemen had annexed royal forests, and were excluding peasants from their customary rights; they were refusing to pay their taxes, so that the poor had to pay more; they were using false measures in their mills; they were defrauding legitimate heirs.[118] More generally:

> A bishop having recognized during a general visitation of all the parishes of his diocese, through the complaints of a great number of poor people, that the rich were abusing the property God had given them to the ruin of the poor, whom they oppressed on the pretext of helping them, by engaging them in numerous contracts and dealings, of which some were totally usurious, others good and just in their original form, but corrupted and spoiled by various injustices, which greed was multiplying day by day: he applied himself to discover the nature of all these kinds of contracts and dealings, to be able to judge their justice or injustice.[119]

This led into a discussion of illicit contracts, many of which concerned the management of flocks and herds, vital to the economy of this largely upland region.[120] Granted the range of abuses considered, and the monotonous regularity with which the verdict was for the refusal of absolution, it is not surprising that there is evidence of resistance. Cases xxii–iv consider situations in which the guilty party tries to force the confessor to explain his action, appears with a notary and two witnesses, cites him before an ecclesiastical or secular court, and pursues a series of appeals up to the level of the archbishop's metropolitan *officialité*.[121] The presence of a good deal of superfluous detail, and the fact that several of these cases can be

cas de conscience touchant la morale et la discipline de l'Eglise (3 vols., Paris 1694–1704), i. 389–478, cases cxii–cxl. For the circumstances of the publication of 1666, see the *avertissement* by his brother Jerome. The book appeared as being the collective work of thirty *curés* and doctors of Paris, whose names were listed at the end.

[117] The only (rather marginal) exceptions are case vi, devoted to gentlemen who fail to perform their judicial obligations, and case viii, concerned with Sunday observance and the protection gentlemen give to innkeepers who break the rules.

[118] Cases i–ii, vii, x.

[119] *Resolutions de plusieurs cas*, 54–5, case xiv.

[120] Ibid., 54–134, cases xiv–xx.

[121] Ibid., 152–73.

linked with individual offenders whom Pavillon confronted, strongly suggests that they were real rather than imaginary examples, which gives them added interest.[122]

Pavillon and those who thought like him were effectively demanding that the church should intervene at the very heart of the social and economic system, defending the peasantry against the exploitation of which they were the constant victims. One can well envisage angry gentlemen and others, faced with such attacks on their customary behaviour, appearing to demand explanations. The confrontation followed quite logically from the serious and sincere effort to give Christian teaching formal expression as a moral code. Working within the most orthodox, impeccable traditions, the catholic moral theologians were producing doctrines almost as potentially subversive of the established social order as were those of the radical sects.[123] If the moral code proclaimed by successive Assemblies of the Clergy and their approved spokesmen had ever been put fully into action, backed up by the coercive power of confession and absolution, then New England ought to have been comfortably upstaged. Not far to the north of the English colonists, indeed, Canada was the scene of one of the most vigorous attempts to turn ideals into reality, eliciting some telling comments from the governor, M. de Frontenac, who remarked of the clergy

> they are all full of much virtue and piety, and if their zeal was not so vehement and a little more moderate, perhaps they would succeed better in what they undertake for the conversion of souls, but to achieve this end they often use methods which are so extraordinary and so little used in the kingdom that they deter most people instead of persuading them.[124]

Two years later he was even more critical:

> almost all the disorders of new France have their source in the ambition of the clergy, who in their wish to combine spiritual

[122] For some of the real offenders, see Broutin, *La Réforme pastorale*, i. 202–3.

[123] There is a valuable collection of texts, with a long introductory essay, by R. Taveneaux, *Jansénisme et prêt à intérêt* (Paris, 1977), which illustrates many of the difficulties which arose from this attempt at economic rigorism.

[124] Plante, *Le Rigorisme*, 131, this comment dates from 1689.

> authority with an absolute power in temporal matters cause all those who are not entirely subject to them to suffer and murmur . . . if I wished to tell all they do to this end by means of the power given them by confession and excommunication, the secret of the confession being impenetrable I make no ⟨ ⟩ on the complaints I have received of the revelations and pressures they employ in that holy tribunal to know the business of everyone . . . [125]

It is noticeable that the governor contrasts practice in the colony with that in France itself; one would expect Canada, with its rather high proportion of clerics and its controlled immigration, to have offered relatively favourable conditions for such a moral crusade, free of many of the accumulated hindrances of the old world. More generally the French church could not hope to escape from the social divisions which were reflected among its own clergy; the bishops and their close advisers were drawn almost exclusively from the nobility and the legal-administrative élite. Only a minority of these men were ever likely to take their professional obligations to the point of questioning the traditional values of their class. For most of the leaders of the church it was a basic assumption that the *curés* would work in concert with local nobles and *notables* to keep the populace in order, not that they should stir up trouble by asserting the rights of the poor. As for the *curé* himself, he must quickly have become aware that his own relationship with the parish was a complex one, and that he could only remain influential so long as he enjoyed some degree of acceptance from the leading members of the community. His ability to raise funds for poor relief, church repairs and similar causes was also at stake, and for these he relied heavily on just those inhabitants who imposed dubious contracts or lent grain at usurious rates. The nature of these problems, however, varied considerably from region to region, and even in neighbouring villages. In areas of sharecropping, and those which were mainly pastoral, the exploiters were more commonly outsiders, seigneurs or town-dwellers; here the *curé* might find little difficulty in sympathizing with a flock who were almost exclusively poor and dependent, yet he had no power to help them.

Although hostility between the ordinary secular clergy and

[125] Ibid., 130–1.

the local gentlemen and *notables* was generally restrained by such factors, it is not surprising that it should have existed. A high proportion of *curés* originated from the towns, particularly in the seventeenth century, so that the classic town–country antagonisms and incomprehension might easily arise.[126] The most clear-cut doctrinal expression of rigorist views, Jansenism, was a religion of lawyers and educated clerics, manifestly out of touch with the feelings and attitudes of both nobles and peasants. Their rigorism found expression in two complementary drives for greater moral control: to eradicate superstition, ignorance, and scandalous behaviour among the people; to control the economic acquisitiveness and conspicuous hedonism of the privileged élites. Here one sees a certain symmetry with the efforts of the royal government and its agents, such as the intendants, to clamp down on much the same range of abuses. It is hardly surprising that such reform by coercion does not appear to have secured more than the most formal external compliance in most cases, yet it remains striking how far criticism of the leaders of society could go. An instance may be found in the work of the Jansenist Louis Habert, generally known as *La Pratique de Verdun*, or by his critics, mockingly, as *La Pratique impratiquable*. Habert provided a model examination for gentlemen and seigneurs, which asked about an impressive list of possible sins, including leaving crimes unpunished, refusal to pay tithes, damaging crops when hunting, using force to prevent pursuit for debts, and imposing excessive *corvées*. He went on:

Have you not involved yourself with the tithes or the billeting of troops, to obtain exemptions for your tenants and other persons, against the ordinance?
Have you not behaved immodestly in church, scandalizing the congregation by your indecent postures and your chatter?
Have you not profaned, or caused the profanation, of Sundays and feast-days by hunting, games, dances, displays, as by the jugglers? How many times?

[126] For the social origins of the clergy see below, p. 370. There is a particularly valuable detailed study by D. Julia and D. McKee, 'Le clergé paroissial dans le diocèse de Reims sous l'épiscopat de Charles-Maurice Le Tellier: origine et carrières', *Revue d'histoire moderne et contemporaine*, 29, (1982), 529–83. The authors note that there was a sudden increase in rural recruitment (among the sons of prosperous *laboureurs*) after 1700.

Have you not treated priests unworthily? Was it seriously?
Have you not been full of pride, idolatrous of yourself, despising others, particularly your vassals?[127]

One can guess how many *curés* dared subject the local seigneur to a formidable inquisition of this kind, even if privately some of them would have agreed with Antoine Godeau that 'there is nothing so dangerous for eternal salvation as prosperity, abundance, reputation, honour, and authority in the world.'[128] More generally, it will be plain by now that the more ambitious and elevated visions of confession as an agency of general moral reform were quite unrealistic when considered on a national scale. Peasants might be more docile than gentlemen—they were unlikely to seek legal remedies if refused absolution—but they had great resources of obstinacy and passive resistance. Perhaps this was as well, for had they flocked to make elaborate confessions and seek spiritual counselling the church could never have mustered sufficient manpower to service their needs. One suspects that under the surface many rigorist authors assumed they were writing for the elect few, and were prepared to be satisfied by a modest show of outward conformity from the mass of the population.

In just one case a rural priest left a detailed and explicit account of the problems he faced as a confessor. Christophe Sauvageon was prior of Sennely-en-Sologne from 1675 to 1710; a well-trained Picard, he drew up a manual of advice to his successor in which he described his flock much as an anthropologist might an aboriginal tribe.[129] The people of Sennely were unwilling even to enter the confessional box, since to do so would have amounted to a public admission that the individual in question had some appalling sin to reveal. Sauvageon had therefore to abandon the new technology, and heard confession elsewhere in the church, while

[127] Habert, *Pratique du sacrement*, 297–9.

[128] A. Godeau, *Instructions et ordonnances synodales et exercices de piété pour la direction des confrères et confréries du très-Saint Sacrement, erigées dans le diocèse de Grasse* (Paris, 1648), 157.

[129] Sauvageon's *livre de raison* was published by E. Huet in *Mémoires de la Société historique et archaéologique de l'Orléanais*, 31–2 (1907–8). His village is the subject of an excellent study by G. Bouchard, *Le Village immobile: Sennely-en-Sologne au 18ᵉ siècle* (Paris, 1972).

people crept up trying to overhear what was being said, a practice which could lead to serious quarrels.[130] Not that the penitents could be relied on to tell the truth, anyway: 'A youth who has abused a girl before marriage waits until death, as she does, before confessing it; the thieves, perjurers, arsonists, and generally all those who have committed shameful and punishable crimes, wait their last hour before admitting them to the priests, giving as their reason that it's no good trusting everybody.'[131]

The prior blamed the natural secretiveness of the people, so fearful of being tricked that they 'never confide their secrets or their intentions, neither to their friends nor to their wives, nor even to their confessors, of whom they are even more distrustful than of the others.'[132] Even when they were not frustrating the system completely by waiting to make deathbed confessions of any worthwhile sins, the Solognots infuriated the prior by their attitude:

There is a deplorable and inveterate custom in this parish of coming quite unprepared to confession. They approach without making the least examination of their conscience, rush or throw themselves into the confessional, almost fighting to be first, yet when they are at the priest's feet they don't even make the sign of the cross unless reminded, hardly ever remember when they last confessed, have not completed their last penitance, have done nothing, accuse themselves of nothing, laugh, recount their poverty and misery, plead their own cause when accused by the priest of some evil he saw them commit, blame their neighbour, accuse everyone else in order to justify themselves, in a word they do everything in the confessional except that which they should, which is to avow all their sins sorrowfully and sincerely. They claim evil to be good, gloss over their faults, mutter their important sins between their teeth for fear that the priest hears them, that is to say that they try to cheat themselves in cheating him, and it is certain that there are very few valid confessions above all from those whose life is not Christian and regular.

The young people don't understand what one means when one speaks to them of the sin of masturbation and pollution, so that it is necessary to explain the matter to them more clearly, and when their age and complexion make one presume that they may have committed these kinds of impurities it is necessary to interrogate them thus:

[130] Sauvageon, fo. 25. [131] Ibid., fo. 218. [132] Ibid., fo. 217.

have you not thought of doing things with women; or indeed have you not been so unfortunate as to make your seed flow.

By dishonest thoughts they only understand thoughts lacking civility and respect, and failing to lift their hat to their superiors, by bad thoughts only giving themselves to the demon or going to steal, burn or do some other wrong to their neighbour, also when examining them on the sixth commandment it is necessary to ask them if they have not been tempted by the sin of the flesh.

The girls never accuse themselves of their lewdnesses until death, although they are very prone to these.

It is also very rare for them to accuse themselves of the sins of sodomy and bestiality except at their death or in times of jubilee. This is why it is necessary to interrogate them on the matter, but it must be done with singular prudence for fear of teaching them sins they have never known nor in consequence thought of committing . . .

When the priest has seen them commit some evil such as swearing, losing their temper and cursing their neighbour, or if he has seen them get drunk, they are under the false impression that in such cases they need not confess, alleging as their reason that they should only tell what the priest does not know.

They think they have the right to employ a kind of compensation against their masters, taking and keeping on their own initiative as recompense grain and other property of these masters when they have performed what they call *troches* for them, that's to say carting, doing other work, or providing food, and they never confess anything of all this.

As the people of the Sologne are naturally perfidious, they cannot believe that their confessors keep their secrets, and this mistrust prevents them accusing themselves of sins whose disclosure they fear might harm them.

There are no people in the world more given to theft than those of the Sologne, yet there is no country where the confessors find fewer thieves.

When they had stolen something belonging to me, such as wood, fruit, linen, and so on, they never made any mention in their confessions, and when I was sure about it and gave them proofs they denied everything, preferring to perjure themselves in the confessional rather than admit themselves guilty.[133]

Sennely-en-Sologne was only one parish out of some 30,000, yet it is hard to resist the impression that the prior's

[133] Ibid., fo. 26.

experience was all too typical, a shattering contrast to the ideals of the moral reformers. This did not mean that the peasants were irreligious, but rather, as Gerard Bouchard puts it, that their religious universe existed on the margins of the official church, and often in opposition to it.[134] They crowded noisily and irreverently into the church on feast days, and loved nothing better than a procession which took them past a healthy number of inns.[135] Their treatment of confession carried no implication of indifference, for Sauvageon noted how absolution still had a vital role to play in settling disputes: 'When they have quarrelled, one must take the precaution of reconciling them before absolving them, otherwise they persist in their hatreds and are never reconciled.'[136] This observation should not be taken to imply that refusal of the sacrament was a powerful coercive weapon, however; it is more likely that the need to save one's soul was a convenient excuse for climbing down within a social ritual which had to observe the proper order. At all periods confession can only have been part of a much wider set of devices for securing reconciliation, in the ultimate interests of community, family, and individual alike, in which the *curé* played a leading part. Far more disputes were settled through his mediation, or that of other local notables, than ever began the perilous and expensive voyage through the courts. A putative reduction in the use of Easter confession in this context need not therefore have been very significant, if the same results were being secured by other means.

The widespread use of missions to evangelize rural France was particularly important here, for the various missionary orders made confession a key part of their operation. If their hell-fire sermons represented 'la pastorale de la peur', they were balanced by a much more sympathetic treatment of those sinners who responded by making a general confession, recounting all the sins of their past life, not merely those committed since they were last absolved.[137] The missionaries

[134] Bouchard, *Le Village*, 293.

[135] Ibid., 289–90, 302–6.

[136] Sauvageon, fo. 27.

[137] This emerges from many accounts, but is particularly well brought out by N. Lemaître, 'Un prédicateur et son public: les sermons du père Lejeune et le Limousin, 1653–72', *Revue d'histoire moderne et contemporaine*, 30 (1983),

saw themselves as mixing fear and consolation; they aimed to build up an emotional climate leading to a kind of catharsis, in which sinners recognized their guilt and obtained remission from its burden. The results were often spectacular, with long queues forming at the confessionals, followed by dramatic public reconciliations and restitutions. Very few institutions of the catholic reform were so successful in harnessing popular feeling, yet there was still an underlying ambiguity. The attitudes of the missionaries were close to those of the local élites who summoned them; their idea of individual reform and salvation was not really in tune with the essentially collective religion of the people, although their activities became a focus for collective displays of fervour. The missionaries gained enormously from their position as outsiders to the community, for as they recognized themselves, people would confess freely to strangers in a way they found impossible with the *curé* they saw virtually every day, and whom they never entirely trusted to keep their secrets. There was also the attraction that the missionaries usually had power to grant indulgences, and to absolve even in reserved cases, those notable crimes which were normally the preserve of higher authority.[138] This system undoubtedly hampered and annoyed the *curés*, since ordinary believers were terrified of being singled out in this way, and often had little idea of exactly which sins were liable to such treatment.[139] Missions clearly had a remarkable impact, but it was bound up with

33–65, and B. Dompnier, 'Pastorale de la peur et pastorale de la séduction: la méthode de conversion des missionaires capucins', in *La Conversion au 17e siècle,* Actes du Colloque du CMR 17 (Marseille, 1983), 257–73; see also Dompnier's article 'Les missions des Capucins et leur empreinte sur la réforme catholique en France', *RHEF* 70 (1984), 127–47. For the subject in general, see the excellent overview by Dompnier, 'Missions et confession au 17e siècle', in Groupe de la Bussière, *Pratiques de la confession*, 201–22. I have not attempted to footnote the individual points at which I have followed these articles.

[138] For this point, see C. Berthelot du Chesnay, *Les Missions de Saint Jean Eudes* (Paris, 1967), 150–2.

[139] The issue rumbled on, and was to become a central point in the arguments over the rights and status of the *curés* in the late eighteenth century, particularly as expressed by the formidable controversialist canon lawyer Maultrot. See E. Préclin, *Les Jansénistes du 18e siècle et la constitution civile du clergé* (Paris, 1928), 336–62.

their status as exceptional events, which were ultimately more in line with popular religiosity than with the austere discipline proclaimed in the manuals. There was an obvious danger that they merely provided an opportunity to wipe the slate clean, so that even their organizers saw the need to back them up by an effective, well-trained parish clergy if their effects were to be durable.[140] At this point all the familiar difficulties reappeared, largely frustrating the best efforts of even the much more diligent *curés* of the end of the century. Ultimately, therefore, missions emphasized the limitations of the system as much as its possibilities.

A parallel can be drawn between missionaries and such aspects of popular belief as shrines and magical healers, all creating a sacred space or time marked off from ordinary life, and outside normal social relations.[141] The widespread perception of annual confession as a judicial exercise would have been much less helpful, in view of the general attitude towards the legal system, pithily described by Sauvageon in the case of his parishioners: 'They hate lawsuits, and still more the judicial officials whom they call the eaters [*des mangeurs*], they are easily brought to an agreement when they have disputes on account of their fear of falling into their hands.'[142] There could be little chance that the priest who applied the system rigorously would escape another peasant attitude recorded by the same sharp-eyed observer: 'They are all the sworn enemies of their masters and their superiors, eternally thinking of ways to do them harm, whether by stealing from them, calumniating them, or revealing the secrets confided in them. All superiority is odious to them. Among themselves they do not preserve any respect for the best qualified people of all ranks.'[143] Clerical superiority must have been precisely what the confessional symbolized for most believers, while those who regularly betrayed secrets were unlikely to relish revealing their own. As an instrument for both social control and personal reformation, then, confession was going to need some

[140] Berthelot du Chesnay, *Les Missions*, 184–8.

[141] For a subtle discussion of this and related points, A. Dupront, 'La religion populaire dans l'histoire de l'Europe occidentale', *RHEF* 64 (1978), 198.

[142] Sauvageon, fo. 216.

[143] Ibid., fo. 217.

kind of coercive backing, for willing co-operation was virtually inconceivable. In this light the demands for the use of delayed absolution make good sense, but apart from the other difficulties already mentioned, the practice would have rapidly lost its effect if employed too often. Significant numbers of excommunicated parishioners, who suffered no apparent social or economic disadvantages from their state, would have advertised both the *curé*'s failure to overawe them and his lack of real power. For all the tremendous stress on the torments of hell, in both sermons and iconography, there is remarkably little evidence that the threat of eternal damnation and punishment affected everyday thought or behaviour. No doubt many of the population were too busy worrying about immediate hardships and dangers to concern themselves much with remote and uncertain ones. Anxiety over the after-life was concentrated on the very specific cases of those thought to be in immediate need, which meant ancestors lingering in purgatory and unbaptized children. Maunoir and his missionaries in Brittany found that it was more effective to invoke the familial responsibility to ancestors than to threaten the living with individual damnation, while most areas of the country had their *sanctuaires à répit*, where dead infants might give sufficient signs of life to be baptized.[144] People did worry about sacrilege, not so much because it was a mortal sin, but because it represented pollution of the community, and might bring punishment in this world rather than the next, a belief which the church's teachings still tended to encourage.

Sin as a threat to the community was a much more promising line if some kind of sanction had to be invoked; this concept was at the heart of protestant discipline as practised by the consistories, and the idea of an interventionist God was strongly entrenched in catholic teaching. Propitiatory rituals sanctioned by the church were regularly practised, while relics and miracles were overwhelmingly concerned with this world rather than the next. The practice of general communal confession at Easter, widespread in Northern France until the early seventeenth century, was intended to purge the community of venial sins, leaving mortal ones for individual

[144] For Maunoir, see above, p. 226. For the *sanctuaries à répit,* Delumeau, *Le Peché et la peur*, 303–14, and J. Gélis, *L'Arbre et le fruit* (Paris, 1984).

confession, but it also represented a collective recognition of sin and request for pardon.[145] The notion of a general culpability for the sins of the impious, which could not be entirely reduced to the individual level, was still referred to, admittedly as a feature of the past, by Archbishop François de Harlay of Rouen in 1651.

There are two consciences to be purged, the individual conscience which everyone purges in the secret of confession, and the public conscience . . . whose confession should be public . . . In earlier times it was even the custom to make open public confession in this form of all the kinds of sin which might be committed by the most wicked and abandoned: and this was very appropriate, to indicate that anyone at all could have committed them and should hold himself guilty of them, for which it is here question of making an honourable reparation for all public, contagious, and scandalous sins.[146]

The disappearance of this potentially valuable observance at the time of the catholic reform emphasizes the general tendency of that movement to redefine religious practice in terms of the individual at the expense of its communal aspects. Public penance now attracted rigorists as a way of shaming particular notorious sinners into reforming their lives, a view defended in a work published under the auspices of the archbishop of Sens in 1673.[147] When he imposed public penance on a few hardened sinners he was accused of reviving a practice which had been abolished, and which was stigmatized as 'dangereuse, scandaleuse, et qui peut faire beaucoup de mal.'[148] Among the bishops whose approbations preceded the book was Pavillon, who argued that most penances were too light to be effective, then went on: 'Besides is it not just that the person who has scandalized the public should for his part do all that depends on him to wipe out the

[145] On this subject see N. Lemaître, 'Confession privée et confession publique dans les paroisses du 16e siècle', *RHEF* 59 (1983), 189–208, and 'Pratique et signification de la confession communitaire dans les paroisses au 16e siècle', in Groupe de la Bussière, *Pratiques de la confession*, 139–64.

[146] Lemaître, 'Confession privée', 199.

[147] [A. L. Varet] *Deffense de la discipline qui s'observe dans le diocese de Sens touchant l'imposition de la Penitence publique pour les pechez publics* (Sens, 1673).

[148] Ibid., 4.

baneful impressions that his crime has left in the mind and in the heart of those who have witnessed it'.[149] As usual, such ringing calls for moral discipline seem to have produced little real action; most bishops would have been all too well aware of the likely reaction among the laity.

Hankerings after more effective sanctions against sinners may have reflected a growing disillusionment with the operation of confession as a means to moral and social control. In one sense clerics were preconditioned to expect failure in this quest, since they generally believed that the elect were a tiny minority. As the Franciscan Colomban Gillotte put it in 1697:

> if the number of the reprobate is incomparably larger than that of the predestined: this is because there are few people who observe his commandments; few who make faithful use of the talent he has freely given them; few who commit the violence against themselves necessary to fight and conquer their passions; and few who having fallen into sin hear with faithful attention both his voice and his urgings by which he invites them to penitence.[150]

Much earlier in the century the Jesuit Jacques Salian had been prepared to quantify the situation; God's hatred for original sin was so great that although the Saviour's merits would have more than sufficed for everyone, they only operated effectively for a minority, while 'the rest gallop furiously to damnation, in punishment of this first sin, and then of the others, which like rotten branches have issued from that diseased root. Is it not a marvellously terrible thing? Men would only punish one of ten criminals, and this great God, father of mercy, will only save one or two out of fifty . . .'[151] Those who held such views could hardly expect a widespread amendment of life through the effects of divine grace, as mediated through the sacrament of penitence, so the only logical conclusion for them to draw would have been that sin must be repressed by some combination of ecclesiastical and secular authority. Something like this was indeed the policy of the *Compagnie du très Saint-Sacrement* and its offshoots, but clerical

149 Ibid., unpaginated approbations at front.

150 C. Gillotte, *Le Directeur des consciences scrupuleuses* (Paris, 1697), 197–8.

151 J. Salian, *L'Ambassade de la Princesse Crainte de Dieu fille de Dieu vivant et de sa justice à la Jeunesse Françoise* (Paris, 1630), 393.

writers do not appear to have followed their own rhetoric through to the point of open advocacy for such a campaign. A more typical response was that of pessimistic rigorists like Treuvé, who advocated the technique of withdrawal from a world incapable of true reformation:

Of all the counsels one can give, this is the most important and most necessary. See few people, speak little, go out rarely. A soul which abandons solitude without an absolute necessity is a soldier who leaves his trenches, and exposes himself quite defenceless to the missiles of his enemies; it is a fish which leaves the water, and in so doing places itself in danger of death. The more you see men, the more your heart becomes empty. The best conversations, when they are too long and too frequent, become pernicious to our soul, because we are so corrupt, that we change everything into poison. There is in all men something contagious, which arises from their natural depravation, which spreads itself imperceptibly, and communicates itself with prodigious ease. One of the ancients said: each time I go among men, I return a little less of a man: a Christian can say in the same way, every time I frequent the Christians of this century, I return less Christian.

And if this is true of the visits which the just pay one another, alas! it is a thousand times more so of those one pays in the world, and of the commerce one has with it.[152]

The rhetoric of 'le mépris du monde', well suited to justify older styles of withdrawal into closed orders, did not fit very easily into the pastoral emphasis of the catholic reform. The strongly Augustinian sympathies of the leading French reformers, shading into outright Jansenism, may help to explain why it nevertheless figures so heavily in the writings of the period. This instance does point up more general worries about the extent to which clerical discourse is a reliable guide to 'la religion vecue', and particularly about the thesis of *culpabilisation* developed by Jean Delumeau. On the basis of the kind of literature cited in this essay, and an extensive survey of sermons, Delumeau builds up a deeply sombre picture of the moral teachings of the French church.[153] He sees its clerics as inculcating a religion of fear, prohibitions, and repression,

[152] [S. M. Treuvé], *Le Directeur spirituel, pour ceux qui n'en ont point* (Paris, 1691), 76–7.

[153] This is most clearly the case in *Le Péché et la peur*.

which left little room for any more positive aspects of the Christian message. This naturally leads him to pose the question whether the excessively harsh pastoral teachings of the church may not have been responsible for the dechristianization of a substantial proportion of the population. There is no doubt that his general assessment of the literature is correct; the issue is rather that of its relevance and applicability at parish level. Unlike Delumeau, I believe that there was a radical discontinuity at precisely this point, with the vast majority of *curés* adopting a far more conciliatory position. It was precisely over confession and absolution that confrontation should have occurred, yet a number of careful diocesan and parish studies have found scarcely any trace of serious conflicts over these matters. Alexandre Dubois, *curé* of Rumegies in the diocese of Tournai, had a marked penchant for Jansenism, and when the Carmelites sought to establish themselves locally he was one of seven *curés* who protested, alleging among other things that:

particulièrement pour le temps de Pâques, dès que les pasteurs auraient quelque méchant paroissien, qu'ils jugeraient indisposé pour le devoir pascal, il serait reçu chez ces Pères, qui sont connus pour être les plus faciles à donner l'absolution indifférement à toute sorte de personnes . . .

[in particular at Easter time, as soon as the pastors had some wicked parishioner, whom they judged unsuitable to perform his Easter duty, he would be received by those fathers, who are known for being the easiest in giving absolution indiscriminately to every kind of person . . .][154]

Yet there is no sign from Dubois's journal that he ever used these powers against his parishioners, for all his keen awareness of their numerous failings. At the end of a decidedly critical summary of these faults, he unexpectedly commented:

Il y a bien pourtant des honnêtes gens et une grande partie de qui on ne saurait dire que du bien, particulièrement de presque toute la paroisse qu'ils ont beaucoup de charité pour les pauvres. Je crois que Dieu les sauvera par ce moyen.

[154] H. Platelle (ed.), *Journal d'un curé de campagne au 17ᵉ siècle* (Paris, 1965), 129.

[Nevertheless there are many honest people, of a large proportion of whom one could only speak well, in particular of nearly all the parish that they have much charity towards the poor. I believe that God will save them by this means.][155]

This *curé* was a well-informed and intelligent man, whose personal commitment to high moral standards is as evident as his shrewd judgement of his flock; it is all the more striking to find that he directly repudiated the rigorist authors on this vital point, and was prepared to believe that salvation was for the many rather than the few.

Visitation records and other sources confirm that in virtually all French parishes only an infinitesimal number of catholics were excluded from Easter communion.[156] This was often a self-imposed situation, where individuals who were living in concubinage or were feuding with others stayed away from both confession and communion; only the exceptional *curé* seems to have made much use of delayed absolution. Parish priests may have distrusted regular clerics, whom they commonly regarded as competitors in any case, because they feared being placed in a humiliating position towards this fringe of public sinners, who might obtain absolution and present themselves at communion. That they themselves should have failed to enforce the moral code is hardly surprising, when one considers the endless confrontations involved, even at the individual level. The *curé* who found himself at loggerheads with any substantial group of his parishioners was in serious trouble; as Timothy Tackett had shown for the diocese of Gap, he might be persecuted to the point where resignation from his benefice was the only real option.[157] Hearing the kind of confessions described by Sauvageon must have been a thankless and wearisome task, so many priests probably reacted by limiting their own involvement. This inference could certainly be drawn from the complaint which

[155] Platelle, *Journal*, 77.

[156] For example L. Pérouas, *Le Diocèse de La Rochelle de 1648 à 1724* (Paris, 1964), 421–2, and J. Ferté, *La Vie religieuse dans les campagnes parisiennes (1622–1695)* (Paris, 1962), 316–22.

[157] T. Tackett, *Priest and Parish in Eighteenth-Century France* (Princeton, 1977), 189–93. The sources of such conflicts are analysed at greater length by P. T. Hoffman, *Church and Community in the Diocese of Lyon, 1500-1789* (London, 1984), 128–61.

bishop de Pressy of Boulogne included in his *statuts synodaux* of 1744:

we learn with sadness that the sacrament of penitence is not frequented in several parishes, because the *curés,* instead of drawing the faithful to this court of mercy where the guilty are absolved, drive them away through their lack of assiduity in hearing confessions; and, when they do hear them, it is in such haste and in so curt and negligent a manner that the penitents are rebuffed and scandalized by it . . .[158]

One is reminded of Bourdoise's famous lament a century earlier:

Depuis trente ans j'ai estimé que la plupart des confessions sont des chansons ou, pour mieux dire, des sacrilèges et non pas des sacrements . . . O routine malheureuse des gens qui ne font autre chose que de se confesser sans changer de vie! O routine damnable d'un très grande nombre de confesseurs, qui semblent tenir boutique du confessional! . . . C'est la question si, en tant de confessions, il y a seulement une seule particule d'attrition, vu qu'apres la confession on poursuit sa miserable vie comme auparavant.

[For thirty years past I have reckoned that most confessions are songs, or, to put it better, sacrileges rather than sacraments . . . O miserable routine of people who do nothing but make confession without changing their life! O damnable routine of a very great number of confessors, who seem to keep a shop of the confessional! . . . The question is whether among so many confessions there is even a single particle of attrition, seeing that after confessing they pursue their miserable life as before.][159]

The assumption made by both men was that the prime responsibility for this state of affairs lay with the clergy. They do not seem to have recognized the virtual impossibility of enforcing an alien morality through the precarious authority of individual *curés.* If confession was to work in this way it could only be through two possible routes. The first was that of direct repression, imposing harsh penances and excluding sinners from communion; the second by instilling feelings of guilt which would deter men from sin. It is hard to see how

[158] A. Playoust-Chaussis, *La Vie religieuse dans le diocèse de Boulogne au 18ᵉ siècle (1725–1790)* (Arras, 1976), 174.

[159] Cited by J. Orcibal, *Jean Duvergier de Hauranne, abbé de Saint-Cyran, et son temps* (Paris, 1947), 534.

either method could work without a significant degree of co-operation from the populace. Peasants were expert in evading unwelcome attentions from outside authorities, and would simply refuse to admit most of their sins, as those of Sennely did, if they knew punishment would follow. As for personal guilt, this is not created merely by telling people they are sinful, but depends on very complex structures of thought; the approach of clerical writers was essentially that of bourgeois individualists, and was highly unlikely to mesh with the essentially communal vision of their social inferiors. It should be noted that the missionaries' primary impact was not through personal guilt of this kind, for they relied mainly on either direct fear of damnation or responsibility for the sufferings of one's relatives in purgatory, techniques developed more by experience than from theory.

Rather than envisaging the rigorist theories as a powerful instrument for the *culpabilisation* of the masses, I think we should see them as expressions of the inner conflicts of the clergy and their class. They were never a practical programme for the mass conversion of the semi-pagan populace clerics saw around them, whom Sauvageon wearily described as 'indisciplinables et enemies de correction'.[160] These doctrines were essentially developed by educated clerics who originated among the urban bourgeoisie; their whole style and imagery reflected the concerns of this élite, and no doubt their unconscious fears and tensions as well. It was to the children of this class that they were most effectively and repressively applied, and if there was any area in which they contributed to dechristianization it was probably here. Elsewhere in the urban context it was a very different story, for there are strong grounds to suspect that the church never really came to grips with the problem of evangelizing the poor in the towns, where parishes were generally too large for effective systems of control. Wherever one looks, in fact, the impression is of well-meaning failure to achieve impossibly ambitious goals. Once they were left isolated with their flocks the clergy naturally responded with a series of makeshift compromises; if they felt sufficiently combative to tackle general sinfulness, they normally chose the route of direct prohibitions on dancing, cabarets, or dubious ceremonies, much more accessible

[160] Bouchard, *Le Village*, 341.

targets than individual sin. Despite the formal teachings of the church, they could not rely on the support of their superiors. As Sauvageon clearly saw, when writing of superstitious practices:

Il n'y a que la seule autorité episcopale qui puisse reformer ces abus, mais je doute fort que cela arrive parceque nosseigneurs les Eveques ont pour maxime detre populaires et de ne point choquer les personnes du monde.

[The episcopal authority is the only one capable of reforming these abuses, but I greatly doubt that this will happen because our lords the bishops hold as their maxim that they should be popular and avoid shocking the worldly.][161]

While this judgement can hardly be extended indiscriminately to all *ancien régime* bishops, it does seem likely that it applied to most of the respectable but courtly prelates chosen by the crown from the later seventeenth century onwards. Divisions between the upper and lower clergy were to be a prominent feature of the pre-revolutionary decades, and here we may see a very real influence from rigorist moral philosophy. Many *curés* showed great sympathy for the economic plight of their humbler parishioners, which led them to criticize not merely the greed of the rich, but also the inequitable distribution of resources within the church. Behind the curious mixture of egalitarianism and moral austerity which emerged as the programme of the *curés* at the outset of the revolution it is not hard to detect the ideas of the seventeenth-century reformers.[162]

These ideas were expressed with particular force and persistence in the literature of moral theology, which naturally centred itself on the sacrament of penitence. The result was a dogged attempt to work out a comprehensive set of rules, which threatened a normative tyranny of a potentially revolutionary kind. Jansenists and others might look back to the primitive church for precedents, but even if their ecclesiastical history was accurate there could be no serious comparison between small self-selected groups of believers and the

[161] Sauvageon, fo. 25.

[162] For this movement see Tackett, *Priest and Parish* 225–86, Hoffman, *Church and Community*, 161–6, and J. McManners, *French Ecclesiastical Society under the Ancien Régime* (Manchester, 1960), 190–254.

unwieldy national churches of the early modern period. It has been the repeated argument of this essay that the catholic reformers vastly overreached their strength, by producing an impossibly demanding manifesto, which all but the most extreme rigorists tacitly abandoned in practice. The idea of using confession to enlist the individual conscience as the supreme agency of moral control, however psychologically brilliant, was completely at odds with the social and mental universe of all but a small segment of the population. It is doubtful how far it had been fully assimilated by the majority of the clergy, for it is evident that the French church continued to offer a religion of ceremonies, relics, pious practices, charity, and communal values to the mass of ordinary believers. In so doing it demonstrated an essential capacity to cover up the unacceptable face of its own doctrines, and to adapt its message to the demands of its hearers. The peasantry successfully resisted Borromeo's moral vision along with his confessional, asserting their own pragmatic styles of life and worship. What might in theory have been the greatest repressive movement in European history disintegrated before their largely mute tenacity. The process was aided by the ambivalence of the clergy themselves, many of whom found it more attractive or simply easier to concentrate on the innumerable external aspects of the faith. Even when they were concerned with the sacrament of penitence, the *curés* were inevitably drawn into concentrating on those public offenders who represented a direct challenge to their authority; only the most determined were likely to persist in seeking out hidden sinners. Confession and absolution therefore remained an agency of external control, in line with the older communal conceptions, while the idea of moral regeneration through the individual conscience was very little applied. It is a moot point whether a more serious attempt to develop this technique would have yielded positive results in any case, since the lack of practical experience among those who formulated the doctrines always threatened to render them quite unrealistic. The more the concept of sin was refined and developed by the moral theologians, the more demanding and unmanageable it became, demonstrating the deep incompatibilities between human life, abstract logic, and absolute moral standards.

8

The catholic puritans

Jansenists and rigorists in France

ENGLISH puritan divines seem to have found no difficulty in dismissing Roman Catholics *en bloc* as superstitious and idolatrous.[1] If they looked across the Channel, their sympathies naturally went to the protestant minority in France, whose churches exemplified a much purer Calvinist organization and discipline than that favoured by Anglican episcopalians. The divisions within the French catholic church were obscured, in protestant eyes, behind the one issue which came near to uniting it, the desire for an early abrogation of the Edict of Nantes and the elimination of 'la religion prétendue réformée'. Puritans and Jansenists could never have understood one another directly in any case, because they belonged to different epochs as well as different faiths. Historians have nevertheless drawn close parallels between the two movements, with their predestinarian, Augustinian theology, their desire to revive the virtues of the primitive church, and their strict moral standards. One particularly tiresome characteristic shared by puritanism and Jansenism is their resistance to any attempt at close definition. Both were terms of opprobrium, often violently denied by those to whom they were applied, changing their meaning with circumstances and periods. This is no good reason for abandoning the terms, for we should not expect to find labels with which we can classify men's beliefs like jam-pots, but it does impose some caution in their use. Officially Jansenism consisted in the Five Propositions censured by the papal bull *Cum occasione* in 1653, yet the inadequacy of such a definition was already recognized by Fénelon when he wrote 'Les cinq propositions

[1] C. Z. Weiner, 'The beleaguered isle: a study of Elizabethan and early Jacobean anti-catholicism', *Past and Present*, 51 (1971); R. Clifton, 'The Popular Fear of Catholics during the English Revolution', ibid., 52 (1971).

ne réalisent nullement le jansénisme, parce que le parti les condamne'.[2] The archbishop of Cambrai's own attempt to mount a comprehensive attack on Jansenism produced the bull *Unigenitus* of 1713, which erred in the opposite direction. The censure of 101 propositions from Quesnel's *Réflexions morales* covered so many aspects of theology and practice, in so ambiguous a fashion, that it merely created greater confusion. One reason for these difficulties, of course, was the insistence of the Jansenists that they were the heirs of the real tradition of the church, whereas their opponents were the innovators. Like the puritans, they stood on their right to remain in full membership of the church, trying to argue away the relevance of any measures taken against them.

Puritan hostility to separatists was reflected in the way many Jansenists denounced protestantism; both reactions must to some extent have been an unconscious defence against dangerously plausible attacks. While it would be a gross caricature to describe Jansenism as a kind of protestantism *manqué,* such accusations were frequent, and not necessarily insincere. Richelieu was plainly telling less than the whole truth when he justified the arrest of Saint-Cyran to Hardouin de Beaumont with the remark: 'On aurait remédié à beaucoup de malheurs et de désordres si l'on avait fait arrêter Luther et Calvin dès qu'ils commencèrent à dogmatiser' ['Many misfortunes and disorders would have been remedied if Luther and Calvin had been arrested as soon as they began to dogmatize'], but the insinuation was typically shrewd.[3] In a later period we find Joachim Colbert, Bishop of Montpellier, complaining in 1717 of the provincial intendant Basville: 'Lui et ses amis travaillent avec succès à me décrier dans le public, et l'on dit hautement que je veux établir une religion nouvelle, détruire les sacrements, faire marier les prêtres et me marier moi-même' ['He and his friends work successfully to discredit me with the people, and they say openly that I want to establish a new religion, destroy the sacrements, make priests marry, and get married myself'].[4] In a more spontaneous

[2] L. Ceyssens, 'Le Jansénisme: considérations historiques préliminaires à sa notion', *Analecta Gregoriana,* 76 (1954), 8.

[3] References for this and similar statements in J. Orcibal, *Jean Duvergier de Hauranne, abbé de Saint-Cyran, et son temps* (Paris, 1947), 573.

[4] V. Durand, *Le Jansénisme au 18e siècle et Joachim Colbert, évêque de Montpellier (1696–1738)* (Toulouse, 1907), 43–4.

episode three years later, the parishioners of Quernes refused the benediction offered by the bishop of Boulogne, shouting 'nous ne voulons point de bénédiction d'un hérétique, d'un appelant . . . '.[5] After a visitation of his parish by the rigorist Bishop Coislin of Orléans in 1682, the prior of Sennely tried to carry out his instruction to remove a statue of S. Antoine, judged 'ridicule et indigne', only to find himself confronted by 'toute la paroisse soulevée jusque là que quelques femmes dirent insolemment que Monsieur l'Evêque n'aimoit pas Saints parce qu'il étoit d'une race d'huguenots . . .'. ['the whole parish up in arms to the point where some women said insolently that Monsieur the bishop did not like the Saints because he came from a family of Huguenots . . .'].[6] Far-fetched though these attacks were, such slurs were still effective enough, so that hostile polemicists never tired of portraying the Jansenists as thinly disguised Calvinists.[7]

Any credibility such claims ever possessed has been destroyed by the work of modern scholars, who have demonstrated abundantly that the Jansenists were an integral part of the great movement for reform within the Gallican church. Practically every one of their individual views can be found in authors of unquestioned orthodoxy, and one of the reasons for the duration of the disputes was the ease with which Jansenist sympathizers could use respected authorities in their own defence. The most important of all was naturally St Augustine, the ultimate authority on grace and predestination. The keen eye of Pierre Bayle saw to the heart of the catholic predicament:

Il est certain que l'engagement où est l'Eglise Romaine de respecter le système de S. Augustin, la jette dans un embarras qui tient beaucoup de ridicule. Il est si manifeste à tout homme qui examine les choses sans préjugé, et avec les lumières nécessaires, que la doctrine de S. Augustin et celle de Jansenius Evêque d'Ipre sont une seule et même doctrine, qu'on ne peut voir sans indignation que la Cour de Rome se soit vantée d'avoir condammé Jansenius, et d'avoir néanmoins conservé à S. Augustin toute son autorité et toute sa gloire. Ce sont deux choses tout à fait impossibles.

[5] J. Carreyre, *Le Jansénisme durant la régence* (Louvain, 1929–33), iii. 50–3.

[6] G. Bouchard, *Le Village immobile* (Paris, 1972), 299.

[7] Examples from the 1640s in A. Adam, *Du mysticisme à la révolte: les Jansénists du 17ᵉ siècle* (Paris, 1968), 175–6.

[It is certain that the Roman Church's obligation to respect the system of St. Augustine places it in an embarrassing position which is often ridiculous. It is so obvious to anyone who examines the question without prejudice, and with the necessary ability, that the doctrine of St. Augustine and that of Jansenius bishop of Ypres are one and the same doctrine, that one cannot see without indignation that the Court of Rome prides itself for having condemned Jansenius, while nevertheless preserving all the authority and glory of St. Augustine. These are two completely impossible things.]

Bayle went on to praise the Arminians for having at least had the honesty to bracket Augustine with Calvin as a *prédestinateur* to be rejected, commenting 'les Jésuites en auroient fait autant sans doute, s'ils avoient osé condamner un Docteur que les Papes et les Conciles on aprouvé' ['the Jesuits would doubtless have done as much, if they had dared to condemn a doctor who had been approved by the popes and the Councils'].[8] Although he may not have been entirely 'sans préjugé', Bayle knew very well that in the decades following the reformation Rome had taken a number of steps, of which the decrees of the Council of Trent and the condemnation of Baius in 1567 were the most important, to assert a 'central' doctrinal position against the extreme Augustinianism of Calvin. The reformer had asserted of Augustine's views on election: 'Il s'accorde si bien en tout et partout avec nous, il est tellement nôtre que, s'il me fallait écrire une confession sur cette matière il me suffirait de la composer des témoignages écrits de ses livres' ['He agrees so well and in every place with us, he is so much one of us that if I were obliged to write a confession on this question it would be enough for me to make it up from the testimony written in his books'].[9] Pascal and others tried to dissociate themselves from Calvin by rejecting the doctrine of 'double predestination', under which the reprobate are as positively chosen by God as the elect, and by maintaining a tenuous element of free will. Although such distinctions are perfectly valid, they are also very fine; paraphrasing Bayle, I think an honest reader must conclude that Calvinists and Jansenists were far closer to one another than they were to the 'new theology' of the Molinists.

[8] P. Bayle, *Dictionnaire historique et critique (*1702 edn.), art. 'Augustin'.

[9] P. Sellier, *Pascal et Saint Augustin* (Paris, 1970), 299.

The Council of Trent had reacted to the protestant challenge by reasserting the value of the sacraments and of good works, but had not succeeded in clarifying the knotty theological problems associated with predestination and grace. The effrontery of the heretics in claiming part of the Christian tradition for their own had not helped, but in any case the Council would have risked tearing the church apart if it had ventured too far. A degree of equivocation had worked well enough in the past; unfortunately new standards of critical scholarship, combined with the effects of printing, were fast making old compromises untenable. A much clearer sense of historical time, something of a mania for legalistic definitions, and the beginnings of textual criticism were all threatening to upset the authority of tradition, less by denying it than by fragmenting it. The protestants had found at least a temporary escape route by emphasizing the Bible, as God's direct message to his people. Although an appeal to Scripture could support some alarming innovations, on a conscious level it was still possible to treat the Bible as the repository of absolute truth, so that in combination with the theory of predestination it could produce a very rigid and seemingly watertight theology. The psychological appeal of such devices was probably enhanced by the growing intellectual uncertainties of early modern Europe, and by the division of Europe into warring confessions. There was no real consumer choice in religion, so that within each creed the characteristic polarities remained: personal faith against institutional practice; moral rigour against the need to win back sinners; divine obligations against secular; individual holiness against the saving power of the sacraments; austerity against ceremonies. As long as men believed that there was some absolute answer to these conundrums, they were bound to generate endless conflicts; in a sense the intensity of these struggles was a sign of the vitality of the churches, and of their involvement with the real problems of their time.

Whereas the protestants had radically simplified the ecclesiastical structure, the catholics had added a whole range of new orders to the complex medieval heritage. Among these the Jesuits were outstanding, for their rapid success, for the violent feelings they aroused, both positive and negative, and for the boldness of their approach to the great problems of the day. Without the Jesuits the Jansenist quarrel seems almost unthinkable, although this is not to make them solely or even primarily responsible for it. Few contemporaries realized the extent of the divisions within the Society, which became more serious as its expansion inevitably slackened.[10] Just as fearful protestants credited the Jesuits with superhuman powers, so their catholic opponents saw deep-laid designs behind what was often mere confusion.[11] It remains true that the Jesuits had produced a distinctive new theology; the doctrines of Molina and Lessius sought to take the sting out of predestination, by detaching it from God's will, to make it merely a statement of divine foreknowledge. This device, by giving man back control over his own fate, made it possible to mount a much more positive defence of catholic doctrines on works, penance, and the sacraments. It reflected the Society's missionary role, for it could also help to justify the teaching of a simplified, even rather formalized religion to the ignorant masses, in the hope of leading them into something better in due course. Immediately denounced as semi-Pelagian, these theories were never formally condemned by Rome; the popes disliked them, but valued the services of the Jesuits too highly to risk humiliating them. The new theology was also very much in keeping with the Tridentine defence of the efficacy of the sacraments *ex opere operando,* while in pastoral terms it was doing little more than make explicit the practice of the medieval church, always potentially at odds with the stricter interpretations of Christian teaching. Saint-Cyran seemed to admit as much when he praised St Bernard as the last defender of true faith, and in some of his comments on the corruption of the church.[12] Many catholics, among them the Jansenists, implicitly rejected any idea of the organic development of tradition, looking instead for an absolute, fixed truth. Pascal wrote that theology consisted simply in knowing 'ce

[10] M. de Certeau, 'Crise sociale et réformisme spirituel au début du 17e siècle: une 'nouvelle spiritualité' chez les Jésuites français', *Revue d'ascétique et de mystique,* 41 (1965).

[11] Wiener, 'Beleaguered isle', 42–4.

[12] Orcibal, *Duvergier de Hauranne,* 533, 655–6; *Lettres inédites de Jean Duvergier de Hauranne,* ed. A. Barnes (Paris, 1962), 105.

que les auteurs ont écrit . . . d'où il est évident que l'on peut en avoir la connaissance entière, et qu'il n'est pas possible d'y rien ajouter'.[13] The differing attitudes of the opponents to the permitted mode of discourse itself made dialogue impossible, and encouraged the contestants to lapse into mutual abuse.

In France the Jesuits had to contend with a powerful indigenous movement for religious reform, attached to the Gallican traditions, whose most prominent leader was perhaps Pierre de Bérulle, the founder of the French Oratory. A convinced Augustinian, and a defender of the rights of the bishops, Bérulle was no friend to the Jesuits. Operating through the networks of friendship and influence which were characteristic of the period, he became a dominant personality among the catholic enthusiasts known as the *dévots*. These clerics and pious laymen wanted the crown to put religion first in making policy decisions, but they were also active in initiating new foundations and good works across the country. Jansenism can only be properly understood, in its first phase at least, as one of the expressions of the *dévot* movement, and as a reaction to the terrible crisis it underwent during the ministry of Cardinal Richelieu. In 1624 Richelieu still had strong links with the *dévots;* he was a protégé of Bérulle and the queen mother, a friend of Saint-Cyran, a bishop known for his reforming zeal. Soon, however, his ambition and his political realism led to a parting of the ways, as he grappled with the major issues of policy in Europe and towards the Huguenots. While the English puritan preachers were calling for intervention in the Thirty Years War, the *dévots* wanted France to remain neutral, leaving the Habsburgs a free hand in Germany and the Netherlands. Both groups saw the struggle as primarily one over religion, in which the fate of Christendom might be decided; the pursuit of secular interest by their respective governments baffled and dismayed them. On this issue at least the *dévots* might hope to make common cause with the Jesuits, themselves leading advocates of a counter-reformation crusade against the heretics. Richelieu himself attributed the two pamphlets *Mysteria politica* and *Admonitio,* which attacked his policy in 1625, to a Jesuit, commenting 'la pernicieuse doctrine qu'ils

[13] Sellier, *Pascal,* 304.

renferment étant la doctrine particulière de leur Ordre'.[14] In practice the cardinal managed the Society so well, by offering protection against its numerous enemies, that he was able to secure its compliance. In his doctrinal works Richelieu had already shown himself to favour a laxist position similar to that of the Jesuits, and adapted to the needs of the ordinary believer, especially in his views on the vexed issue of attrition and contrition. Now the apologists for his foreign policy put forward theories the *dévots* could only regard as Machiavellian exercises in sophistry, bearing a strange resemblance to the morality of the casuists. While their opponents appealed to arguments from analogy, assuming the existence of an absolute, God-given natural order, the cardinal's publicists used secular rationalism to draw endless distinctions for use in the real world.[15]

The political defeat of the *dévots* in the great crisis of the 'Day of Dupes' (November 1630) threw the leaders of the catholic reform into disarray. Many, like Vincent de Paul and Condren, felt that they must preserve their own orders and foundations, and that to oppose Richelieu directly would be disastrous for them and for the whole reform movement. Their submission, and the servility of most of the bishops, left the way open for other leaders who might crystallize the widespread disquiet of the traditionalists. This was the role in which Saint-Cyran found himself cast, probably quite against his will, for nothing in his career suggests personal or political ambition. His reputation for learning and piety, his known refusal of favours from Richelieu, and his association with the abbey of Port-Royal were enough to make him the object of grave suspicion, even if he had been less given to imprudent sallies and paradoxes. His friend Cornelius Jansen was the author of the *Mars Gallicus,* a bitter attack on Richelieu's declaration of war in 1635. M. Orcibal has shown how Saint-Cyran's arrest and imprisonment in May 1638 was the culmination of a series of episodes in which he had been identified as a potential centre of disaffection.[16] Ironically, one of the most crucial was the imprudence of a Jesuit, Father

[14] E. Thuau, *Raison d'état et pensée politique à l'époque de Richelieu* (Paris, 1966), 110.

[15] Ibid., 148.

[16] Orcibal, *Duvergier de Hauranne,* 435–594.

Caussin, dismissed from the key post of royal confessor in November 1637 for trying to make the king's policies a matter of conscience. Another was the appearance of the *solitaires* attached to Port-Royal; the government's reaction to these apparently harmless religious enthusiasts ceases to be puzzling when it is realized how far they defied social conventions. The alienation of property, the rejection of careers in the service of the state, and the willing assumption of manual labour—regarded as humiliating and degrading by *les gens aisés*—all seemed a direct defiance of the social order.[17] The simple existence of the *solitaires* was a kind of blueprint for a new order based on religion, which the French government treated much as the English parliament would the Diggers.

Like Richelieu's investigators, the historian must admit that Saint-Cyran was largely innocent of the charges brought against him. He held an extreme version of the Bérullian position on the way to individual salvation, regarding retreat from the world as the ideal, to which even those who remained in the world should approach as nearly as possible. In his defence of Gallicanism, under the pseudonym of Petrus Aurelius, he had already objected to the works of the Jesuit casuists, in which he saw a prime example of the errors introduced by the excessive rationalization of faith.[18] He thought that the true value of the sacraments was being lost, because 'le monde est devenu corrompu . . . on a mis presque toute la piété de la Religion Chrétienne dans les seules choses extérieures, comme sont les confessions et les communions . . .' ['the world has become corrupt . . . we have located almost all the piety of the Christian religion solely in external things, such as confessions and communions . . .'], whereas in contrast 'Dieu a réduit toute la Religion à une simple adoration intérieure faite en esprit et en vérité. . . .' ['God has reduced the whole of religion to a simple interior adoration made in mind and in truth . . .'].[19] In such passages from his letters Saint-Cyran may strike a note reminiscent of many puritans, and he went even further with remarks such as 'on

[17] L. Cognet, 'Le mépris du monde à Port-Royal et dans le Jansénisme', *Revue d'ascétique et de mystique*, 41 (1965).

[18] Orcibal, *Duvergier de Hauranne*, 348.

[19] Barnes, *Lettres inédites*, 247, 298.

ne se sauve que rarement et avec mille difficultés dans le mariage . . . et dans les grandes conditions' ['one only saves oneself rarely and with a thousand difficulties in marriage . . . and in positions of high rank'.][20] In practice, however, he seems to have been a relatively moderate *directeur de conscience*, and to have accepted the corruption of the world as beyond repair. His arbitrary imprisonment was a matter of politics, not of theology; it did, however, set the stage for the Jansenist quarrel by preparing a small group of his sympathizers centred on Port-Royal to employ any weapon they could find against the enemies of the 'truth'.

Saint-Cyran's experience foreshadowed a repeated pattern, which would see the Jansenist movement created, as an independent entity, by the efforts of anti-Jansenists. The posthumous publication of Jansen's *Augustinus* in 1640 led to another attempt to find heresies, which trawled up so pathetically small a catch that it only enraged the Augustinians; the blatant partisanship of Jansen's critics made them determined to defend his book. Saint-Cyran's own most partisan act was in response to Isaac Habert's violently hostile sermons, which led him to suggest that Antoine Arnauld should denounce the trivialization of the sacraments and 'la dévotion facile'. The celebrated *De la fréquente communion* (1643) was the result, setting off another round of polemic which did more to popularize the book than anything else could have done. Later in the same year Arnauld repeated the trick with his *Théologie morale des Jésuites*, whose criticisms of propositions drawn from the casuists elicited yet more unwise replies from his adversaries. By raising major issues of commonplace religious practice, which affected laymen as well as clerics, Arnauld had shifted the revolt of the traditionalists towards a more fruitful ground than that of the abstruse debates over grace and predestination. The conflicts of the 1640s and 1650s would be all the more intense as a result, widening the divisions within the church, and creating parties which threatened to fight one another interminably.

The formal persecution of Jansenism began with the papal condemnation of the Five Propositions in 1653, the expulsion

[20] J. Orcibal, *La Spiritualité de Saint-Cyran* (Paris, 1962), 45.

of Arnauld from the Sorbonne in 1655, and the imposition of the Formulary in 1656. Throughout the complex manœuvres leading to these events the popes themselves seem to have remained consistently ill informed about French affairs. Preoccupied with defending their rather limited sovereignty over the Gallican church, they repeatedly misunderstood the likely results of their interventions, falling easy victims to a variety of misrepresentations. On the other hand, Cardinal Mazarin exploited the dispute with cynical skill; he kept the pot boiling, to give him another ploy for use in his elaborate diplomatic game.[21] Identified with opposition to the government from the start, the Jansenists had seen their position worsened as a result of the Fronde, and their marginal but highly damaging association with the fugitive Cardinal de Retz. Memories of the Fronde would remain a major element in Louis XIV's thinking, and half a century later his confessor reported:

J'ai souvent entendu dire à Sa Majesté que cette cabale de gens révoltés contre les décisions et les sentiments de l'Église n'était pas moins nuisible à l'État qu'à la Religion; qu'il savait, par une fâcheuse expérience, qu'un certain esprit d'indépendance se glissait partout où ce parti de novateurs trouvait quelque accès et que, généralement parlant, cette secte était ennemie de toute domination, tant spirituelle que temporelle.

[I have often heard His Majesty say that this cabal of people in revolt against the decisions and the opinions of the church was no less damaging to the state than to religion; that he knew, from tiresome experience, that a certain spirit of independence spread everywhere that this party of innovators found some access and that, speaking generally, this sect was the enemy of all lordship, as much spiritual as temporal.][22]

Those who put religion before political expediency could expect a rough ride in seventeenth-century France, except when it suited the Most Christian King to agree with them. The only effective defence against the hostility of the crown lay in an appeal to public opinion, the tactic brilliantly

[21] P. Jansen, *Le Cardinal Mazarin et le mouvement Janséniste français* (Paris, 1967).

[22] G. Gutton, *Le Père de la Chaize* (2 vols., Paris, 1959), ii. 206.

exploited by Pascal. The *Lettres provinciales* did not create mistrust of Jesuit casuistry, but they exploited it to devastating effect. In 1694 the Jesuit Gabriel Daniel admitted that Pascal had 'given the Jesuits' reputation a terrible wound . . . That book alone has made more Jansenists than all Mr Arnauld's books together.' Equally bad, it had also created a third party in France, 'almost as numerous as the other two', of those who submitted to the church, but reproved Jesuit laxity.[23]

The existence of such a party was already evident in the same Assembly of the Clergy which decided to enforce the Formulary, when in February 1657 it approved the printing and circulation of a French translation of Borromeo's *Instructions aux confesseurs,* to serve as a barrier 'pour arrêter le cours des opinions nouvelles, qui vont à la destruction de la Morale Chrétienne'.[24] The whole tone of Borromeo's crisp and practical guide was very much in line with Jansenist thinking; he advocated public penitence, full restitution of wrongs, the turning-away of those who appeared improperly dressed, and the deferment of absolution until there had been a genuine amendment of life. Such ideas must have found a ready audience among the large group of Parisian *curés* who joined in the hunting of the casuits, much aided by the appearance of Pirot's *Apologie pour les casuistes,* a damningly honest defence of the extreme Jesuit position. In the article on Loyola in his *Dictionnaire historique et critique,* Bayle would later remark: 'Ceux qui ont lu le livre du Père Pirot m'avouëront qu'il est plus aisé de le censurer, et de sentir qu'il contient une mauvaise doctrine, que de résoudre ses objections' ['Those who have read Father Pirot's book will admit to me that it is easier to censure it, and to feel that it contains bad doctrine, than to meet his objections']. To a modern eye Pirot often seems to take a thoroughly sensible position, pointing out that Jansenism could discourage men from making any effort, that such impossibly high standards would mean denuding the church of priests, and that the kind of cases discussed by the casuists would have the ordinary parish priest in great diffi-

[23] [G. Daniel], *The Discourses of Cleander and Eudoxus on the Provincial Letters* (Cullen, 1694), 13–14 (French original same year).

[24] Procès-verbal de l'Assemblée, cited in contemporary editions.

culties.[25] Hardened sinners who were refused absolution risked being deprived of it to the end of their lives, when the grace of the sacraments could have helped them to overcome their temptations.[26] Less sympathetically, although in an equally common-sense fashion, Pirot attacked extreme views on almsgiving, for:

les conditions et le partage des biens, ont esté introduits par le droit des gens, afin de rendre les particuliers laborieux, car si toutes choses estoient communes, personne ne voudroit travailler: La maxime des Jansenistes fomente cette fainéantise, parce que personne ne se soucieroit d'aquerir du superflu; si les riches estoient obligez de donner à ceux qui sont en grande nécessité, tout leur superflu . . .

[ranks and the disposition of property have been introduced by human law, in order to make individuals hard-working, for if everything were held in common, no one would wish to work: the maxim of the Jansenists breeds such laziness, because no one would be troubled to acquire a superfluity; if the rich were obliged to give all their surplus to those who are in great need . . .][27]

Although Pirot's *Apologie* is often maladroit or unfair, it is not the absurd work it is often assumed to be; behind it lies the vision of a pastoral method adapted to the needs of the ordinary parishioner. The strict Augustinians could never admit the kind of compromises this involved, and remained blind to the psychological insight the Jesuits had shown in developing their missionary techniques. The majority of the French church seems to have agreed that absolute standards had to be enforced at every level, an attitude which must make the Jesuit doctrines suspect.

The theology of the Jansenists might have been condemned, but their rigorous conception of Christian morality was on the way to becoming the official position of the French church. There were good reasons for this apparent illogicality on the part of the church's leaders, which was far from being the direct result of the *Lettres provinciales*, and might well have happened without them. The dispute between regulars and seculars still smouldered on, and nowhere were the bishops

[25] *Apologie* (1659 edn.), 5, 13, 73.
[26] Ibid., 49. [27] Ibid., 59.

more jealous of their authority than over control of the confessional. The whole century resounded with the complaints of the *curés* against the interference of the various orders, always ready to tempt penitents away with the promise of easy absolution. The accusations brought against a congregation of Penitents in the diocese of Paris in 1672 are typical:

tous indifférement entendent les confessions de tous ceux qui se présentent et mesme donnent l'absolution à ceux dont on a différé l'absolution pour des cas de conséquence, comme pour des scandales, pour ne pas vouloir restituer le bien de l'Eglise, en ayant le pouvoir et autres, ce qui fait un grand désordre à la Pasque.

[they all indiscriminately hear the confession of all those who present themselves, and even give absolution to those who have had it deferred for important reasons, as for causing scandal, not wishing to make restitution of church property although they have power to do so, and others, which causes a great disorder at Easter.][28]

The Jesuits' policy, virtually admitting that the ends might justify the means, naturally included the toleration, even the encouragement, of the multifarious forms of popular devotion. Most of the bishops, sharing the prejudices common to the educated classes of their time, were coming to see popular customs as unwanted accretions, offensive to morality, good taste, and reason. This trend was to culminate in the great compilations of popular superstitions by the *curé* Jean-Baptiste Thiers and the Oratorian Father Le Brun.[29] The widespread sympathy for such censorious views was sufficient to ensure that the Jesuits, although they continued to play an important part in the great movement for pastoral reform, would never dominate it as they had once hoped. Many reforming bishops, even if they were not Jansenists, would look for a more rigorist approach, aimed at producing a revolution in both the outer and the inner man. It was probably an appreciation of the extent to which such ideas were winning acceptance that led some Augustinians to break with the leaders of 'le parti Janséniste'. When the intransigence of the latter frustrated an attempt at a compromise settlement in

[28] J. Ferté, *La Vie religieuse dans les campagnes parisiennes (1622–1695)* (Paris, 1962).

[29] J.-B. Thiers, *Traité des superstitions selon l'Ecriture sainte* . . . (Paris, 1679); P. Le Brun, *Histoire critique des pratiques superstitieuses* . . . (Paris, 1702).

1662, Jacques de Sainte-Beuve, whose support for Arnauld had earlier lost him his chair at the Sorbonne, commented despairingly: 'Le feu est aux quatre coins de l'Eglise, et au lieu de l'éteindre, on y jette toujours plus d'huile. Ils ne peuvent s'empêcher d'écrire' ['The fire has reached the four corners of the church, and instead of putting it out, they are always throwing more oil on it. They cannot stop themselves from writing'].[30] The reforms which Sainte-Beuve thought more important than the doctrinal issues depended crucially on the parish clergy, so the first essential was to create or expand the seminaries necessary to train them properly. A great deal would depend on who ran these insitutions, and what they taught.

The organization of a seminary was one of the many responsibilities of a bishop, who was free to invite any suitable order or group of clerics to run it. There was always a small group of avowed Jansenists among the bishops, for the obscurity, not to say unreality, of the distinction between orthodoxy and heresy was such that the government periodically miscalculated in its selection of prelates. More important for the general tone of the church, however, were the numerous bishops who belonged to no party, while looking for high moral standards and strict discipline. Neither group was likely to call on the Jesuits to run their seminaries; often their eye fell on the Oratorians, if they did not simply appoint individual clerics of whom they approved. Despite repeated attempts to purge it of Jansenists, the Oratory remained Augustinian in its basic attitudes, so that the clerics it trained were likely to feel a certain sympathy for the Jansenist position. In 1678 and again in 1684 the government considered the suppression of all Oratorian colleges and seminaries, as the result of a number of individual imprudences.[31] The congregation had a dangerous knack of producing troublesome and independent minds; apart from Quesnel and the pioneering biblical scholar Richard Simon (both eventually expelled), there was the Cartesian Bernard Lamy, disgraced and exiled by *lettre de cachet* in 1675. In his moral theology course he had dared to

[30] Adam, *Mysticisme*, 250.

[31] P. Lallemand, *Essai sur l'histoire de l'éducation dans l'ancien Oratoire de France* (Paris, 1887), 135–6.

suggest disapproval of the social order, for 'dans l'état d'innocence, il n'y aurait point eu d'inégalité des conditions: c'est par une suite du péché qu'il y a maintenant une différence parmi les hommes, dont les uns commandent et les autres obéissent' ['in the state of innocence there would not have been any inequality of rank: it is as a result of sin that there is now a difference among men, of whom some command and others obey'].[32] Despite these various embarrassments, in the early eighteenth century the Oratory still ran sixteen seminaries, while another eight were under control of small orders holding similar opinions. Allowing for seminaries not directed by a specified order, perhaps a quarter of these institutions were in the hands of clerics with a strong Augustinian bias.[33] Before 1700, when a more tolerant attitude towards such doctrine prevailed, the number had been greater. At La Rochelle, for example, the seminary was directed between 1667 and 1694 by Michel Bourdaille, author of a *Théologie morale de S. Augustin*; after his death the new bishop called in the Jesuits, who introduced major changes in the teaching.[34] The opposing parties had become increasingly aware of the need to woo young priests, and in 1682 the king's confessor, Pére de La Chaize, persuaded the Jesuits to relax the rule under which they would only direct seminaries in towns where they already had a college. He argued that although seminarists did not spend very long under training,

L'expérience cependant, nous apprend que, pendant ce peu de temps, les ecclésiastiques qui nous étaient les plus attachés et qui n'avaient qu'une saine doctrine changent entièrement de sentiment dans ces séminaires où nous n'avons point d'accès.

[Nevertheless experience teaches us that, during that short time, the clerics who were the most attached to us and followed only a sound doctrine change their sentiments entirely in those seminaries where we have no access.][35]

[32] Lallemand, *Essai*, 126–7.

[33] E. Préclin, *Les Jansénistes du 18e siècle et la constitution civile du clergé* (Paris, 1928), 84–8.

[34] L. Pérouas, *Le Diocèse de La Rochelle de 1648 à 1724* (Paris, 1964), 260–1, 365–6, 378–9.

[35] Gutton, *La Chaize*, ii. 171–2.

In many respects the personal rule of Louis XIV (1661–1715) was a decisive period for the pastoral orientation of the French church. Obligatory training in the seminaries, periodic visitations, a flood of episcopal orders and publications, all helped to transform the parish clergy into a relatively orderly and respectable group. The attitudes and ideas which became current during these decades of reform would subsequently be hard to change completely, even if they were liable to some slow evolution. The *dévot* movement might have lost the political battle, but at the parish level, where the government's power was so much less, it had gained a degree of compensation. The imposition of strict moral standards on the people had an obvious appeal, not least as a method of social control; the government itself subscribed to something of a double standard, wishing to deprive its subjects of the moral liberty it claimed for itself. One result of the doctrinal quarrels over Jansenism was to make anything that hinted at laxism suspect to a great number of people; even the Jesuits became highly sensitive to possible attacks, so that from the 1660s there are far fewer instances of 'scandalous' propositions being advanced by their casuists. The austere Pope Innocent XI condemned sixty-five propositions tending towards moral laxity in 1677, as a result of a new Jansenist campaign which had begun with the publication of *La Morale pratique des Jésuites*,[36] while in 1700, at the prompting of Bossuet and others, the Assembly of the Clergy censured no fewer than 127 similar propositions. Helped by the influence of Mme de Maintenon over ecclesiastical appointments, a party of *dévot* rigorists had become the most powerful force in the church; their success was signalled by the appointment of Cardinal Noailles as archbishop of Paris in 1695. Strongly Gallican and Augustinian in their sympathies, the leaders of this group were behind the attempt to discredit the Jesuits through accusations that their missionaries in China had tolerated paganism. Although they were themselves quite clear in their stand against Jansenism as defined by the previous condemnations, they commonly promoted clerics and approved books whose orthodoxy was more disputable. The

[36] First vol. Cologne, 1669, by Nicolas Perrault and others.

support many of them had given to Quesnel was a primary reason for the choice of his *Réflexions morales* as the target for the bull *Unigenitus*. Catechisms were among the most important instruments of pastoral reform, and here too the rigorist school seems to have had the upper hand. While the variety was immense, Bossuet's catechism of Méaux and Colbert's catechism of Montpellier were very popular; as has been remarked, these manuals had a striking tendency to emphasize the all-present, all-seeing nature of God as judge, creating a negative image of a church based on prohibition.[37]

The association between the *dévot* movement and moral repressiveness was inevitable, and had already found expression in the *Compagnie du Très-Saint Sacrement*, a secret association of pious laymen and clerics founded in 1627, only to be proscribed by Mazarin in 1660.[38] The dissolution of the national organization did not, however, mean the end of the local groups it had encouraged, which continued with their well-meant but odious work of searching out abuses for denunciation to the authorities. The minute-book of the Company in Marseille reveals a good deal of genuinely charitable work, alongside an apparently continuous battle to eliminate abuses from processsions, dancing and drinking on Sundays and feast-days, and the misbehaviour of travelling players or the scandalous poor. Even though the bishop and the municipality probably agreed in disapproving of mixed bathing, 'ceste grande immodestie des nudités des femmes', or the use of images of saints as tavern signs, it was clearly an uphill battle to achieve more than a temporary result.[39] A cosmopolitan port like Marseille was not very promising ground for such endeavours, but there was no lack of resistance in the remoter dioceses where individual bishops tried to introduce the new spirit. Caulet's episcopate at Pamiers (1645–80) saw a great improvement in the standards of the clergy, but this was partly at the cost of keeping their numbers down, for when the brothers Foreau visited Caulet's diocese in 1669 they reported him to be short of thirty 'bons ouvriers', so that the people 'crient sans cesse après lui pour avoir des

[37] J. C. Dhôtel, *Les Origines du catéchisme moderne* (Paris, 1967).

[38] R. Allier, *La Cabale des dévots* (Paris, 1903).

[39] Idem, *La Compagnie du Très-Saint-Sacrement de l'autel à Marseille* (Paris, 1909), 68–9 and *passim*.

prêtres et que même on lui chante injure'.[40] Prohibitions against dancing resulted in repeated gestures of defiance; despite a campaign of preaching and missions at Foix, Caulet himself was serenaded by revellers for three nights, from St John's Eve 1663 to the night after the feast-day.[41] The *curés* were encouraged to impose public penances, while delaying absolution for habitual sinners until there was evidence of a genuine change of life, and in an attenuated form such practices are reported to have persisted in the Ariège well into the nineteenth century.[42] There is a certain irony in the way the population responded to such ferocious methods of evangelization; instead of reforming their conduct, they treated Caulet as a thaumaturge—once he was safely dead. Despite the hostility of the authorities, his tomb became a centre for miraculous cures and pilgrimages, which continued at least thirty years after his death.[43] Caulet's neighbour, Bishop Pavillon of Alet (1639–1677), ordered his *curés* to keep registers recording 'les qualités et les défauts de chacun des paroissiens', backing up the confessional itself with an annual visit to each household to interrogate both inmates and neighbours.[44] One can only agree with another rigorist, Cardinal Le Camus of Grenoble, who judged Pavillon's policy 'sèche et peu propre à convertir le monde, et il [Pavillon] n'a aucune ouverture pour les expédients et les tempéraments nécessaires. Il y a donc dans sa discipline quelque chose du temps des Goths' ['harsh and ill suited to convert the worldly, and he has no awareness of the expedients and the modifications that are necessary. There is therefore in his discipline something of the time of the Goths'].[45]

Le Camus nevertheless described Pavillon as 'un saint sur terre', while he had earlier lamented of his own diocese:

> En vérité je suis bien combattu entre les règles que je lis dans tous les Pères et la nécessité dans la pratique, que l'expérience nous fait voir, ou qu'il faut tout rompre ou qu'il faut des condescendances qui passent l'imagination.

[40] J.-M. Vidal, *François-Étienne de Caulet, évêque de Pamiers (1610–80)* (Paris, 1939), 106.
[41] Ibid., 149–51.
[42] Ibid., 285, 292. See Chapter 7 above for a fuller discussion.
[43] Ibid., 601.
[44] Ibid., 288–9.
[45] B. Neveu, 'Le Camus et les Jansénistes français' in *Le Cardinal des montagnes*, ed. J. Godel (Grenoble, 1974), 102.

[In truth I am very torn between the rules which I read in all the Fathers and the necessity in practice, as experience teaches us, either that everything must be shattered or to resort to compromises which go beyond anything imaginable.][46]

In his own corner of the Alps he fought a similar, if less impossibly idealistic, battle between 1671 and 1707. While he seems to have achieved a considerable reform of clerical morals, the visitation records of his last years still emphasize the high level of sexual misconduct, profanation of the Sabbath, drunkenness, and other disorders. It can hardly have been worth the cardinal's while to condemn so mild a deviation as winter sports for women, even 'attendu l'indécence et le danger de cette sorte de divertissements'.[47] Another bishop who had to combat the popular attitudes of a mountainous diocese was Soanen of Senez (1695–1740), the victim of the Council of Embrun in 1727. An isolated and exceptional figure in the region, he evidently had little permanent effect on the deeply rooted culture and life-style of its inhabitants, failing to create a wide following even among the clergy. A study of wills demonstrates the modest and ephemeral nature of his influence over an aspect of religious practice which was more easily influenced than most, because it was relatively uncontaminated by secular concerns. Requests for funeral processions of penitents declined sharply, while those for simple funerals rose, but both trends were quickly reversed once Soanen had been confined to a distant abbey.[48] Similar limitations emerge from the journals of two *curés* in very different areas of France, Christophe Sauvageon at Sennely-en-Sologne in the diocese of Orléans (1675–1710), and Alexandre Dubois at Rumegies, in the northern frontier diocese of Tournai (1686–1739). Both were confronted by parishioners who were assiduous in their attendance at church, and charitable towards the poor, but saw no reason to alter their life-style in any other respect. Troublesome over the tithes, they

[46] Neveu, 'Le Camus . . .', 102.

[47] J. Solé, 'La crise morale du clergé du diocèse' in *Le Cardinal des montagnes*, 204 and *passim*.

[48] R. Collier, *La Vie en Haute-Provence de 1600 à 1850* (Digne, 1973), chs. vi–vii; M. Vovelle, *Piété baroque et déchristianisation en Provence au 18ᵉ siècle* (Paris, 1973), 463–97.

were subject to all the normal human frailties; at Sennely their sexual morality was particularly dubious, while at Rumegies personal violence ran high. The *curés* had to yield ground everywhere, using their journals to record their dismay, and to warn their successors of the problems they would face.[49] In practice there is little doubt that the aims of the rigorist bishops and seminary teachers disintegrated at parish level, because they were so remote from the behaviour and beliefs of ordinary people. Many *curés* managed some kind of compromise, but it was hardly in the best interests of the church for much of its painfully constructed system of clerical education to be inculcating notions which could only prove counter-productive.

Resistance was not confined to the lower classes or the uneducated; superstition may have been declining in polite society, but the other great sticking points were at least as much in evidence. The theatre, dancing, concupiscence, gambling, and usury were the targets of repeated denunciations, with the most minimal effect. Despite numerous threats from the newly devout Louis XIV, and Noailles's efforts to set up a kind of moral policing, Parisian society moved steadily towards the free and easy Deism which would characterize the eighteenth century.[50] The church was treading on particularly delicate ground when it intervened over the economic relations between rich and poor; one of Pavillon's most outrageous breaches of convention occurred when he collected a series of problems of casuistry which had arisen in his diocese. After obtaining opinions from the celebrated Parisian experts Sainte-Beuve and Porcher he published them, apparently without permission.[51] Most of the twenty-nine cases were concerned with economic misbehaviour by the local landowners, including depriving peasants of their customary rights, evading taxes, using false measures, imposing unjust sharecropping contracts, and making usurious loans of grain.[52] The rigorist doctors faithfully condemned a

[49] G. Bouchard, *Le Village;* H. Platelle, *Journal d'un curé de campagne au 17ᵉ siècle* (Paris, 1965). For Sauvageon in particular see Chapter 7 above.

[50] J. de Saint-Germain, *La Reynie et la police au 17ᵉ siècle* (Paris, 1962); O. A. Ranum, *Paris in the Age of Absolutism* (New York, 1968).

[51] *Résolutions de plusieurs cas importants pour la morale et pour la discipline ecclésiastique* (Paris, 1666).

[52] Ibid., esp. cases i, ii, vii, xiv–xvi, xviii.

vast range of abuses, demanding restitution in almost every case. Had such standards ever been successfully imposed, they would have effected a revolution in economic life; one is hardly surprised to find that in cases xxii–xxiv the doctors had to consider situations in which these offenders, refused absolution, appealed to secular courts, or appeared with a notary and witnesses to demand a public statement of the *curé's* reasons for excluding them from the sacraments. After Sainte-Beuve's death his brother collected hundreds of his individual consultations, which appeared in a monumental three-volume work, forming a kind of encyclopedia of rigorist casuistry to set against the laxist manuals of the Jesuits.[53] The church had certainly got into a frightful tangle over the question of usury, where established doctrine and general practice clashed head-on. The Jesuits were uncertain themselves; in 1677 La Chaize intervened to obtain the suppression of a book by Jacques Tiran, which virtually legitimized lending at interest for fear of 'troubles graves', presumably protests over laxism.[54] The kind of embarrassed justification offered by Gabriel Daniel was not going to convince many critics:

> But if these learned men, who are but Schoolmen, Canonists, or Casuists, be deceived in a matter, so subject to error, what shall become of an infinite number of Christians, of all estates, conditions, and employments, who daily either give, or take money, at this kind of interest? . . . But in a word, when once any doctrine makes an universal trouble and disorder in civil affairs, that very disorder ought to pass for a moral demonstration that it is false, and that it is against the order of Divine Providence, and, by consequence, against the eternal verities, of which common sense and reason are but the expressions which make part of the rules, God has given us for our conduct.[55]

Many dioceses were apparently unaffected by Jansenism as such, yet here too the rigorist current often left its mark, while all reforming bishops seem to have aimed at a more personalized religion, which should have manifested itself in changed life-styles. Some Jesuits became remarkably strict,

[53] J. de Sainte-Beuve, *Résolutions de plusieurs cas de conscience touchant la morale et la discipline de l'Eglise* (Paris, 1689–1704).

[54] Gutton, *La Chaize*, ii. 120–3.

[55] Daniel, *Cleander and Eudoxus*, 114.

like Father Paul-Gabriel Antoine, whose *Théologie morale* of 1726 advocated delaying absolution, opposed probabilism, and won the approval of rigorist bishops.[56] Some of his colleagues in Lorraine, however, under the patronage of Stanislas of Poland, embarked on a missionary campaign after 1739 which used all the methods of 'la dévotion facile', stirring up widespread clerical opposition.[57] Such missions were far more popular than the attempts of the rigorists to repress popular festivities and superstitions, which led to endless struggles, rarely successful outside the towns.[58] Faced with the people's desire to extend the faith, as a matter of communal practice, the élites reacted by seeking to limit and restrict it. This did not exclude aiming at a high pitch of exaltation; for the Jansenists this was to be obtained through love of a God one was condemned to ignore, with no assurance that the affection was reciprocal. It also involved a rigorous self-discipline which limited life as well as faith. The extension of such high standards to the people, under the cover of a return to tradition, may have been partly the unconscious imposition of class-determined moral views; in this sense rigorism perhaps represented the subjection of the poor and unruly to bourgeois values. Inevitably this fell short of real success, and detailed research has cast doubt on some of the assumptions made earlier about the church's effectiveness in tightening up sexual morality.[59] The preoccupation of some Jansenists with sexual pollution is certainly striking, though few went so far as Pavillon's catechism, which warned that a simple conversation between two people of opposite sex was never without sin.[60] Confessional manuals like those of Habert and Treuvé worried obsessively about the attachments that might develop between priest and penitent, the 'chutes effroyables' caused by immodest dress, and the illicit habits of the married.[61] Such

[56] R. Taveneaux, *Le Jansénisme en Lorraine* (Paris, 1960), 674–5.

[57] Ibid., 687–90.

[58] Y.-M. Bercé, *Fête et révolte* (Paris, 1976), chs. iv–v.

[59] J.-L. Flandrin, *les Amours paysannes (16ᵉ–19ᵉ siècle)* (Paris, 1975), and *Familles* (Paris, 1976).

[60] Vidal, *Caulet*, 284.

[61] Louis Habert, *Pratique du sacrement de penitence, ou methode pour l'administrer utilement* (Paris, 1691); Simon Michel Treuvé, *Instructions sur les dispositions qu'on doit apporter aux sacrements de penitence et d'eucharistie* (Paris, 1676).

dangers loomed larger as the church moved from providing a mechanism for reconciling disputes (once the dominant function of the confessional) to trying to reinforce and intensify man's personal relationship with God.[62]

Although Jansenism and rigorism shared many characteristics with English puritanism, their rejection of the world seems to have been more sweeping and more absolute. Unlike the puritans, they never indulged in dreams of an 'elect nation', preferring a radical pessimism which even excluded millenarian hopes, until the belated appearance of the abbé Vaillant, 'figurism', and the *convulsionnaires*.[63] These deviations only appeared after Fleury had made *Unigenitus* a law of the state in 1730, following this up with a renewed persecution. More positive elements in Jansenism, like the opening of the Bible to the faithful in such translations as that of Mons, ran into determined obstruction from the authorities. In this area, as in the case of attacks on 'superstitious' devotion to the Virgin and the saints, the likenesses between Jansenism and protestantism proved uncomfortably obvious even at the time.[64] The retreat into Richérism, with its assertion of the apostolic role of the *curés*, is also reminiscent of presbyterianism, especially when one finds it being pursued through the meetings of *curés* in the regular *conférences ecclésiastiques*.[65] Jansenism never succeeded in becoming a popular religious movement; the evidence is rather that it was disliked by the majority of those subjected to attempts to enforce its teachings. The doctrine of the elect and the reprobate might rationalize this situation, but it was no basis for a sustained pastoral campaign. The narrower forms of Jansenism were too highly intellectualized to relate closely to the demands of the everyday world, dominated as they were by a vision of the primitive church as a small community of true believers. Yet somehow Jansenism did come to incorporate many of the positive elements in the catholic reform, so that the attempts to eradicate it proved highly damaging to the church. Louis XIV's crude persecution allowed the movement to acquire the prestige attaching

[62] J. Bossy, 'The social history of confession in the age of the reformation', *Transactions of the Royal Historical Society*, 5th ser., 25 (1975).

[63] Préclin, *Jansénistes*, 112 et seq.

[64] See esp. Adrien Baillet, *De la dévotion à la Sainte Vierge* (Paris, 1694).

[65] Taveneaux, *Lorraine*, 703–4; Platelle, *Journal d'un Curé*, 140–1.

to a doctrine of opposition, while no Laudian supporter of dumb dogs could have improved on Massillon's complaint against the Jansenists:

'd'avoir mis dans la bouche des femmes et des simples laïques les points les plus relevés et les plus incompréhensibles de nos mystères et d'en avoir fait un sujet de conversation et de dispute. C'est ce qui a répandu l'irréligion, et il n'y a pas loin pour les laïques de la dispute au doute et du doute à l'incrédulité.

[that they have placed in the mouths of women and ordinary laity the most remote and most incomprehensible points of our mysteries, which they have made subjects of discussion and dispute. This is what has spread irreligiousness, and for the laity it is not far from disputing to doubting, and from doubting to unbelief.][66]

Such defeatism was hardly much of an alternative.

[66] Letter of 1724 in E.-A. Blampignon, *L'Épiscopat de Massillon . . . suivi de sa correspondance* (Paris, 1884), 256–7.

9

Idées and *mentalités*

The case of the catholic reform movement in France

dans les matières où l'on recherche seulement de savoir ce que les auteurs ont écrit, comme dans l'histoire, dans la géographie, dans la jurisprudence, dans les langues et surtout dans la théologie, et enfin dans toutes celles qui ont pour principe, ou le fait simple, ou l'institution divine ou humain, il faut nécessairement recourir à leurs livres, puisque tout ce que l'on en peut savoir y est contenu, d'où il est évident que l'on peut en avoir la connaissance entière, et qu'il n'est pas possible d'y rien ajouter . . . . Il n'en est pas de même des sujets qui tombent sous le sens ou sous le raisonnement: l'autorité y est inutile; la raison seule a lieu d'en connaître. . . . L'éclaircissement de cette différence doit nous faire plaindre l'aveuglement de ceux qui apportent la seule autorité pour preuve dans les matières physiques, au lieu du raisonnement ou des expériences, et nous donner de l'horreur pour la malice des autres, qui emploient le raisonnement seul dans la théologie au lieu de l'autorité de l'Écriture et des Pères. Il faut relever le courage de ces timides qui n'osent rien inventer en physique, et confondre l'insolence de ces téméraires qui produisent des nouvautés en théologie. Cependant le malheur du siècle est tel, qu'on voit beaucoup d'opinions nouvelles en théologie, inconnues à toute l'antiquité, soutenues avec obstination et reçues avec applaudissement; au lieu que celles qu'on produit dans la physique, quoique en petit nombre, semblent devoir être convaincues de fausseté dès qu'elles choquent tant soit peu les opinions reçues: comme si le respect qu'on a pour les anciens philosophes était de devoir, et que celui que l'on porte aux plus anciens des Pères était seulement de bienséance!

[in those matters where one is simply concerned to know what authors have written, as in history, in geography, in jurisprudence, in languages, and above all in theology, and generally in all those which have as their principle, either simple facts, or what is established by men or by God, one must of necessity turn to their books, since all that one can know is contained therein, from which it is evident that one can have complete knowledge of them, and that it is impossible to add anything to this . . . It is not the same in those subjects which are subject to the senses and to reasoning: in these authority is useless; reason alone is fitted to learn about them . . . In the light thrown by this difference we should bemoan the blindness of those who bring authority alone for proof in physical questions, in the place of reasoning or experiments, and feel horror for the malice of those others, who rely solely on reasoning in theology in the stead of the authority of the Scripture and the Fathers. We must raise up the courage of those timid persons who do not dare to invent anything in physics, and confound the insolence of those rash persons who produce novelties in theology. However the misfortune of our times is such, that we see many new opinions in theology, quite unknown in the ancient world, obstinately upheld and greeted with acclaim; on the other hand those which have been produced in physics, although few in number, seem destined to be condemned as false as soon as they give the slightest shock to received opinion; as if the respect we have for the ancient philosophers was a matter of duty, while that we have for the most ancient of the Fathers was a mere question of politeness!][1]

THIS text from one of the greatest and most original of seventeenth-century thinkers reminds us how often the ideas of the period confound our expectations. It also illustrates an attitude to religion which forms a central theme of this paper, and which became more firmly established as the seventeenth century wore on, invoking authority to support a highly ascetic and demanding conception of true religion. A parallel strand of thought appears in a text sponsored by Henri de Pardaillan de Gondrin, archbishop of Sens, a contemporary who shared Pascal's Jansenist predilections. In his synodal statutes of 1658 we find a passage which might have come from almost any French diocese of the time:

Nous ordonnons que nos Archidiacres & Doyens Ruraux s'informeront diligemment dans leurs visites de tous les abus et Superstitions dans les Paroisses, tant des Villes que de la Campagne, comme sont les brandons, conjurations de fievres, chancres, feu volage, avives & autres maux, par certaines paroles, billets ou ligatures, & en quelqu'autre maniere que ce puisse être, consultations de Devins, preferences ineptes de certains jours ou certains mois, soit pour les Mariages, soit pour les autres affaires, comme si les uns étoient heureux, les autres malheureux, & autres de quelque espec qu'elles

[1] B. Pascal, *Préface pour le traité du vide, in Œuvres complètes,* ed. J. Chevalier (Paris, 1954), 530.

soient, sous pretexte de quelque coutume ou experience que ce soit, afin d'y pourvoir selon l'exigence du cas, instruction & pouvoir à ceux que nous jugerons à propos, pour les deraciner et en desabuser les Fideles, exhortant cependant les Curez à remontrer à leur peuple, que ces Superstitions ne sont autre chose que les restes de Paganisme & des inventions du Demon; par lesquelles il tache de les tromper & de les détourner de l'obligation qu'ils ont dans leurs adversitez de recourir à Dieu.

[We order that our archdeacons and rural deans should inform themselves diligently on their visits of all the abuses and superstitions in the parishes, in both town and country, such as are the Lenten torches, spells for fevers, cancers, burning illnesses, wounds and other ills, by certain words, notes or ligatures, and in any other manner there may be, consultations of wizards, unsuitable preferences for certain days or months, whether it be for marriages or other business, as if some were lucky others unlucky, and others of whatever kind they may be, in order to provide, according to the needs of the case, instruction and power to those whom we judge appropriate, to root them out and disabuse the faithful, meanwhile exhorting the *curés* to point out to their people, that these superstitions are nothing but the relics of paganism and inventions of the demon; by which he seeks to trick them and deflect them from their obligation to turn to God in their misfortunes.][2]

In their different ways both views could be characterized as austere and even repressive; although there were numerous alternatives available, they were probably characteristic rather than exceptional in the French church at the height of the catholic reform.

The multiple impulses which were working to transform the French church should offer a notably good prospect for investigating the connections and contrasts between *idées* and *mentalités*. On the one hand we have the enormous formal superstructure of ideas or dogma, expressed through a vast body of literature ranging from very complex theology to simple manuals for ordinary believers. Religious books dominated seventeenth-century French publishing, in terms of both titles and print runs; much may have been dead wood, derivative and unread, but the phenomenon cannot be dismis-

[2] J.-B. Thiers, *Traité des superstitions (*1st edn., 1679), reprinted in *Superstitions anciennes et modernes: préjugés vulgaires qui ont induit les peuples à des usages et à des pratiques contraires à la religion* (Amsterdam, 1734), i. 15.

sed.[3] Much of the serious thought about the nature of man, his relationship to the world and to his fellows, came from clerics writing what were avowedly religious works. These divines set out programmes for moral and disciplinary reform, normally adopting highly critical attitudes towards the beliefs of humble people, which they aimed to change. The structure of this popular belief has been one of the main preoccupations of the *histoire des mentalités*, not least because it can be investigated with the help of intellectual tools developed by anthropologists concerned with 'primitive' cultures in our own time. Here one may hope to detect clashes between élite and popular cultures, which to some extent overlap with distinctions between *idées* and *mentalités*. Such conflicts might be expected to set up dynamic relationships between the antithetical elements which caused them. At the crudest level, one might ask how far the educated minority (who produced the sophisticated texts on which we must usually rely) were influenced by their hostile vision of popular culture into formulating a religious stance which excluded many positive features of this supposedly debased popular belief. For the people, pressure from above might force the abandonment of some traditional practices, or at a deeper level alter their perceptions of both religion and the church. Historians cannot easily confine themselves to analysing the structure of beliefs, taken in a static context, for both their training and the nature of historical evidence pushes them towards the study of change. This is probably fortunate, since many of the methodological problems of this kind of history seem less acute when one concentrates on isolating the causes and constituents of change, rather than pursuing the chimera of a comprehensive map of a society's mental structures at some fixed point in time.

A static view is most defensible in the case of popular religion, which can plausibly be portrayed as a constant phenomenon over many centuries. Such a picture would stress its nature as an essentially pre-literate, animistic, and semi magical style of thought, whose fluidity could accommodate

[3] For the predominance of religious books, see H.-J. Martin, *Livre, pouvoirs et société à Paris au 17e siècle* (2 vols.; Geneva, 1969), and H.-J. Martin and R. Chartier (eds.), *Histoire de l'édition française* (Paris, 1983–4), vols. i and ii.

repeated changes in detail without any real modification of structure. I would not argue with this so far as it goes, but there is a danger that in concentrating on the style and structure of popular religion, the great constants, we will underestimate the changes that were possible within them, and their significance for the lives of both communities and individuals. There is a paradoxical sense in which such a *mentalité* can endure precisely because its structures are so loose and non-limiting, a characteristic which must render it liable to change directed by more self-conscious and systematic outsiders. The movement of religious ideas in the sixteenth and seventeenth centuries carried within it many forces liable to promote such intervention. The period saw an unprecedented expansion in the elaborated religious culture of the élites, to a level never surpassed since. This was associated with the appearance of a much more marked divide between élite and popular religion. Consciousness of this growing separation was precisely one of the factors impelling the clergy and the devout towards a missionary campaign directed at their social and intellectual inferiors.

The authoritarian overtones of this campaign are no surprise, when one remembers the tremendous emphasis on authority found in nearly all the religious discourse of the age. This emphasis on the authority of the Word (as by Pascal) and of the church as an institution (as by his opponents) must have an ironic ring to modern historians, aware of the numerous disputes which reference to both kinds of authority served only to exacerbate. To modern eyes this attitude may well appear to be as irrational, even as harmful, as any popular superstition. It is also evident that this nominally conservative ideology was not automatically hostile to change; self-proclaimed novelties were disfavoured, but the ambiguities of authority provided an abundance of fig-leaves for covering up such shame. The camouflage worked as much to resist self-awareness as to deceive others; men generally advocated change without wishing to admit it to themselves. This strongly implies that change was not the product of any positive and self-conscious striving or demand, so that it must have arisen from factors which were not very evident to the participants themselves.

The model of change might be summarized along the following lines: stretching back through the middle ages and even before, we can perceive numerous localized or short-term campaigns against paganism and superstition, or to revive religious fervour. The terminology and declared aims of many of these movements differed very little from those of reforming clerics, both catholic and protestant, in early modern Europe, and they commonly appealed in similar fashion for a return to some idealized past state. The difference seems to be partly one of scale, with such impulses appearing more frequently and in more places simultaneously. This is manifest in the reformation itself, but also in the catholic reaction to it, whose tendency to move down parallel paths is strong evidence for common pressures at work. While it is tempting to use the metaphor of a catalytic effect, this risks being misleading, since no single cause could possibly be found. The kind of factors involved will be familiar to everyone: printing and the spread of both clerical and lay education; the development of urban centres, with their commercial and legal élites; the growing power of nation states and the administrators needed to run them. What may be slightly less obvious is the way in which religious fervour, even fanaticism, was itself a powerful force for change in society and politics, yet was in its turn being inflected by the very developments it helped to create. The result was the coexistence of a congeries of interdependent forces for change, which for a time at least tended to reinforce one another. One of the determinants of the speed and nature of change was the extent to which these factors worked in the same direction. If they diverged, or one underwent a temporary blockage, becoming static or even regressing, this would affect the whole process by which they jostled one another forward. Even movements which built up a powerful inner dynamism must eventually have been limited by the *conjoncture* within which they had to operate.

Within this model it seems that ideas previously confined to small circles of clerics developed wide-ranging significance in early modern Europe, extending their appeal to large sections of the ruling groups, themselves often expanding in numbers and power. There was a further self-reinforcing effect here, since the church recruited most of its personnel from among these same élites, who might be expected to bring in what were, in many ways, socially determined ideas about the nature and role of the church.[4] Earlier reform movements had been overdependent on individual saintly leaders, or on essentially short-lived bursts of popular enthusiasm. Now there seems to have been something more like a continuum, with a steady pressure for the extension of Christian morality and the religious life into new areas, extending throughout lay society. With due caution I would suggest that French reforming catholicism was essentially a 'bourgeois' religion, exhibiting an increasingly prudential, rational, and self-consistent approach which would have come naturally to clerics from the appropriate social background. The attitude of the traditional nobility looks more ambiguous, but between the growing power of secular rulers on the one hand, and 'bourgeois' officials and town rulers on the other, these aristocratic hesitations long remained marginal, not least because they lacked any coherent means of expression. Such crude equivalences between social status and religious attitudes are not intended to disguise the enormous variations of individual response, nor to impose some ultimate reductionism on spiritual experience. Instead they draw our attention to the central role of bishops, diocesan officials, and leaders of religious orders who came from the 'robe' nobility, from other legal and administrative families, or (to a lesser extent) from the commercial élites. These men and women brought with them assumptions and styles of thought, we might even dare to say *mentalités*, which coalesced with traditional Christian dogma to generate a reforming drive of unprecedented vigour. Within this movement it is hard to see that individuals could do more than encourage and articulate the broad social and intellectual forces which were at work. The enormous scale of the church as an institution, its communal provision for education and for the formulation of doctrine, and its intricate connection with the hierarchical structures of lay society, all limited the possibilities for any individual to alter its direction;

[4] The social origins of the *curés* are discussed in numerous local studies. Excellent examples may be found in P. Goubert, *Beauvais et le Beauvaisis de 1600 à 1730* (2 vols.; Paris, 1960), 198–200, and P. T. Hoffman, *Church and Community in the Diocese of Lyon, 1500–1789* (New Haven, Conn. 1984), 156–8.

individual efforts were most effective precisely where they worked with the tide.

It could well be argued that many central doctrines of the Christian church had been waiting until this moment to find a society that could respond fully to them. The intellectual materials had been there all along, but the personnel to implement their message had been lacking, in both numbers and quality. The force with which the reformation and counter-reformation struck the societies of early modern Europe is obvious. At the level of the educated, they might arouse an intense search for personal spiritual satisfaction or certainty; across a much wider social range they inspired civil war, violence, and dramatic action extending even to 'tyrannicide'. They also initiated a longer-term drive towards the imposition of a new style of religion on the community as a whole. This movement was in part negative, with the prohibition of many traditional features of popular culture, along the lines so well described by Peter Burke.[5] While resistance to this intended emasculation of popular culture was substantial, so that the movement for purification fell far short of total success, it was still the easiest part of the programme to implement. Far harder to achieve was the positive aim of turning the ordinary practitioner of religion into a heartfelt Christian believer in the sense conceived by his or her betters. Both the nature of this reform movement, and the reasons for its partial failure have been extensively discussed, but there is still room for an attempt to relate the catholic reform to both ideas and *mentalités*.[6]

Despite the enormous literature on doctrinal problems connected with Jansenism, the characteristic writings of the catholic reformers were not theological in the narrow sense, but concentrated on encouraging 'true' religion and inculcating

[5] P. Burke, *Popular Culture in Early Modern Europe* (London, 1978). The repressive moves against festivals and similar activities in France are particularly well described in Y.-M. Bercé, *Fête et révolte* (Paris, 1976), chs. 4 and 5.

[6] For France alone the bibliography is enormous; good general accounts include J. Delumeau, *Le Catholicisme entre Luther et Voltaire* (Paris, 1971), and F. Lebrun (ed.), *Histoire des catholiques en France* (Toulouse, 1980). A recent overview for Europe as a whole, with a very full bibliography, can be found in A. D. Wright, *The Counter-Reformation* (London, 1982).

a deeper Christian morality for everyday living. In the latter area a kind of moral absolutism established itself, becoming dominant by the middle decades of the seventeenth century. Generalizing very dangerously, one might claim that hitherto sacred and profane had coexisted without too much difficulty; since boundaries were not clearly drawn, the ordinary believer could move between the two fairly freely, without being aware of any incongruity or inconsistency. The new style brought a trend towards legalistic definitions of the same distinction, combined with an increasing demand for the pre-eminence of the sacred, which was ideally supposed to permeate every level of experience, giving all aspects of life a religious dimension. The elevated scale of personal commitment required by such a faith went far beyond anything previously demanded, and was manifestly beyond the capabilities of most of the population. These theories are a very misleading guide to everyday practice, by clerics, the educated, or still more the peasantry. In fact it is likely that they created a widening gulf in just that area between theory and practice, so that the official prescriptions of reforming bishops and theologians became increasingly remote from reality. There was a kind of ratchet effect in operation; since it was impossible to challenge the extreme assertions of both divine demands and human potentialities head on, they tended to become more extreme with each restatement. This was particularly dangerous because it seems to have removed the possibility of any effective self-righting mechanism, whereby the church could moderate its demands in the light of experience, or adapt them to human realities. There was a special type of holy rhetoric at work, whose inbuilt tendency was to raise the stakes beyond the edge of possibility. These writers and preachers were often more concerned to present themselves as holy men, even as prophets, and to forward their careers in the church, than to propose practical solutions to practical problems. They vied with one another in painting a dark picture of the devout few struggling against the evils of '*le siècle*', which almost necessitated the elaboration of a suitably unattainable religious ideal.[7] Whether this can be classified as

[7] Two examples of many will have to serve here. J.-F. Senault, *L'Homme criminel* (Paris, 1653; 1st edn. 1644), 223, asserts that because man has

a peculiar clerical *mentalité* it is difficult to say; it appears to represent a strange mix of *déformation professionnelle* and imprisonment within a closed system of ideas.

A natural consequence of this situation was that French clerical writers came to assert with surprising accord that the great majority of the population were damned, sunk irreparably in paganism, idolatry, and concupiscence.[8] In order to combat this depressing situation, and at the very least to impose an outward conformity which might avert more direct manifestations of divine wrath, various new techniques of religious instruction were developed. These included the catechism, a range of popular devotional literature, more serious use of the confessional, and the establishment of itinerant missionaries to supplement the work of the parish clergy. Their introduction revealed some of the detailed problems inherent to any attempt to graft élite ideas on to popular religion. The language of the catechisms, tending always to dry abstractions, must have sounded very strange to peasant children, who naturally resorted to rote learning of these alien texts.[9] Despite such barriers, the long-term result must have been positive, in the sense that essential Christian dogma was better known; one may doubt, however, to what extent people changed their lives, or actually abandoned more traditional beliefs. The case of the missionaries is more complex, for here the church seems at times to have achieved considerable success precisely by harnessing features of popular belief and affectivity, thus reintroducing many of the ambiguities which

become a slave to concupiscence, and acts in obedience to it, 'la pluspart de ses bonnes œuvres son des pechez, et ses actions partant d'un mauvais principe, ne scauroient estre que criminelles'. According to Antoine Paccori, in his anonymously published *Règles pour vivre chétiennement dans l'engagement du mariage et dans la conduite d'une famille* (Paris, 1726), 13–15, writing of the father in the household, 'Fidèle à Dieu dans les tentations du démon, de la chair et du monde, il se fera une sainte violence, pour renoncer à son propre sens, et à ses propres volontés, et pour suivre dans sa conduite les lumieres de la foi et la volonté de Dieu, commme les seules regles de ses actions.'

[8] The point is made, with numerous references, by J. Delumeau in *Le Péché et la peur* (Paris, 1983), 457–69; he cites cases (pp. 460–2) in which members of the clerical élite doubted not the doctrine itself but the wisdom of preaching it to the people.

[9] For the catechism in general, see J.-C. Dhôtel, *Les Origines du catéchisme moderne* (Paris, 1967).

had been the original target of the enemies of 'paganism'. Within the emotional fervour created by processions, ceremonies, and hell-fire sermons, villages might confess their secret sins to a stranger (as they would not to their own *curé*), make ostentatious reconciliations with their enemies, and even offer restitution for past wrongs. The missionaries themselves were often enraptured by their apparent success, and such exercises certainly had a genuine spiritual value, yet for most participants they represented a periodic bout of religiosity, which symbolically wiped the slate clean, and might go some way to counterbalance a persistently sinful existence.[10]

At their most powerful, as operated by Maunoir and his Jesuits in Brittany, the missions instilled guilt into peasants who could not comply with the moral demands of the church, yet now understood them well enough to comprehend something of their own failure. Paradoxically, the result was often the intensification of ostentatious devotional practices which at least bordered on the superstitious, in the hope of redeeming the weaknesses of the flesh. More even than for their own souls, their concern seems to have been for those of their relatives and ancestors, for the missionaries dwelt shrewdly on the sufferings these risked undergoing for the sins of the living. An element of ancestor-worship might thus be encouraged, quite against any overt intention of the reformers.[11] The church did often display an obsessional concern with death, the ultimate moment of truth when those who had not heeded its message would discover their error. This was the repeated message of the most popular (or at any rate most widely used) missionary pamphlet of the seventeenth century, *Pensez-y bien*, while an edifying death, preferably after lengthy suffering, was the favourite theme of hagiographers, fascinated with the odours of sanctity in which their subjects

[10] For the missions, see *inter alia* the correspondence of S. Vincent de Paul, C. Berthelot du Chesnay, *Les Missions de S. Jean Eudes* (Paris, 1967), and two special journal numbers: *17ᵉ siècle*, 41 (1958), and *Annales de Bretagne et des pays de l'Ouest*, 81 (1974).

[11] Maunoir's work in Brittany is described in A. Croix, *La Bretagne aux 16ᵉ et 17ᵉ siècles* (Paris, 1981), ii. 1183–246. The connection between the doctrine of purgatory and fear for one's ancestors is made by Delumeau, *Le Péché et la peur*, 427–46.

passed away.[12] The logic which helped to create such an emphasis was, like so much in the thinking of the period, irrefutable in its own terms. The church could hardly be expected to undersell its crucial role as the bridge between transient earthly existence and eternity, yet there was a risk that excessive concentration on the necessity and difficulty of winning salvation would pervert and damage the essential Christian message. The more repressive visions of religion virtually sacrificed life to death, so that some of their extreme exponents arrived at a fantasy world which was repugnant even to educated society; crude overemphasis on the prospect of eternal torment ultimately helped bring the whole notion into doubt.[13] The very ideas which inspired one personality type were in any case liable to repel others, a problem which may have been particularly acute with a style of faith which generated so much of its emotional force from the glorification of various kinds of renunciation. Few reformers seem ever to have realized that, by making religion so heavily dependent on personal commitment, they were creating the possibility that men would refuse such commitment or be weighed down by a sense of their failure to live up to it; either attitude might eventually develop into general scepticism in the face of what were perceived as unattainable standards.

Such an evolution towards scepticism was of course to be characteristic of the very élites originally associated with the new faith; as the eighteenth century wore on this left clerics (who now seemed traditional or even reactionary) in an increasingly exposed position. The attitudes of the educated and prosperous had always been tinged with ambiguity, and they had often balked at the more ambitious claims of religious writers. To the moral credit of the *dévots,* they had spent more time denouncing the worldliness and irreligion of polite society than they had the superstitions of peasants, but this

[12] For *Pensez-y bien* and similar works, see Delumeau, *Le Péché et la peur,* pp. 389–415. An interesting description of attitudes to death and suffering within a religious order is by G. Drillat, 'La Vie spirituelle des Visitandines', in J. Delumeau (ed), *La Mort des pays de Cocagne* (Paris, 1976), 194–205.

[13] See D. P. Walker, *The Decline of Hell* (London, 1964), and J. McManners, *Death and the Enlightenment* (Oxford, 1981), esp. chs. 5 and 6.

had generally proved counter-productive.[14] Even those local rulers who wanted the church to help in establishing good order and moral control were irritated by zealots whose campaigning against popular recreations actually tended to disturb the public peace. Another reason why neither the bourgeoisie nor indeed the mass of the clergy were ever solidly behind the more logical reformers was the manner in which their own mental worlds usually retained strong traces of earlier and less systematic ways of thought. The ideas of the anthropologist C. R. Hallpike may be helpful here: following Piaget, he suggests that there is a developmental pattern common to all human beings, which sees thought develop in a logically determined sequence.[15] This proceeds from the pre-operatory thought characteristic of young children (in which reasoning moves from the particular to the particular, but cannot go from particular to general and back) to the highly developed logical and mathematical thinking which requires literacy and long schooling, and is unknown in primitive cultures. At the same time, the theory emphasizes how we all retain the ability to revert to earlier stages of cognitive development and think in very different ways in different circumstances. This is particularly true of persons under some kind of stress, or when strongly affective elements are involved, so that even those highly skilled in formal thought often fail to employ it. It does seem clear that religious faith, and a mode of living according to religious teachings, must take account of this basic thought structure; they cannot hope to succeed through exclusive reliance on elaborated formal constructs, nor by a crude combination of these with threats. It is also possible for intellectuals to develop an almost fetishistic attachment to abstract reasoning to the exclusion of other kinds, a mental puritanism perhaps illustrated by Port Royal's concern with the teaching of logic.[16] A curious

[14] This emerges clearly from most of the encyclopaedic collections of casuistry, such as that by J. Pontas, *Dictionnaire des cas de conscience* (3 vols.; Paris, 1714). Most clerical moralists would have agreed with Paccori (*Règles pour vivre,* p. 32) that salvation was easier for the poor, whereas 'les richesses y mettent des obstacles presque insurmontables'.

[15] C. R. Hallpike, *The Foundations of Primitive Thought* (Oxford, 1979).

[16] Exemplified in the famous textbook for the schools at Port Royal by A. Arnauld and P. Nicole, *La Logique ou l'art de penser* (1st edn., 1662).

situation could develop, in which the internal dynamics of an over-intellectualized system pulled it away out of sympathy even with those groups which had done most to create it.

The rulers of lay society commonly remained attached to religious practices and forms of devotion which risked attracting the censure of rigorist clerics. When Bishop Noailles (nephew of the Cardinal) visited the relic of the Saint Nombril at Notre Dame de Vaux in Châlons in 1707, he had the reliquary opened to discover only some pieces of gravel, which a local surgeon declared could never have been part of Christ's umbilical cord. The bishop removed the fraudulent relic, but the protest he received from the canons and parishioners is a remarkable document, signed by an impressive list of local notables—*conseillers* in the *présidial*, *trésoriers de France*, notaries and other lesser legal practitioners, and local merchants and tradesmen. Although much embarrassed by Noailles's discovery, they alleged the long tradition concerning the provenance of the relic, the veneration proper to anything pertaining to the Saviour (recalling the miraculous blinding of a bishop of Arras who had opened the vessel containing the 'sainte Manne'), the daily cures performed by the relic, and the bishop's alleged failure to follow the proper legal procedures in such a case. Although not explicitly stated, one can easily detect their concern for the loss of honour and glory to their church, and for the decline in revenue from pilgrims. The anonymous cleric whose account was printed by Le Brun ironized on the '*clameurs*' of the parishioners, whose opinions came '*des siècles d'ignorance et de grossierté*' whereas the bishop was concerned to uphold '*l'honneur de la pure Religion*'.[17] In such respects, and also when they attacked usurious contracts and other forms of economic exploitation, reforming clerics were going too fast and too far even for those educated believers who wanted decency in church ceremonies and the repression of obvious sinfulness among the ignorant masses.

[17] This account appears in *Superstitions anciennes et modernes*, ii. 75–82. The editor remarks that Bishop Noailles 'faillit à être lapidé par le Peuple superstitieux & toujours avide de fables & de pratiques prétendues-religieuses', and also comments 'il faut avouer qu'on ne sauroit rien lire de plus pitoyable que cette Requête, qui prétend justifier la vérité de la Relique.' Voltaire referred twice to this incident; see R. Pomeau, *La Religion de Voltaire* (2nd edn., Paris, 1974), 26.

Perhaps fortunately for the church, the actual execution of reform was usually in the hands of people whose own attitudes were far more mixed. *Curés* who had spent a few months in the diocesan seminary, Capucins and Lazarists recruited among the lesser bourgeoisie, humble girls working in the teaching and nursing orders; few of these were adherents of '*la pure Religion*' as Noailles might have envisaged it. Their pastoral experience was likely to separate them still further from the ideology I have been describing, to the point of giving them sympathy for the sufferings and the style of belief they found so widespread among those entrusted to their charge. This may help to explain the apparent paradox that a period which declared itself so fundamentally hostile to 'superstition' saw the creation of many new devotions and local rituals. The function of the ordinary clergy, as conceived by the reformers, was to close the gap between orthodox Christianity and the dubious beliefs of the people. While this objective was partly achieved, there were many ambiguities left, and not a few awkward side-effects created. One, as just suggested, was that the new 'purified' faith was contaminated afresh by popular or superstitious elements in the very operation of extending it to the whole population; this was almost certainly in the long-term interests of the church, but it was not what the new ideology demanded. Another problem stemmed from a recognition of the incompatibilities between time-honoured practices and 'correct' dogma; the resulting self-consciousness was a mixed blessing which, once acquired, could never be lost again.

The implications of this complex history for the relationship between history of ideas and *histoire de mentalités* are hard to summarize. In a sense it can be claimed that the would-be reformers had their own *mentalités,* clerical, bourgeois, legal and others; they were certainly not free agents developing ideas unconditioned by their social and professional circumstances. This area has so far been less discussed by historians than that of popular *mentalités,* not just for reasons of historical fashion. There is a certain coherence about popular *mentalités,* perhaps because, in Hallpike's terms, they never advance beyond the level of 'operative' cognition to that of formal thought. Among the literate and educated classes we find no such restraints; the coexistence of different available levels of cognition allowed much greater freedom for individual choice, rendering generalization very hazardous. The fact that these groups had access to a much more fully elaborated world of ideas was crucial in opening the possibilities for change, and for forceful action to secure that change. In the area of religion, however, the ideas were narrowly circumscribed by the permitted modes of discourse, so that they evolved dangerously towards the unrealistic or the irrelevant. In so doing, they risked not only relative failure in the assault on popular superstition, but also the loss of active support from the élites. This may well be a case where ideas evolved themselves into a cul-de-sac, just the situation envisaged by Hallpike when he remarks 'since the greater flexibility and analytic power of formal thought is directly dependent on its freedom from concrete imagery and the constraints of the real world, it is by that very fact also given powers of intellectual self-deception denied to those whose thinking remains at the level of pre-operatory or concrete thought, who must be content with the illusions of imaginative fantasy and the deceptions of the phenomenal appearances of things'.[18]

Much later evidence exists to suggest that popular belief retained its essential characteristics long after the repressive drive against it had waned. Only after the trauma of the Revolution would the church, as an official body, begin to see more virtue in the popular piety on which it had always, *de facto,* rested. Yet one may note that the potential for this *rapprochement* had always been present, for even in the works of such forthright critics of popular superstitions as the Abbé Thiers and the Père Le Brun one can readily perceive the difficulties these authors faced in distinguishing superstitious popular abuse from the lawful practices of the church.[19]

[18] Hallpike, *The Foundations of Primitive Thought*, 35–6.

[19] See for example Thiers in *Superstitions anciennes et modernes*, i. 95–7, discussing exorcisms and similar practices. He reaches the rather lame conclusion 'Ainsi afin que les Exorcismes, les Benedictions & les Oraisons soient dans l'ordre de l'Eglise, & qu'on ne puisse les soupçonner de Superstition, elles doivent se faire par des personnes que l'Eglise autorise pour cela, & avoir elles memes l'approbation de l'Eglise. Sans ces deux conditions elles sont illicites, & il y a de la Superstition à s'en servir.' In Le Brun's *Histoire*

Magical elements remained strong in catholic, and even some protestant, rituals, while the line drawn between licit and illicit devotion often depended less on any logical distinction than on some past pronouncement by authority. Perhaps it is not surprising that in the religious sphere, above all, ideas had such difficulty in breaking free from the tenacious grip of *mentalités*, for not even the most rational clerics could really hope to separate belief from emotion and feeling.

critique des pratiques superstitieuses (first published 1702), reprinted in the same volume, ch. VII of Book I (pp. 134–6) demonstrates the problems a Cartesian theologian faced: 'De milieu qu'il faut garder entre la trop grande credulité & l'incrédulité, ou l'obstination à ne rien croire d'extraordinaire & de merveilleux. Reflexions sur la manière de discerner si ces faits extraordinaires sont vrais.'

Conclusion

Religion, repression, and popular culture

THE essays in this book have all been concerned with processes of change within a specific period of time, the 'long' seventeenth century of the first three Bourbon kings of France. This was the period of the 'great repression', during which the monarchy and the church joined forces to impose order and obedience on the mass of the population. Sexual licence, rituals of inversion, popular festivities, and local saints were among the most prominent targets for this offensive. In such a traditional society, which habitually looked to the past for validations, it was predictable that attempts to innovate should frequently be obstructed or diverted; the surprise is rather that these processes operated as powerfully as they did. While the explanations I have advanced here have assumed that the social and economic developments since the later middle ages were a necessary precondition for change on this scale, this should not lead to any facile assumption that they were a sufficient cause. The expansion and self-assertion of a powerful class of urban rulers, who provided the main body of recruits for both the royal administration and the church, did not produce a completely new ideology, moulded in their own image. In the catholic Europe of the time, with its elaborate systems for establishing and maintaining orthodoxy, this would have been inconceivable. What took place was rather a selective development of the intellectual legacy of antiquity and the middle ages, in a dynamic relationship which allowed ideas themselves to be both vehicles and causes of change. As suggested in the final essay, for the first time the more ambitious formal doctrines of the church found themselves in conjunction with a social and political situation in which they could be effective on a general scale.

Religion indeed provided the only fully developed set of concepts through which men could rethink their relationship to one another, to society and its institutions, and to the physical world. It is notable how radicals in both politics and science began by borrowing its language and by using religious arguments to justify attacking other forms of authority. This protean capacity to adapt was in a sense a common feature of Christianity and popular culture, but it operated in very different modes. Popular beliefs and practices rather lacked any internal dynamic, tending to respond in a passive way to particular social demands, and undergoing a constant process of small-scale internal mutation. The formal doctrines of official religion, on the other hand, were capable of generating self-sustaining and powerful ideologies which could be very resistant to modification. It is important to recognize that this is not comparing like with like; the distinction is between *la religion vécue* and the written codes maintained by the church. The latter did not determine how even the intellectuals experienced religious feelings within themselves, while as we have seen, they could divide the élites among themselves almost as much as they did the élites from the illiterate or semi-literate masses. The latter distinction, however, was the most characteristic feature of the seventeenth century, for both contemporaries and modern historians, growing more profound with each decade. The educated minority made a series of very self-conscious distinctions of this kind, precisely because they experienced the period as one of transition, in which their identity and status required such boundaries to be drawn with much greater determination. For many of the families which had risen into the élites it was also a time of increased concern about maintaining their position as economic growth slowed or stopped. In the later sixteenth century a shift in the social composition of the secular clergy became apparent, with larger numbers of young bourgeois colonizing the church as part of a family strategy for distributing assets in each generation. It was no accident that this period saw the general introduction of property qualifications for entry to the priesthood, virtually excluding the great mass of the population. The slow process by which the social and educational standards of the clergy were raised was an indispensable element in the catholic reform movement.

It is only in recent years that historians have begun to confront this complex amalgam of economic, social, religious, and cultural change. Two French historians, Jean Delumeau and Robert Muchembled, have offered challenging interpretations of the process as a whole. Delumeau writes from the position of a liberal catholic, concerned to understand the deeper causes of 'dechristianization' and to challenge comfortable myths about the 'age of faith' before the Revolution. In a series of general books and erudite studies he has built up a dark picture of 'la pastorale de la peur' as applied to a semi-pagan peasant world by the catholic reformers.[1] The extreme nature of these teachings is explained by a coincidence between a long-established pessimistic style of preaching and the sequence of collective misfortunes which struck Europe between the Black Death and the Wars of Religion. These experiences confirmed the picture of a vengeful and demanding father-figure, punishing his children for their sins, and consigning all but the most virtuous to eternal damnation and torment. Once substantial groups among the élites had adopted this outlook they naturally saw it as their responsibility to repress sin throughout society, in the hope of appeasing divine anger. The result was a perversion of Christian teaching, with an overemphasis on ascetic virtues and an excessive rejection of human values, which eventually undermined the church's own position. At the same time Delumeau makes it clear that he sees popular faith as more superstitious than Christian, so that for him the catholic reformers were justified in trying to encourage a more personal and internalized religion.

While Delumeau has performed an invaluable service in raising fundamental questions about the catholic reform movement, and in exposing its ambiguities, his interpretation is open to serious objections. It is notoriously difficult to prove that one age is more fearful or pessimistic than another, and he has not convincingly done so; other specialists might regard his account of reactions to bubonic plague (for example) as very exaggerated. A similar problem is that of the relationship between doctrines and their application; a mountain of citations from sermons and devotional literature

[1] J. Delumeau, *Catholicism between Luther and Voltaire* (Paris, 1977); *Un chemin d'histoire; chrétienté et christianisation* (Paris, 1981); *La Peur en Occident, 14–18e siècles* (Paris, 1978); *Le Péché et la peur: la culpabilisation en Occident, 13–18e siècles* (Paris, 1983).

does not necessarily reveal very much about what happened between the *curé* and his flock. This is only one of the areas in which there is an alarming discrepancy between the great quantity of evidence adduced and the rather elementary theoretical structure in which it is placed. Delumeau's view of popular culture is also disturbingly external, not to say ethnocentric, while he tends to underplay the ritual and semi-magical elements which persisted in official catholicism. For him true religious experience essentially consists in individual reflection, and the social fabric of religion is only valued as it provides stimulus and support for this activity. This does lead to serious distortions when describing a religion which most believers experienced as a sequence of ceremonies and formal observances, activities which many clerics also found more congenial than preaching and hearing confessions. What Delumeau ultimately gives us is an important and helpful account of clerical discourse, rather than a rounded picture of what clerics did. The rather exaggerated and one-sided quality of the literature does indeed need explaining, and he offers a sensitive analysis of the intellectual filiations involved. As a general interpretation of the catholic reform, however, this is not really adequate, above all because it needs integrating into a larger and more balanced picture.

A structure of this kind is just what Robert Muchembled has set out to provide, in a striking analysis which appears basically Marxist in inspiration.[2] For him the centralizing absolutist state is the vital agency of change through which the culture of the ruling groups was imposed as the only valid model. The development of urban society, the transition from feudalism to capitalism, and the pressures of royal fiscality all helped to break down an older world of essentially local political and cultural units. The insistence on a single culture was extended to the rural world through the church and through the popular literature of the *bibliothèque bleue*; this latter was in fact produced in the towns, and its general effect was to

[2] R. Muchembled, *Culture populaire et culture des élites* (Paris, 1978). For his work on witchcraft, see the references in Chapter 1 above. There is a critique of the 'acculturation' thesis by J. Wirth, with a restatement by Muchembled, in *Religion and Society in Early Modern Europe, 1500–1800*, ed. K. von Greyerz (London, 1984), 56–78.

advocate subservience to official values. The crucial group in peasant society was that of the *laboureurs*, increasingly marked off from the mass of villagers by education and attitudes as well as by wealth. As they became dominant in local political life, so they came to ape their superiors, and showed an exaggerated hostility towards older forms of culture in order to demonstrate their own emancipation from them. They and the *curés* therefore became the representatives of an alien culture in the villages, with literacy as a crucial dividing line. The community was split, losing its cohesiveness, and the same happened to the popular culture which had been a crucial expression of the older localized society. Hierarchical authority structures were imposed with new force, to secure the domination of the privileged minority, and to destroy any vestiges of opposition. The process was at its most intense between 1650 and 1750, after which popular culture had been reduced to a dislocated set of folklore practices, so that the élites were prepared to extend a certain scornful tolerance to these survivals, no longer seen as dangerous.

This interpretation has great apparent attractions; with respect to the motivation behind the reforming movement it probably contains a substantial element of truth. Yet Muchembled's account of the relationship between the two cultures tends towards massive oversimplification. It has a certain resemblance to Cartesian physics, as a daring reconstruction which runs way ahead of the evidence, supported by a chain of risky hypotheses. Bold assumptions are made without any factual support, such as the assertion that peasant marriages were normally arranged by parents, that there was no tenderness towards spouses or children, or that popular religion was no more than thinly disguised paganism. The claim that rural popular culture had been stripped of most of its coherence and power by the middle of the eighteenth century is hard to understand or accept, since all the main features described for earlier centuries would still seem to be present in force at that date. The whole attitude to this culture (which is essentially similar to that of Delumeau) is particularly worrying. Traditional rituals are depicted as an inadequate substitute for effective technology, a desperate attempt by peasants, faced with a series of traumatic disasters which induced deep insecurity, to equip themselves with a reassuring mental structure. The attribution of such intentionality to social groups is a highly dubious procedure, which becomes plainly invalid when none of those concerned could possibly have interpreted their own actions in such a way. Psychological explanations of this kind are not only impossible to verify; they lead for example to the reduction of the richness and subtlety of popular festivals to a crude pattern of tension and release in an unmanageable world. The key role ascribed to the *laboureurs* is hardly adequately sustained by the semi-literary case of Rétif de la Bretonne's *La Vie de mon père*; in any case this group only dominated the rural communites in perhaps 25 to 30 per cent of France. Even the link with absolutism looks less convincing when we consider Peter Burke's demonstration of parallel trends in other parts of Europe with rather different political systems.[3] Some of the weaknesses of Muchembled's interpretation of witchcraft, very much a central part of his thesis, and the only one for which he produces much detailed evidence, have been suggested above.[4]

There is a dangerous tendency here not just to overstate actual polarities and oppositions, but to argue by analogy in order to depict what appears to be a coherent pattern. Such intellectual operations can be suggestive and valuable, but they need to be controlled with the greatest care, and the hypotheses they generate must be tested. This is particularly true of vital causal links, such as Muchembled's view of the roles of the *laboureurs* and the *bibliothèque bleue*. In the latter case it does stretch credibility to see this essentially commercial operation, in which publishers printed cheap (and often simplified) editions of existing works in line with their vision of popular taste, as a major instrument of acculturation. Compared with Robert Darnton's subtle investigation of folktales, bringing out their mixture of subversiveness and acceptance, this seems unacceptably crude and restrictive.[5] It also exemplifies the dangers of writing as if all the products of

[3] P. Burke, *Popular Culture in Early Modern Europe* (Paris, 1978), *passim* and esp. 207–43.

[4] Chapter 1, pp. 53–7.

[5] R. Darnton, 'Peasants tell tales: the meaning of Mother Goose', in *The Great Cat Massacre and other Episodes in French Cultural History* (Harmondsworth, 1985), 17–78.

urban literate society were somehow united by a common class interest, a frequent fallacy in Marxist interpretations.[6] While such identifications are often correct, they cannot merely be assumed without proof, any more than can the existence of the class interests themselves. The identification of urban, literate, and authoritarian values with the ruling élites is acceptable in most contexts as a form of historical shorthand; when it is the central plank of a theory it needs to be analysed in detail. The social groups concerned were far from having a common view on everything, any more than all French rural communites resembled those in the north-east of the country. The whole conception of the élite as a single identifiable group is a highly dangerous one, which can easily become as much of a historical *passe-partout* as the rise of the middle class.

Unlike Muchembled, then, I do not believe that French popular culture suffered mortal wounds in the seventeenth and eighteenth centuries, at least in the rural world. It had certainly come under prolonged and unprecedented attack, but it had proved astonishingly resilient and durable under this pressure. The growing tolerance of the later eighteenth century is more plausibly seen as deriving from the internal development of élite thought than as a sign of victory, for in the short run this was a battle the crown and the church had generally lost. In the very long run, stretching down to the present century, peasant culture was doomed to slow erosion and fragmentation, but numerous historians have testified to its enduring strength right up to the present day.[7] These processes of decline would seem to be virtually identical with the dissolution of peasant society itself; the village community has been losing its culture and its social identity simultaneously.[8] Where matters certainly went differently was in the towns, where we should be wary of seeing the various

[6] For an acute discussion of this and other problems arising from Marx's notion of class, see J. Elster, *Making Sense of Marx* (Cambridge, 1985), 318–97.

[7] For two recent examples see E. Weber, *Peasants into Frenchmen: the Modernization of Rural France 1870–1914* (London, 1977), and J. Devlin, *The Superstitious Mind* (New Haven, Conn., 1987).

[8] The same point is made by M. Vovelle, 'Histoire des mentalités—histoire des résistances ou les prisons de la longue durée', *History of European Ideas* 2 (1982), 10.

traditions of irreverence and festival as static. They were certainly capable of developing into expressions of social dissent, and the urban oligarchs and clergy who suppressed many of them were well aware of this danger. Exceptional though the Carnival of Romans may have been, the possibilities of disorder were very obvious, and are quite sufficient to explain the attitude of the authorities.[9] As Peter Burke has pointed out, the 'safety-valve' theory of licensed inversion which actually confirms the established order, plausible though it generally is, cannot be universally applied.[10] It does not necessarily conform to the perceptions of seventeenth-century *échevins* and *consuls*, all too conscious of the depth of popular resentment at some aspects of their self-interested rule. This very real cultural repression in the towns must not be confused with the more general issue of how peasant culture was treated, using examples from the first as if they were generally valid. Inevitably a great deal of the evidence has to do with the urban campaign, which has left much fuller documentary traces, and which loomed much larger in the minds of the élites since it took place on their very doorsteps and at their direct instigation. The same warning against regarding popular culture as static should be extended into the rural world, however, building on an acute observation made by Gunther Löttes with reference to sixteenth-century Germany. Having pointed out how the village community had a democratic surface structure, but an oligarchic deep structure based on the growing dominance of the rich peasants, so that while the village operated very effectively against outside challengers it offered no mechanism for resolving internal confrontations, Löttes goes on to say:

> Protest appears to have transformed time-honoured village customs into occasions for social blackmail, in order to enforce traditions of mutuality and generosity that the have-nots felt were disappearing, or into occasions for popular justice of the charivari kind, in order to accuse and punish infringements of what they held to be the norms governing the life of the village.[11]

[9] For the Carnival of Romans see the references given on p. 111 above.

[10] Burke, *Popular Culture*, 201–4.

[11] G. Löttes, 'Popular culture and the early modern state in sixteenth-century Germany', in *Understanding Popular Culture*, ed. S. L. Kaplan (Berlin, 1984), 147–88. The citation is from p. 158.

To the extent that the *laboureurs* did co-operate in repressing the activities of their poorer neighbours, it makes much better sense to see their motivation in terms of a natural reaction to such developing criticism of their own behaviour, rather than as the product of a mysterious and ill-attested process of acculturation.

Both French historians tend towards different versions of reductionism, identifying external reasons for changes in Christian practice, themselves described in broad and suspiciously homogeneous terms. John Bossy has adopted what is in many ways the opposite approach to the same phenomena, taking an original and idiosyncratic viewpoint.[12] Working outwards from the ritual and social acts which lay at the heart of everyday religious life, he seeks to identify their meanings, and the ways in which these codes were changed. There is a danger here of producing an ideal type of pre-reformation Christianity, which Bossy does not wholly avoid, but there are many valuable insights to be gained by this technique. His work emphasizes the variety of response found within western Christendom on almost every important issue, and provides a welcome reminder that there was an immensely complex intellectual superstructure of religion, which many highly talented individuals devoted their lives to understanding, expounding, and developing. What Bossy offers is primarily an 'internal' account of change, describing what happened in terms which at least some of the participants might have understood, yet which are not the formal categories of theologians. His particular sensitivity to the meanings contained in rituals often brings out the hidden implications of what might be thought fairly modest changes. It is impossible to summarize the results adequately, not least because Bossy makes virtually no attempt to formulate general explanations. What does come through, however, is that between the fifteenth and the seventeenth centuries the fragmentation of Christendom went far beyond the obvious divisions caused by the reformation. In both catholic and protestant Europe a whole range of time-honoured practices, which had given

[12] J. Bossy, *Christianity in the West, 1400–1700* (Oxford, 1985); 'The counter-reformation and the people of catholic Europe', *Past and Present*, 47 (1970); 'The Mass as a social institution, 1200–1700', *Past and Present*, 100 (1983); and other articles.

western Christianity its characteristic colour, were being destroyed or radically transformed. The effect was to diminish the sense of the church as a community, to restrict the notion of charity, and to limit the more social aspects of the sacraments. The old divide between priests and laity had been replaced by a much deeper one between the godly, both clerical and lay, and the majority of the population who clung to older ways. Many key concepts were redefined, along the lines suggested by Foucault, with the result that the vision of an essentially harmonious universe was undermined by a new stress on difference and distinction, while analogical reasoning was dethroned by a new analytical style.[13]

The ultimate result of Bossy's interpretation is to raise a new series of questions about causes and effects, while giving only oblique hints about how they could be resolved. His discussions do suggest that the causal links involved are extremely complex, to a point where their full reconstruction is probably an impossible task. Symbols, beliefs and rituals emerge in many cases as deeply ambivalent, capable of developing in very different directions according to circumstances. Christianity is thus allowed a powerful dynamic of its own, but one which often appears haphazard or capricious, capable of being deflected by individual theologians as well as by social pressures. While all this is very salutary, there is a decidely gnomic quality in Bossy's treatment; he deals with many crucial issues by ignoring or dismissing them, aided by his aversion to certain types of generalization. At the risk of impertinence and inaccuracy, I would draw certain conclusions from which he prudently refrains. Firstly, although the communally-based religion practised by most Europeans possessed enormous resilience in itself, its relationship to official doctrines was always rather precarious. The disturbing factor here was far less the behaviour of the people than the tendency for clerics to exploit particular features of belief for economic or status advantage. The development of masses for the dead, of indulgences, images, and healing shrines, represented supply as much as demand. Whereas local religion was relatively economical, discarding outworn

[13] M. Foucault, *Les Mots et les choses* (Paris, 1966).

saints as new favourites emerged, and allowing practices to wax and wane in response to particular needs, the official church, with its propensity to create lasting structures, tended to accumulate and institutionalize. Devotional practices which seemed natural and inoffensive at village level became contentious, even sometimes obviously abusive, when manipulated by the clergy or developed on a large scale. The whole development of 'la comptabilité de l'au-delà', so well described by Jacques Chiffoleau, was a particularly important example of these dangers.[14] Indeed Bossy has described the concern of late medieval religion with helping the dead to escape from purgatory as 'this enormous construction, under whose weight the medieval church collapsed'.[15] One might almost say that popular religion was too successful in penetrating the pre-reformation church, to the point where more austere theologians were driven to react against what they considered a perversion of the Christian message.

At this point the more general patterns of causation described in the last essay can be related to these largely internal evolutions of the church. Their interaction gave far greater force to the reforming campaigns which now emerged, expressing themselves in three main forms. The first was a drive for doctrinal purity, often linked with a special emphasis on strict Augustinian views, and a reassertion of the unbridgeable gap between God and man. The second was a movement to discipline and educate the clergy themselves, notably the secular priesthood, in order that they should propagate this purified faith, rather than accommodating themselves to popular superstitions. Thirdly came the general repression of scandalous and irreverent practices and behaviour, to create a new kind of godliness, or in the minds of some to resurrect the standards of the primitive church. In protestant Europe these concerns were completely dominant for a time; where catholicism remained the official faith the picture was inevitably more complicated, but there can be no mistaking the very powerful operation of similar trends within the catholic reform movement in France. They did not escape

[14] J. Chiffoleau, *La Comptabilité de l'au-delà* (Rome, 1980).

[15] Bossy, 'The Mass as a social institution', 42.

the observant eye of Joseph Addison, who pointed up the contrast with a very different style he found in southern Italy.

I must confess, tho' I had liv'd above a year in a Roman Catholick Country, I was surpris'd to see so many Ceremonies and Superstitions in Naples, as are not so much as thought of in France. But as it is certain there has been a kind of Secret Reformation made, tho' not publickly own'd, in the Roman Catholick Church, since the spreading of the Protestant Religion, so we find the several Nations are recover'd out of their Ignorance, in proportion as they converse more or less with those of the Reform'd Churches. For this reason the French are much more enlighten'd than the Spaniards or Italians, on occasion of their frequent Controversies with the Huguenots; and we find many of the Roman Catholick Gentlemen of our own Country, who will not stick to laugh at the Superstitions they sometimes meet with in other Nations.[16]

The very appearance of the reformation and the collapse of Christian unity intensified the process, for as rulers and ruling classes acquired bitter experience of the disruptive potential of religious zealotry, so they came to favour strict orthodoxy and the repression of all kinds of deviance. They had an additional motive in the possibilities of national churches as reinforcing elements in the structure of monarchical power, which could only be fully realized if they were free of major internal divisions, while the clergy themselves could use the relationship to induce repressive action by the state. The resulting persecutions were sometimes effective in the short run; in the longer term, however, they stimulated reactions which did much to advance the cause of toleration and pluralism. Doctrinal unity within the churches proved unobtainable, as efforts at definition created new divisions instead of ending arguments. A reformed and well-trained clergy proved less docile than expected, particularly towards their ecclesiastical superiors, and tended to sympathize with their parishioners on many matters; the failure of much of the assault on popular culture has already been suggested. When writing on such themes I have tried to bring out the numerous inconsistencies which help to explain these relative defeats, despite the huge effort mounted by the catholic reformers. Almost all the

[16] J. Lough, 'Two more British travellers in the France of Louis XIV', *The Seventeenth Century*, 1 (1986), 162–3.

agencies through which the catholic clergy, at least, tried to operate were ambiguous, so that 'acculturation' itself is a two-edged concept. New devotions such as those to the Holy Sacrament and the Rosary were interpreted differently, according to the prior attitudes of the participants. As Roger Chartier perceptively observes, 'the search for a specific and exclusively popular culture, often a disappointing quest, must be replaced by the search for the differentiated ways in which common material was used'.[17]

The dubious nature of any single division between popular and élite culture is emphasized if we replace the writings of moral theologians in the wider context of baroque catholicism as a whole. The writers of mystical and devotional books, particularly in the first half of the seventeenth century, used a highly coloured and emotive imagery, invoking supernatural signs without any obvious inhibitions. Educated clerics and laymen alike often conceived the world in terms just as 'animistic' as those commonly ascribed to any benighted peasant; this was not just a question of miracles and divine intervention in the world, but of widespread beliefs in astrology, alchemy, and magic. The normal reaction of the authorities to natural disasters was to organize processions and displays of relics, designed to appease the divine anger which must have been the cause. Similar practices proved effective in calming popular revolts, as at Agen in 1635.[18] Holy men were treated with veneration, and the crowds which gathered around them were not just composed of the poor; a mass of unreadable seventeenth-century hagiography bears witness to the readiness of educated clerics to publicize their miraculous powers. Mère Jeanne des Anges, flaunting her dubious stigmata, was made welcome at the royal court in the course of a triumphant progress around France.[19] The world and its history were still largely perceived through the Christian myths, so that in the 1670s the Jesuit Kircher found many admirers for his attempts to provide biblical sources for

[17] R. Chartier, 'Culture as appropriation: popular cultural uses in early modern France', in *Understanding Popular Culture,* ed. Kaplan, 229–53. Citation from p. 235.

[18] For this episode see Bercé, *Histoire des croquants,* i. 355.

[19] M. de Certeau, *La possession de Loudun* (Paris, 1970), 314–18.

Egyptian hieroglyphs and Chinese ideograms.[20] The natural tendency of modern scholars has been to concentrate on those thinkers whose ideas culminated in the 'scientific revolution' and the Enlightenment, unintentionally giving a very skewed picture of the intellectual world of the seventeenth century. Religion remained for virtually everyone the experience of powerful affective practices and symbols, which included rituals of protection and sanctification. It involved defining certain times, places, and objects as sacred, thereby providing a common framework for social life and relationships. There was also a persistent tendency to seek direct authentication through signs from heaven; the Jansenists themselves were quick to use the miracle of the *sainte épine* to prove the justice of their cause, although in the 1730s wiser heads were not so sure about the *convulsionnaires de Saint-Médard*.[21]

By this date there had been a real sea-change in élite attitudes, but one which came at a time when the catholic reform was already losing momentum. To explain it one may look at the very telling suggestions made by Jacques Revel, who identifies three criteria around which the oppositions between popular and official cultures were set up.[22] Firstly there was the position of theologians such as Thiers, using scholarship to distinguish between true and false knowledge with reference to past authority. Secondly there was the rational approach, contrasting practices and attitudes that were coherent and understandable with those that were not, such as were the products of ignorance or the nefarious passions. The third line of discrimination was that of convention, a more or less explicit social code which determined what was culturally acceptable. All three functioned simultaneously in the period 1650–1750, but gradually the third ousted the first as the predominant one; popular beliefs became evidence

[20] For Kircher see J. Solé, *Les Mythes chrétiens de la Renaissance aux Lumières* (Paris, 1979), 117. Although this book is not very reliable on detail, its general thesis is interesting and often persuasive.

[21] For the *sainte épine* see A. Adam, *Du mysticisme à la révolte* (Paris, 1968), 232; for the *convulsionnaires* there is a full account in B. R. Kreiser, *Miracles, Convulsions, and Ecclesiastical Politics in Early Eighteenth-Century Paris* (Princeton, NJ, 1978).

[22] J. Revel, 'Forms of expertise: intellectuals and "Popular" culture in France (1650–1800)', in *Understanding Popular Culture*, ed. Kaplan, 255–73.

of social and cultural archaism. It is characteristic of human society that this loose set of unthinking attitudes should be so much more effective than the rationality and logical argument everyone claims to admire, but virtually no one actually follows. The establishment of such conventions, which rapidly became the model which all those who aspired to high status tried to emulate, led to the attitude of disdain correctly identified by Muchembled, but misattributed by him to a sense of triumph which would have been quite inappropriate. Popular culture survived very effectively as a nexus of attitudes and practices, although some of its more radical tendencies had been curbed; among the élites conventions had proved more powerful than intellectual positions, so that the elaborate theocentric constructs of the seventeenth century rapidly became obsolete and largely irrelevant.

Chronology alone would make one suspect that witchcraft persecution must be connected with the changes in religion, and historians have generally taken this view. Muchembled sees the persecution as a central element in the attack on popular belief, while Christina Larner described it as part of the repression of deviants characteristic of an age of faith.[23] These approaches have genuine value, for they help to explain why certain groups or individuals proved highly susceptible to demonological theories. There could hardly be a clearer embodiment of such links than Jean Bodin, theorist of the Renaissance monarchy and most formidable of all demonologists. On another level, there was obviously a connection between changing conceptions of the devil and the theories of witchcraft; with his usual acute sense for such changes Bossy explains how the devil had ceased to be 'the mirror-image of Christ, the personified principle of the hatred of one's neighbour', to become instead 'the mirror-image of the Father, the focus of idolatry, and hence of uncleanness and rebellion'.[24] These distinctions are the more convincing because they correspond so closely to those between official and popular conceptions of witchcraft itself.

A much greater concern with the devil and sin must have a

[23] C. Larner, *Witchcraft and Religion: the Politics of Popular Belief* (Oxford, 1984), 113–39.

[24] Bossy, *Christianity in the West*, 138.

place in any explanation of the persecution, but as the chapters on witchcraft in this book make plain, it is only one of a number of interlocking causes, and was more important at the level of demonology than that of village belief and action. Very much the same can be said of the interpretation of the persecution in terms of the identification and punishment of deviant elements within society. Persuasive though Christina Larner's arguments were, she ran into insuperable difficulties when she tried to make this into a primary cause of the whole process. In the course of attacking those historians who attached great importance to social strain she claimed that this theory broke down when applied to Continental examples because of 'the controlled and unequivocal way in which the ruling élite controlled and manipulated the demand for and supply of witchcraft suspects.'[25] This may well have force in the Scottish case she knew best, or for such figures as the bishops of Bamberg and Würzburg, but it is singularly remote from the picture which emerges when one looks in detail at the evidence from France, Lorraine, or Spain. The attitudes of vital élite groups such as the French *parlementaires* and the Spanish inquisitors are very well described as highly equivocal, even at the peak period of witch-hunting.[26] Social and economic strains are very obvious in many cases, as is the basically local and popular instigation of legal action.

Social anthropologists have employed the study of witchcraft with great success as a means to the deeper understanding of social and cultural patterns, a procedure which can also be applied in a very rewarding way to early modern Europe. While there is still a long way to go with such an analysis, even a preliminary attempt emphasizes the complexity of responses and attitudes among all the groups concerned, and the multiple social and cultural interactions at work. It is not inherently impossible to understand all the processes involved, but the kind of multi-factorial analysis required is extremely complex, and it excludes simple answers. An outline model for explaining the European persecution has

[25] Larner, *Witchcraft and Religion*, 52.

[26] For France see the works by Mandrou and Soman cited in Chapter 1 above; for Spain see G. Henningsen, *The Witches' Advocate: Basque Witchcraft and the Spanish Inquisition (1609–1614)* (Reno, 1980).

already been offered in Chapter 1; something more should however be said about the 'age of faith' concept advanced by Christina Larner when she suggested that the emerging nation states demanded overt allegiance to their preferred versions of Christianity because 'like all new regimes they demanded both ideological conformity and moral cleaning'.[27] This ingenious formulation is more provocative than convincing; Renaissance states were mostly old monarchies, not self-consciously new political organizations. There does not seem to be any need for such a theory to explain their desire for religious uniformity and obedience to constituted authority, although it may apply quite well to the brief attempts at 'godly rule' in England and Scotland, or to the programme advanced by some members of the Catholic League in France. Nor did the monarchs of the period have the power to launch serious campaigns for a new type of religious commitment, unless they enjoyed the voluntary support of important social groups. The notion works much better if it is inverted, and we see small knots of enthusiastic reformers trying to involve the apparatus of the state in their cause, promising exceptionally zealous service to both God and king as the potential consequence. Some such individuals were also active in denouncing witches; for many Parisian Leaguers these servants of the devil included King Henri III himself, whom they saw as the ultimate traitor and source of pollution in the kingdom.[28]

Religion and the state could never be separated because they were jointly involved in the exercise of power, for which both claimed supernatural justification. Although both church and state succeeded in extending their authority significantly in early modern France, they were far from having imposed it throughout society. This was a world full of struggles over power, whose political and normative systems provided no recognized ways of expressing or resolving such conflicts; the assumption was rather that communal unity should prevail at all levels. For both the people and the nobility this meant that revolt, itself so often proclaimed as being in

[27] Larner, *Witchcraft and Religion*, 124.

[28] This was the theme of several pamphlets, most specifically *Les Sorceleries de Henry de Valois et les oblations qu'il faisoit au diable dans le bois de Vincennes* (Paris, 1589).

defence of communal rights and feelings, was often the only way in which they could find political expression at all. The response to such direct challenges, and to the widespread factionalism of local society, was to place still greater stress on authority, as embodied by natural rulers who stretched from God through the king and the bishops down to the father as head of the family. The shortcomings of this approach became more evident over time, for it placed an almost unbearable burden on the human embodiments of authority, who were expected to satisfy all their constituents, and to maintain a quite unrealistic level of harmony. The church too lacked satisfactory mechanisms for resolving its internal disputes over doctrine, or for defusing the innumerable greater and smaller power struggles which persisted within its sprawling institutional structure. Human relationships so generally contain latent tensions over power, with its implications for access to certain freedoms and privileges, that culture must provide some means of regulating them or legitimizing the existing patterns. In this period the model of the authoritarian father-figure was developed to extreme lengths in response to those needs, only to prove rather unsatisfactory and inflexible. An analysis of society and ideology in terms of power is a necessary part of any attempt to study culture in the wider sense, for it is a crucial part of the set of human needs and problems which underlie cultural structures, and which may disrupt them and threaten their viability. The theme has appeared in all these essays to some extent, being particularly central to those on popular revolt and church and state. I have also argued that witchcraft beliefs and persecutions were inextricably involved with questions of power, both communal and personal. The objection to popular culture, in the form of superstitious practices, was often not that it was ineffective, but that the results could only be obtained through diabolical help. Behind this lay an implicit contest over the control of access to divine assistance, with the church asserting monopoly rights in this area, and stigmatizing competitors as users of illegitimate 'black' magic.

Another repeated theme has been the need to avoid sharp polarities in describing the complex relationship between social and cultural structures. For many centuries all

Europeans, with the possible exception of a handful of learned clerics, had shared a common culture, although it was not used in identical ways by all the participants. There did come a stage at which a distinctive élite culture partly defined itself by excluding the lower orders; this gulf was becoming serious by the second half of the seventeenth century, and its establishment was a major cultural shift which is still part of our lives today. Its long-term importance, however, should not deceive us into supposing that a series of binary oppositions were established between learned and popular, urban and rural, literate and oral, individual and communal, scientific and animistic. Whilst such polarizing concepts are very useful for identifying trends, they must be treated as ideal types, and not confused with the real behaviour of individuals or social groups. As has already been stressed, there were ambiguities everywhere, so that we should rather think of people occupying points along a continuum, and not even being internally consistent in their own attitudes. Having made these vital disclaimers, I would like to emphasize some of the ways in which these distinctions can help to describe and explain what was happening. French society was differentiated by space as well as by education and culture; the segregation of the élites in the towns encouraged them to develop greater self-awareness and cohesion, but facilitated the coexistence of different cultural patterns. Town–country hostility was a major factor in popular revolts, because of the association of the towns with the hated taxation and legal systems. Both of these required people to displace themselves, with personal contact necessary at every stage. Tax assessment and collection both produced potential confrontations, as did the pursuit of a legal action. The modern world has taken much of the hostility out of similar processes, which have been depersonalized by the use of formalized written or printed documents, but this is only possible in a society where literacy is taken for granted.

The frontiers of literacy are particularly hard to locate, as much recent work has emphasized. The only measure available to the historian, the ability to sign one's name, need not indicate a functional command of writing for other purposes, yet the practice of elementary schooling meant that many children acquired basic reading knowledge without learning to write.[29] Technical literacy of either kind would have made little difference unless used, and little evidence survives to suggest that even the *laboureurs* did much writing; if they did it was to keep accounts. Peasants or artisans who wanted to write a letter or a legal document would probably have gone to a professional scribe or a notary and dictated it. We therefore have to allow for a situation in which the numbers who could read may have been considerably larger than previously thought, but those of the effectively literate rather smaller. Writing was of course another instrument of power, capable of being exploited to the disadvantage of the poor, the means by which their debts and obligations were recorded and enforced. The lack of enthusiasm for educating the masses reflected an awareness of this fact, and a desire to keep them in subjugation.[30] There is some doubt about how far books had penetrated the rural world by the end of the seventeenth century; as late as 1686 royal legislation on the book trade prohibited the rural pedlars from selling books of any kind. Although this measure was clearly unenforceable, one acute recent study of the *bibliothèque bleue* suggests that its products were mainly sold in the towns until the eighteenth century, and that before this date there are very few mentions of pedlars trading books, or of their being read at the *veillées*.[31] We must conclude that rural communities were essentially oral societies, in which literacy played only a very marginal role, and this is in line with everything else we know about them. Some crucial features of oral culture have been pointed out by Walter Ong, such as the way thought can only be preserved in such forms as proverbs, and that the need to repeat past knowledge frequently in order to preserve it at all 'establishes a highly traditionalist or conservative set of mind'.[32] He also cites the fascinating work

[29] There is an excellent basic discussion of the question in R. Chartier, M. M. Compère, and D. Julia, *L'Éducation en France du 16ᵉ au 18ᵉ siècle* (Paris, 1976), 87–109. For greater detail see F. Furet and J. Ozouf, *Lire et écrire: l'alphabétisation des français de Calvin à Jules Ferry* (2 vols.; Paris, 1977). There are some acute and sceptical remarks by P. M. Jones in *Politics and rural society: the southern Massif Central c.1750–1880* (Cambridge, 1985), 125–8.

[30] Chartier, Compère, and Julia, *L'Éducation*, 37–41.

[31] J.-L. Marais, 'Littérature et culture "populaires" aux 17ᵉ et 18ᵉ siècles: réponses et questions', *Annales de Bretagne et des pays de l'ouest*, 87 (1980), 65–105. The royal policy is discussed on 69–70.

[32] W. J. Ong, *Orality and Literacy* (London, 1982), 41.

of the Russian psychologist A. R. Luria, interviewing illiterate and semi-literate peasants in the early 1930s. Luria found that his illiterate subjects, although by no means stupid or illogical in their own terms, could not operate with formal deductive procedures, which seemed to them pointless and irrelevant to the real world. Moreover, they found great difficulty in any kind of self-analysis, referring instead to the opinion others held of them; in such societies 'judgement bears in on the individual from outside, not from within'.[33] This finding correlates very well with the behaviour of suspected witches, who so often took themselves at other people's valuation. It also helps to explain the popular attitudes to confession, with their obstinate resistance to the internalized examination of conscience. As Ong remarks; 'the influence of writing and print on Christian asceticism cries for study', and must be related to 'a world dominated by writing with its drive toward carefully itemized introspection and elaborately worked out analyses of inner states of soul and their inwardly structured sequential relationships'.[34]

Such considerations emphasize the fact that Muchembled has directed attention to a very important issue by stressing the dynamic role of literacy in breaking down traditional culture. His analysis needs severe modification in several respects, however, since he oversimplifies and antedates the processes which are necessary for this to happen. The illiterate cannot be effectively 'acculturated' by their literate neighbours, or by participation in the legal process, since it is only by the actual exercise of writing that individuals learn to structure their thoughts differently. The practice of reading from texts modifies them into a form of oral communication; in this context we may note that the church turned most of its written knowledge into oral form for popular consumption, whether this was in sermons or catechisms. The catholic attitude to the bible, a forbidden text preferably to be kept in Latin, and by implication a source of supernatural power, is also highly significant. To the peasant this must have seemed to assimilate it to the *grimoires,* the notebooks of secret charms and remedies kept by magical healers, even to those scraps of paper with magical prayers on them through which relief might

[33] Ibid. 49–55. [34] Ibid. 152–3.

be sought. Closer attention to the contents of the *bibliothèque bleue* also casts doubt on its role, for Marais argues that it bears witness to the traditional culture formed out of medieval traditions at the beginning of the print era, before the internal transformations of the post-reformation era. Its religious sensibility, its taste for the supernatural and for laughter, its practical wisdom, are all characteristic of this culture. Although the new individualistic religion, with the believer encouraged to ensure his salvation and prepare himself for death, did start to appear from the presses of Troyes in the seventeenth century, in works with such titles as *Pratiques d'un chrétien* and *Pensées chrétiennes,* the older sensibility based on the Passion and the saints as intercessors remained dominant. Since these books catered for a varied audience it is likely that the more 'modern' works of piety were directed to those urban groups of lesser bourgeois who were far the most likely converts to the personalized faith of the reformers.[35]

The distinction between true religion and superstition, particularly in the discrimination against practices of supposedly pagan origin, was one that only the higher levels of the literate élite could be expected to draw. Peasants had no such conception of historical time, but fused all devotional activities into a socially sanctioned whole, a syncretized cultural system shaped to meet their basic religious needs. In this sense their religion can be subjected to the same kind of functional analysis that social anthropologists regularly employ. It gave meaning to the world, by providing a symbolic system which was the basis for ritual and social acts, which explained otherwise mysterious phenomena, and which alleviated the fear of death. Rituals were performed with some enthusiasm because peasants believed they offered protection to themselves, their animals and crops, and their ancestors, and because they reaffirmed the norms governing their relations with their neighbours. Their success in these functions depended largely on their ability to create powerful moods and motivations in the performers.[36] Such an analysis

[35] Marais, 'Littérature et culture "populaires" ', *passim.*

[36] The anthropological literature on this subject is very large. Particularly helpful general discussions are those by E. E. Evans-Pritchard, *Theories of Primitive Religion* (Oxford, 1965) and the essays by C. Geertz and M. E. Spiro in *Anthropological Approaches to the Study of Religion* ed. M. Banton (London, 1966).

does not imply that rituals should be seen as desperate attempts to compensate for inadequate technology; in many of the areas to which they were applied they supplemented very considerable practical skills.[37] In others, most notably the crucial variable of the weather, they made admirable sense within the whole belief system, and were shared by the élites, for the preachers still treated natural disasters as divine judgements, while clerical and lay authorities organized processions and prayers to placate the master of the universe. The catholic reform movement displayed considerable instinctive understanding of these aspects of faith; missions are one example, as are such emotive new devotions as the *quarante heures*. The importance given to ceremony and ritual can be seen in visitation records, which reveal great concern with the state of churches, the provision of suitable vestments and silverware, and the seemliness of divine service. Where the reform movement did bring a change of tone was in expecting religious activity to be solemn and reverential, a sharp contrast with the traditional culture which had encouraged a much wider range of physical actions and affective states, and saw no harm in laughter and drinking in the course of a procession or the meetings of a confraternity. Hostility to these livelier expressions of popular culture led to a very serious split between the village community and the church in one specific respect. Much of the group activity and ritual behaviour within the village was organized by the *jeunesse*, essentially the young unmarried males. The boisterous irreverence they naturally patronized was so much at odds with the gravity now required by the clergy that they were excluded from any recognized role in local culture, their formal organization often being disbanded. This was a notably unsuccessful measure, for the *jeunesse* continued to function, while becoming increasingly a centre of opposition to clerical puritanism; conflicts over dancing and many customary rituals were the common expression of this long-term struggle, which may also help to explain masculine disaffection with the church.[38]

[37] This is one of a number of important points made by Stuart Clark, 'French historians and early modern popular culture', *Past and Present*, 100 (1983), 62–99.

[38] For this valuable insight, see J.-P. Gutton, 'Confraternities, *curés* and Communities in rural areas of the diocese of Lyons under the *ancien régime*',

Culture is not just an amalgam of rituals, beliefs, and forms of group behaviour, but in its wider sense includes the corporate knowledge of a society. Here too we should not underestimate the capacities of peasants and artisans, who were often very skilful in applying practical knowledge to everyday problems, and whose methods of learning by apprenticeship were highly effective. Even their conservatism can be regarded as quite rational, a recognition of the very limited margins within which they had to operate, which magnified the risks inherent in innovation. The failure of French agriculture to deliver better yields should not be simply put down to incompetence; traditional methods were capable of good results, and it seems likely that the major constraints were socio-economic ones. This is certainly the implication of Hugues Neveux's work on agrarian productivity in the Cambrésis, where the failure to equal the high performance of the fourteenth century in the early modern period plainly has nothing to do with technical backwardness resulting from ignorance.[39] Nor should the use of folk medicine be seen as an indicator of primitivism, for it was generally far preferable to the bungling and painful 'learned' medicine inflicted on the rich, frequently with lethal consequences. The product of a healthy empiricism, it had found many genuine remedies, besides obtaining numerous cures through the placebo effect. Peasants and artisans normally had recourse to magic either as a supplementary option, or if they could not identify the disease, or when all else had been tried and had failed. Too much historical writing implies a vision of the common people as ignorant, brutal, and frequently panic stricken, versions of the sub-human creatures glimpsed by La Bruyère. They did indeed inhabit a hard world, living in conditions that the privileged inhabitants of the modern West would find intolerable and which no doubt contributed to a high level of

in *Religion and Society*, ed. von Greyerz, esp. 207–10. Gutton has also discussed aspects of this conflict in *Villages du Lyonnais sous la monarchie (16ᵉ–18ᵉ siècles)* (Lyon, 1978), 80–86, and in 'Reinages, abbayes de jeunesse et confréries dans les villages de l'ancienne France', *Cahiers d'histoire*, 20 (1975), 443–53.

[39] H. Neveux, *Vie et déclin d'une structure économique: les grains du Cambrésis (fin du 14ᵉ–début du 17ᵉ siècle)* (Paris, 1980).

violence, also related to the style of insults which seems to characterize oral society. Yet they managed their environment with considerable resource and determination, while it is very implausible to portray their personal relationships as almost totally lacking in affect. As Jean-Louis Flandrin has rightly emphasized, there was a whole language of gesture in courtship whose meanings are easily misunderstood or ignored.[40] Proverbs and folktales, the oral 'literature' which brings us closest to popular ideas, reveal a quite sophisticated and varied understanding of human relations.[41] In his collection of popular superstitions Thiers reported an impressive range of practices to do with the choice of a spouse, a demonstration of the seriousness with which young people took the matter, and of the importance of the concept of love as part of their thinking.[42] The history of child-rearing practices is peculiarly difficult to reconstruct, but some recent discussions have gone to ridiculous lengths in denying parental affection for children; on this point I have found a surprising amount of evidence in the depositions at witchcraft trials, all pointing to genuine care and concern. The lack of privacy and the strength of communal views may also have protected children to some degree from the extremes of abuse they can suffer in a more individualistic society.

The traditional culture and society deserve our respect and understanding, as much more successful human environments than many observers have been willing to allow. This does not mean bathing them in some kind of sentimental glow, as the embodiment of warm communal values to be contrasted with the lonely and selfish individualism of post-industrial man.

[40] J.-L. Flandrin, *Les Amours paysannes (16ᵉ–19ᵉ siècles)* (Paris, 1975); *Le Sexe et l'Occident* (Paris, 1981), *passim* and 286–91.

[41] For proverbs, see Flandrin, *Le Sexe et l'Occident*, 217–45, and the collection by A.-J.-V. Le Roux de Lincey, *Le Livre des proverbes français* (Paris, 1842). Folktales are discussed by Darnton, 'Peasants tell tales', while some aspects of their relationship to culture and society are explored by E. Le Roy Ladurie in *L'Argent, l'amour et la mort en pays d'oc* (Paris, 1980); The standard source is P. Delarue and M.-L. Tenèze, *Le Conte populaire français* (Paris, 1957–76).

[42] On Thiers see F. Lebrun, 'Le "Traité des superstitions" de Jean-Baptiste Thiers; contribution à l'éthnographie de la France du 17ᵉ siècle', *Annales de Bretagne et des pays de l'ouest*, 83 (1976), 443–65.

Such simplistic oppositions are just what I have repeatedly been arguing against, and although peasants were often charitable to one another, other aspects of their social relations were harsh, sometimes brutally coercive. Few of us would have liked to live in this society, with its lack of privacy and denial of many forms of opportunity and individuality. Nor did popular religion represent some deeper instinctual understanding of the essential Christian message, although it did contain elements which the official church failed to appreciate. These are sufficiently important to make one think that there was something unwise and exaggerated in the attitudes of the catholic reformers. Popular religion can only be classified as semi-pagan if we insist that Christianity is a religion of the book, of clear-cut theological positions, and of individualized commitment; such a position is perfectly defensible, but it means asserting that the mass of the population were virtually incapable of becoming Christians. The implicit position of the bishops who insisted on knowledge of basic dogma as a precondition for admission to the sacraments came very close to this, and would have been disastrous had it ever been seriously applied. Such harsh views now seem as much the product of a particular historical process and movement as do the varied forms of popular religion. Most of their leading exponents were men who had elaborated this ideology without any real experience of the people they hoped to reform; the gulf between the clerical establishment and the parish clergy was crucial in allowing ambition to outrun reality in this way. The divide between literacy and orality was also expressed in this case, since it was easy to be severe and logical on paper in a style which could not possibly be sustained in the endless face-to-face contacts of everyday life, unless the *curé* was prepared to alienate his parishioners completely. The parish clergy were inevitably drawn into a series of compromises which blunted the edge of the reform, and thereby aided the survival of alternative visions of the world. Their continuation also reflected the process of appropriation so deftly identified by Roger Chartier, with the faithful adapting and extending even the new spiritual models to meet their own needs, so that their internal logic (by no means perfect in any case) was rapidly weakened or destroyed as they turned into new expressions of essentially popular religious feeling.[43]

The 'great repression' may have proved unenforceable and counter-productive in many of its aspects. Nevertheless it plainly mobilized large bodies of influential opinion behind it, and had significant effects throughout French society. It marks the passage from what had essentially been a shame culture, in which norms were communally agreed and enforced, to a guilt culture, which placed far greater stress on the individual conscience as an instrument for moral order. This did not mean that shame ceased to be a factor, for the change was to a mixed and self-reinforcing normative system, in which guilt feelings operated to make shame much more powerful, and to extend its claims. If doctrines of *politesse* and *honnêteté* were the formalized expression of such tendencies in French society, the catholic reform adopted them enthusiastically in its own sphere. In Freudian terms this represented a major strengthening of the superego, leading to those conflicts between ego and superego which generate guilt.[44] Norbert Elias, in his highly original study of the civilizing process, identifies a strong shift in self-control, from interpersonal external compulsion to individual internal compulsion, so that spontaneous and emotional impulses could only be expressed with the permission of the 'moral conscience'. The self came to be 'divided by an invisible wall from what happens "outside" '.[45] Elias makes out a powerful case for connecting this shift with the phenomenon of state formation, a necessary concomitant to a much more highly differentiated and interdependent society. He rightly stresses that pyschological structures, far from being common to mankind at all times and places, are socially determined to a very large extent—a point which had also been made by Freud himself.[46] The

[43] Chartier, 'Culture as appropriation'.

[44] Freud's views on guilt are very clearly stated in ch. VII of *Civilisation and its Discontents*, conveniently available in *Sigmund Freud*, xii: *Civilization, Society and Religion*, ed. A. Dickson, (Harmondsworth, 1985), 315–26.

[45] N. Elias, *The Civilizing Process*, i: *The History of Manners* (Oxford, 1978), ii: *State Formation and Civilization* (Oxford, 1982). Some of the themes are developed by Elias in *The Court Society* (Oxford, 1983). The passage cited is from *The History of Manners*, 257.

[46] For example in *The Future of an Illusion;* Dickson, *Sigmund Freud* xii, 190.

analysis by Elias is far too extended and complex to be adequately summarized here, but I would suggest that by concentrating on manners and political behaviour, with relatively little attention to religion, he has presented a rather incomplete picture. While he does point up very considerable differences between noble and bourgeois attitudes, these become even more apparent and significant when explored through their relationship to the religious drive for reform.

The aristocracy possessed a long-established and distinctive moral code, which revolved around such notions as honour, gallantry, generosity, and display. From the viewpoint of a severe clerical moralist these values, however admirable in themselves, retained little merit if they expressed themselves primarily though worldliness, sexual licence, duelling, gambling, and the oppression of the poor.[47] Although Elias is surely right to see a kind of 'courtly rationality' developing, for most nobles this does not seem to have implied any renunciation of hedonism in their general conduct.[48] Indeed, in their concern to establish frontiers between themselves and other social groups, in order to reinforce their special status, they elaborated a style of fashionable behaviour based on these same worldly values. If the progressive refinement of manners required the extension of superego controls, this was only on a very limited basis. The aristocracy never became a devout class, although the shift towards greater self-control probably made it easier for individuals to respond and adopt a life-style which implicitly challenged that of polite society in general. The marked tendency towards role-playing must also have contributed to ensuring that most aristocratic bishops after the middle of the seventeenth century were much more

[47] The particular case of duelling has been subjected to a very thorough analysis by F. Billaçois, *Le Duel dans la société française des 16ᵉ–17ᵉ siècles: essai de psychosociologie historique* (Paris, 1986). He shows how the church mounted an offensive against the practice, particularly after a peak of duelling in 1614–15 (pp. 140–6), and also discusses the attitudes of the casuists (pp. 175–82). In a very interesting passage he explains how Pascal represented a new style of thought, and could not understand the 'baroque sensibility' of the nobles (pp. 238–41); elsewhere he suggests that in a civilization increasingly built on the written word the duel was 'le rituel d'un peuple sans écriture' (pp. 218–19).

[48] For 'courtly rationality', Elias, *State Formation and Civilization*, 281–4, 289–91.

respectable and assiduous than their predecessors. It is hardly surprising, however, to find that most of the leading Jansenist or neo-Jansenist bishops and clerics did not come from the traditional aristocracy, or that it was to this class that Louis XIV and his successors turned for safe appointees who would not rock too many boats. In order to find the groups who were primarily associated with the development of strong superego controls we must turn to the office-holding, financial, and mercantile élites, the families who dominated both urban society and the church. Here there was no such established model of class behaviour as obtained for the aristocracy, leaving the way open for a more dynamic and innovatory response to the specific problems of their class and time. There is a certain irony in the fact that most members of these groups still saw the aristocracy as occupying the pinnacle of society, and aspired to see their families rise to the ultimate consecration of noble status and acceptance; the desire to complete this social ascent was one of the major reasons why they developed a distinctive psychology of their own, which seems to have persisted even among those who acquired the formal titles of nobility.

Whereas aristocrats measured their status by birth and connections, assuming that if necessary they could trade these off against royal pensions or marriage with an heiress of lesser lineage, the other élites were much more conscious of their vulnerability. Their positions depended essentially on their wealth, which in turn called for prudent management over the generations and a willingness to put family interest above individual choices. Only a sustained effort of this kind could bring about an ultimate triumph through admission to the reserved enclosures at the closely guarded centre of the court society, or to the various lesser but still privileged enclaves in provinces and towns. Most families after the middle of the seventeenth century found it difficult enough to maintain their existing status, for which skill and good fortune were also necessary, albeit on a lesser scale.[49] The rapid enrichment and

[49] These patterns are very helpfully discussed by Colin Lucas, 'Nobles, bourgeois and the origins of the French Revolution', *Past and Present*, 60, (1973), 84–126, although he tends to overestimate the amount of social mobility in the seventeenth century.

growth of the élites in the sixteenth century had been accompanied by the pauperization of large sections of the population, leading to a series of neo-Malthusian demographic and economic blockages.[50] The process shows up dramatically in calculations of real wages, which reached their lowest known point in European history at the end of the sixteenth century, and made only a modest recovery during the seventeenth.[51] Hardship among the peasantry was reflected in extensive sales of land, which largely passed into the hands of outsiders living in the towns, and in the constitution of a credit network which imposed a growing burden of debt.[52] The same polarization of society between the haves and have-nots which had helped in the expansion of the propertied élites made the prospect of descending the status ladder frightening as well as humiliating, and such fears operated at every level of society. The crucial danger for a family came at each generational change, for an extravagant or feckless heir could quickly undermine the work of a whole line of prudent ancestors, ending up with capital dispersed, offices and land sold off. Even more than on themselves, it was on their children that the élites sought to impose a system of repression and guilt based on the superego, with the recognition that instinctual gratification must be controlled by prudential calculation. It was not only the devout who behaved in this way; those parents who had been unable to resist temptation themselves were perhaps even more conscious of the dangers it held, and of its insidious power. If this was indeed the case, and there is a good deal of evidence to suggest that it was, then here we have a very important clue to the links between the social and psychologi-

[50] The exact nature of these processes is of course highly controversial. For a good introduction to recent debates see T. H. Aston and C. H. E. Philpin (eds.), *The Brenner Debate: Agrarian Class Structure and Economic development in Pre-industrial Europe* (London, 1985).

[51] The best wage series for France is that for building workers in Paris, calculated by M. Baulant, 'Le salaire des ouvriers du bâtiment à Paris de 1400 à 1726', *Annales E.S.C.* 26 (1971), 463–83. Wage levels are helpfully discussed by P. Goubert, *Beauvais et le Beauvaisis de 1600 à 1730* (Paris, 1960), i, pp. 547–76 and by E. Le Roy Ladurie, *Les Paysans de Languedoc* (Paris, 1966), i, pp. 263–80.

[52] This emerges from the books by Goubert and Le Roy Ladurie mentioned above, and from numerous other regional studies, notably J. Jacquart, *La Crise rurale en Île-de-France, 1500–1670* (Paris, 1974).

cal roots of the 'great repression'. The renunciation of instinct becomes the dynamic source of conscience, and in the general context of early modern society and its belief structures it was likely to produce a great surge of enthusiasm for the imposition of a reformed and purified Christianity. Saintliness had become the duty of the many rather than the few, the failure to display it the mark of damnation. It was inevitable that the application of these doctrines to the mass of the population should be as much symbolic as real, and that they should be greeted by incomprehension and evasion on a massive scale. Among the élites, however, the shift was ultimately independent of religion and its normative system, so that it represents the creation of a recognizably modern set of social and psychological structures, which were crucial to the emergence of the culture in which we now live. Such a wide-ranging change could hardly come about rapidly, and indeed for a long time internalized self-control was little more than a thin veneer even in the case of the élites, maintained primarily by social conventions. A powerful superego was probably still the exception rather than the norm among the members of the urban ruling groups at the end of the seventeenth century; the political, social, and cultural contexts of the time, however, were generally favourable to the minority who had developed it, giving them an influence out of proportion to their numbers.

Since so much of this book has been concerned to chart the negative aspects of the catholic reform, its final words should try to redress the balance a little. A concentration on official doctrines, which can hardly be avoided in view of the limitations of the evidence, is liable to give a grossly misleading impression of social and religious life as it was really experienced. The same can be true of assumptions based on economic data; these too can tempt historians into painting an unduly black picture. As a corrective it is worth citing some comments made by two British travellers towards the end of Louis XIV's reign. Joseph Addison spent some fifteen months in France in 1699–1700, including a long stay at Blois. Having visited Versailles en route he wrote:

I never thought there had been in the world such an Excessive Magnificence or poverty as I have met with in both together. One can scarce conceive the pomp that appears in every thing about the King, but at the same time it makes his subjects go Bare-foot. The people are however the happiest in the world and enjoy from̄ the Benefit of their Climate and natural Constitution such a perpetual Mirth and Easiness of temper as even Liberty and Plenty cannot bestow on those of other Nations.

In another letter from Blois Addison wrote of the French as 'the Happiest nation in the World. 'Tis not in the pow'r of Want or Slavery to make 'em miserable. There is nothing to be met with in the Country but Mirth and Poverty. Every one sings, laughs and starves . . .'[53]

After the next peace treaty the Scottish schoolmaster James Hume visited France in 1714, and what he observed in the same area produced a strikingly similar comment

Just before we entered Blois we saw a company of countrey people together, where, for want of a Fidler, one of the company sang a minuet while others danced in their wooden shoes. Upon which one cannot but observe, how the natural Gayety and cheerful temper of the French supports, and renders them easy, under such pressures as would drive a people of a more saturnine complexion, and a spirit more erected to Liberty, to rebellion and madness, or sink them into inconsolable melancholy and Discontent.[54]

A similar spirit helped to keep religious practice far livelier and more flexible than rigorist reformers would have made it. At the same time I do not think it makes any sense to see the catholic reform movement simply as a disastrous mistake, a deviation from moderate Christian values which led to 'dechristianization' and disenchantement with the faith. To some extent the reform did have such effects, yet without the energy and commitment which it injected the church would surely have found itself in an even worse position. It would have been virtually impossible not to have responded to the social and intellectual pressures of the age; such a failure would furthermore have demonstrated a fatal case of institutional sclerosis. The attack on popular culture, whose success was in any case very limited, should be balanced against the impressive and dedicated pastoral work of the *curés,* whose

[53] Citied by Lough, 'Two more British travellers', 160, from Addison's *Letters,* ed. W. Graham (Oxford, 1941), 12–15.

[54] Ibid., 173.

sympathy with their flocks emerged so strongly as they played their vital part in the events of 1789. It was the old reefs represented by involvement with the monarchy and the papacy on which the church would be shipwrecked during the revolutionary years, problems which long antedated the catholic reform. Nor should repression be regarded as a bad thing, for as Freud himself never tired of emphasizing, it is the necessary condition for a higher culture and a civilized life.[55] Given the place the church occupied in the cultural and social life of *ancien régime* France, it was the predestined vehicle of this process; despite its own deep commitment to many communal values it thereby fostered the individualism which ultimately subverted them. The saintly clerics of the seventeenth century would have been horrified by the society in which we now live, but the historian cannot deny their part in creating it; the best he can do is to use their own concepts, and absolve them on the grounds that one cannot sin without intention.

[55] For two examples of Freud's views, see *The Ego and the Id*, in *Sigmund Freud*, xi: *On Metapsychology* ed. A. Richards (Harmondsworth, 1984), 375–7; and *Civilization and Its Discontents*, in *Sigmund Freud*, XII, ed. A. Dickson, pp. 284–6.

Index